A HISTORY OF RAILWAYS IN 100 MAPS

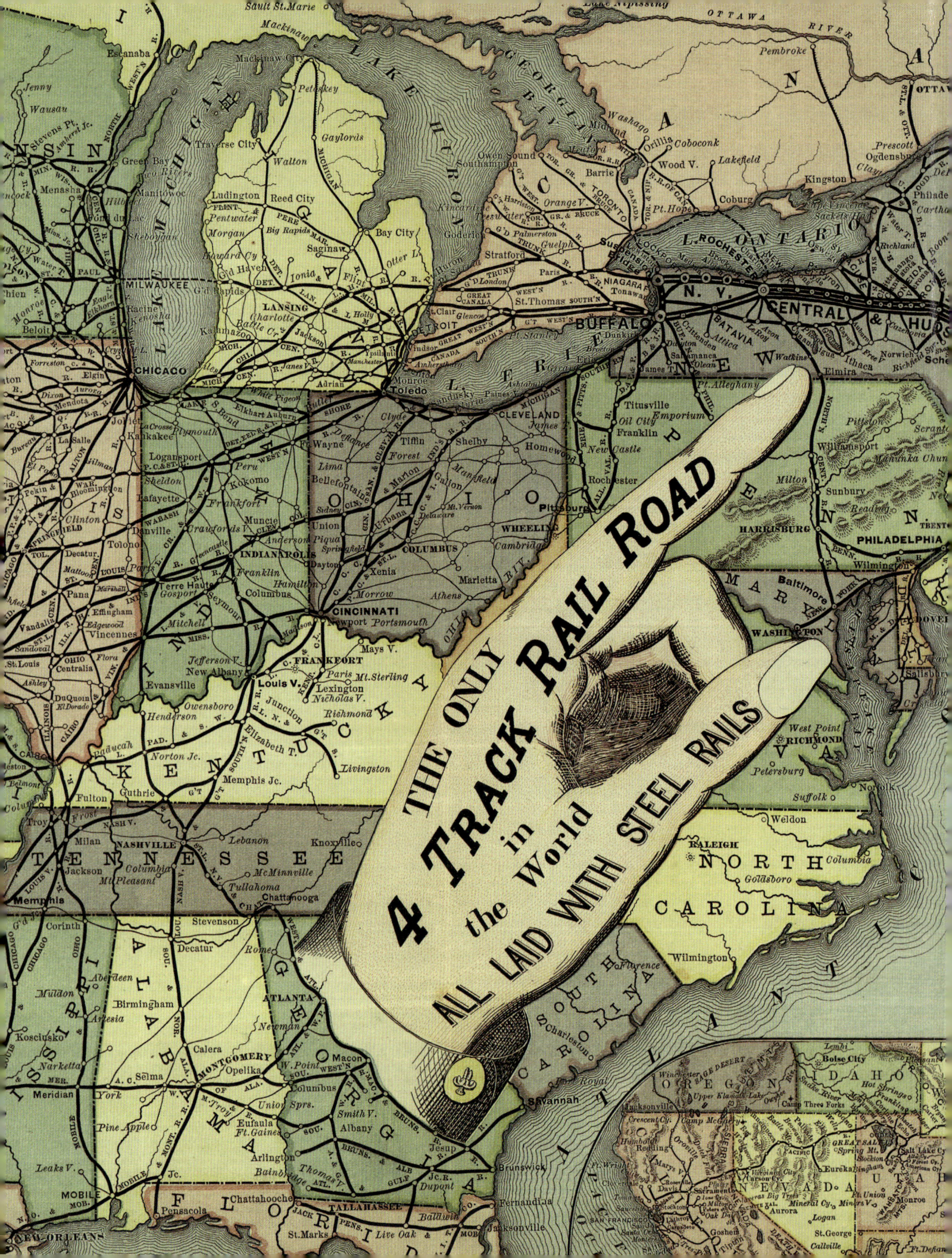

A HISTORY OF RAILWAYS IN 100 MAPS

JEREMY BLACK

For Fred Black

Front endpaper: The GWR Engine House, Swindon, England. An engraving by J.C. Bourne first published in 1846.

Rear endpaper: The celebrated meeting of locomotives from the Union and Central Pacific Lines marking the completion of the United States Pacific Railroad in 1869.

Page 1: Art Nouveau-style poster for the Gotthard Railway Company, designed by Gabriele Chiattone, 1902.

Page 2: Map of the New York Central and Hudson River Railroad and its principal connections, published by Rand McNally, 1876.

Pages 6–7: Claude Monet, *Arrival of the Normandy Train, Gare Saint-Lazare*, 1877. Monet completed 12 paintings of the station and its modern, glass-and-iron train shed, where he found that the light penetrating the glazed roof and the trapped steam rising from the locomotives made for an appealing combination of effects.

First published in 2024 by
The British Library
96 Euston Road
London NW1 2DB
www.bl.uk

ISBN 978 0 7123 5501 8

Text copyright © Jeremy Black 2024
Images copyright © The British Library Board and other named copyright holders 2024
Volume copyright © The British Library Board 2024

The author asserts the moral right to be identified as the author of this work.

All rights reserved. No part of this book may be reproduced or transmitted in any form or by any means, electronic or mechanical including photocopying, recording or by any information storage and retrieval system without the prior permission of the publishers in writing.

Every reasonable effort has been made to trace copyright holders of material reproduced in this book, but if any have been inadvertently overlooked the publishers would be glad to hear from them.

British Library Cataloguing in Publication Data
A catalogue record for this publication is available from the British Library

Edited by Christopher Westhorp
Picture Research by Sally Nicholls
Designed by Karin Fremer

Printed in the Czech Republic by Finidr

CONTENTS

Introduction 8

Chapter 1. **Origins** 14

Chapter 2. **Spreading the New Age 1860–85** 42

Chapter 3. **Geopolitics and War 1885–1918** 80

Chapter 4. **Alongside Road 1919–39** 156

Chapter 5. **War and the Air Challenge 1939–70** 190

Chapter 6. **Rail Developments 1970–Present** 226

Chapter 7. **Facing the Future** 252

Chapter 8. **Rail and the Collective Imagination** 262

> Further Reading 280
> List of Maps 281
> Index 284
> Picture Credits and Acknowledgements 288

INTRODUCTION

THE HAUNTING, DISTANT SOUND OF THE WHISTLE; the trembling in the air as the train nears; its speedy, dominant bulk as it towers past; the hiss of brakes and the rush of this man-made wind – all are physical sensations we can associate with trains. So also with the jolts of crossing the tracks. Each of us will have a different memory, and all contribute to the history of rail.

This book adds to that a particular kind of this history – that offered by maps. The maps produced on railways are fascinating in their own right, and also very significant for the story of trains. The needs of train companies and users, especially passengers but also freight companies, were one of the major drivers for the development of mapping and for the use of maps from the 1820s until railways were partly superseded by the requirements of road a century later. Even if the road network generally now comes foremost in modern-day mapping, the needs of rail transport remain very important for mapping and therefore the understanding of maps. This is notably so in the major urban areas where much of the world's population now lives and where rapid transit systems frequently focus on rail, with the relating networks, which are generally very dense, creating major challenges for mapmakers.

So mapping rail is in part a history of maps and, in that, a history of map production, map use and map perception. All will be discussed in this book. At every stage, it is important to remember that what is interesting about the maps is not only what they tell us about rail, but also the choice that is made about what to show us and how the result tells a message, not least that choice involves what to omit/exclude as well as what to show/include. No map can include everything, and the contrasts in the mapping of railways that can be seen in this book indicate the variety of what can be shown. This is true not only of the railways themselves but also of their context.

CONTENT AND PURPOSE

The choices to be made in relation to maps run from projection and perspective, to content, scale, colour, key and title. For any particular facet of the map, most obviously the coverage of the subject, there is the question of how far it is appropriate solely to focus on the train and how far other elements and features should be included. These range from terrain to towns, borders to roads. There is the interplay of the physical geographies across which the train moves and the human geographies that explain its goals and use.

There is also the question of purpose for specific maps. It is a mistake to suggest that a specific map of a train route must serve a functional purpose. Sometimes the map of the rail route can appear redundant or somehow unnecessary. Thus, to illustrate an article on 'The new murder mystery experience on Britain's poshest train', *The Sunday Times* on 12 March 2023 included three photographs and a map, produced specially for the newspaper, that showed the route, even though the author made it clear that that was of scant interest to her, and indeed had scant consequence for the story: '...poor old Kent doesn't get much of a look-in ... I rather neglect the views ... I can get graffiti and Tescos closer to home.' The same weekend, there was a reminder that maps have become part of the visual wallpaper of rail, as in *The Daily Telegraph* of the previous day that, alongside more prominent photography, had a map to illustrate a luxury train service opened in Sri Lanka in 2022. So any suggestion that a specific map of a train route may not serve a functional purpose can be both mistaken and underrate the wider value of maps for rail.

The value of rail for mapping is also pertinent. The nineteenth century saw much expansion in new forms of technology, but most did not involve any wide need for maps nor any expansion of map culture. In essence, this point was true of the communication of messages, whether by mail services, telegraph or telephone, which successively grew very greatly, with the first two linked to a degree to rail services. Just as more recently with the Internet, there were mail, telegraph and telephone routes that were mapped, but there was nothing like the scale or intensity of mapping for, and about, rail. More particularly, consumers had far less interest in the routes by which the messages passed, whereas rail mapping was very much centred on user-information.

The era's other major use of steam technology for transport was the steamship. As with railways, steam navigation carried passengers and freight. Moreover, on internal routes, both rivers, such as the Mississippi, and canals could either be in direct competition with rail or could lessen the attraction and/or profitability of particular routes for rail. Yet in terms of maps the steam navigation routes were much less mapped. In part, this was a matter of the absence of track and the possibility therefore of following very precise routes at sea; too many to map. There was an alternative mapping need for riverine navigation, that of shallow and therefore hazardous waters, but this aspect was not particular to steam, having already been necessary for sail- or row-powered ships. Moreover, unlike areas close to harbours, such mapping was not needed for deeper waters.

THE TRANSFORMING POWER OF STEAM
Rail was different. It was new. There was a specific route affirmed through a track, and with construction, maintenance and operating consequences according to how that route was selected. For a rail company to operate a specific route necessitated having that route mapped. Indeed, well before a company built a project and operated a route there were pioneer developers and speculators, many of them hustlers, who found that maps were essential in order to suggest, debate and plan potential rail links. And once railways were established there was also the need for passengers and freight consignors to understand the rail options and what they meant. This was particularly so as railways joined together and intersected in order to be part of a system.

The 'scream of steam' for long held generations in awe, as steam-powered locomotives thundered through cuttings and tunnels and over embankments, putting the march of travel onto a new level of scale and speed. The excitement of rail, the energy of builders, the trains cutting through the landscape and the transformations they brought, or made possible, to economies, societies and countries can all be seen readily through maps.

From the outset rail was instrumental to the expansion of coal mining – and coal made the modern world. Coal, with its higher calorific value, transformed the metallurgical industry and took the world from an essentially wooden age to a new iron age. Coal also meant that land could be used for agriculture instead of the forestry necessary to produce wood, and this change, given that cleared land was the major input in agriculture, helped to contain food prices to the benefit of labour, ensuring that workers could afford more for the consumption of goods and services. That affordability was also helped by the movement of food by steamship and rail.

MAPS AS INDICATORS OF CHANGE AND ADAPTABILITY
Railway maps have been a key form of mapmaking and of cartographic use, whether governmental, business or public, since the early nineteenth century. That has remained the case to the present. As such, railway maps have had a history in cartography that has lasted as long as that of most government mapping agencies, and

The renaming of the London Overground in 2024 is an important instance of how history is reinterpreted for rail, and how maps record this process. Initially, the London Overground was established in 2007 to take over the Silverlink Metro Services, which operated from 1997 in North London, and 'Overground' was a 'branding' of the suburban rail network managed by Transport for London (TfL). The London Overground served 113 stations on six lines and was designated on maps by orange lines. In 2024 it was decided that each line would have a name and a colour, and the lines became Lioness, Mildmay, Windrush, Weaver, Suffragette and Liberty. The line names were deliberately not geographical – as in North London line. However, those chosen did not necessarily resonate widely – the Mildmay Line was named after a hospital about which little was known, and the Liberty Line celebrates the Royal Liberty of Havering, the old name for the borough of Havering. The lines had no reference to railway history, such as the navvies who built the railways, railway works or marshalling yards. Moreover, the Weaver Line was a somewhat strange reflection on an urban economy whose history and present are known for much else. Yet, the very controversy in 2024 reflected widespread interest in railway maps and their naming.

longer than very many. Moreover, in terms of the interaction of mapping technology and business, mapping for rail has been a sustained type of cartography. This mapping, furthermore, had to engage with the public: rail was not only about freight. As passengers, the public were readers and scanners of guides and maps, using them as the basis for planning journeys and not merely admiring them as decoration at stations. Indeed, for many people timetables were heavily dependent on maps, the latter serving to establish the options that were then pursued through the timetables.

Railway maps are important, both in themselves, as guides to the development of railways, and also for what they show about developments across a range of spheres and at a number of levels. As this book illustrates, these include economic, social, political, military and geopolitical developments, and they ranged from the local to the continental. Each reflected the impact of rail, both in terms of practicalities and also with reference to mental mapping.

Railway maps show the constant adaptability of cartography and how graphics reveal their flexibility to meet differing needs and requirements, whether that be for practical effectiveness or simply aesthetic appeal, as well as the impact of specific mapping technologies. Adaptability will remain important into the future, reflecting challenges both in what has to be covered and new ideas of how best to do so.

INDUSTRIALISED PRODUCTION

Railway mapping came to the fore in a period in which the production of mapping was becoming easier. Mechanised papermaking became commercially viable in the 1800s, leading to the steam-powered production of plentiful quantities of inexpensive paper. The steam-powered printing press developed in the same period, while lithography made a major impact on mapping from the 1820s. Entrepreneurs, moreover, experimented with a variety of production techniques. Controlling cost was important as publishers sought to expand their market and this factor needs to be borne in mind when looking at the maps in this volume. The choice of map, and notably of map content and appearance, owed much to cost. This need for economy was central to functionality, and not an add-on factor. Thus, in considering maps, there is the central unknown of the budget permitted. For example, in America, cerography, or wax-engraving, was developed in the 1830s and it provided an inexpensive way to produce relatively detailed maps. Cerographic maps were to play a major role in nineteenth- and early twentieth-century American cartography, distinguishing it from that of Europe, where such maps were regarded as inferior to lithography, both aesthetically and in terms of precision.

Colour printing came to play a more prominent role in mapping from the nineteenth century, and it was seen as both a commercial opportunity and a challenge. Although multicolour printing from more than one plate had become possible with the advent of engraving, it was never common. Map-colouring ceased to be a manual process in the nineteenth century, and was transformed by the developments in printing. Colour served to enhance the aesthetic quality of maps. The use of colour also increased the density and complexity of information that could be conveyed, and thus made maps into more powerful explanatory devices. The average map now had more information that had to be assimilated, not least through a process of separating out the components, then integrating them in a comprehensible form. In this complexity and depth of meaning lay much of their interest. In rail terms, colour printing provided the opportunity to distinguish between a number of lines where they were in close proximity.

DATA AND DIFFERENTIATION

The mapping of transport history is greatly dependent on data availability, and quantifiable information also throws light on the qualitative nature of the system. Different indices of density of usage can be mapped: numbers of trains or passengers per day, freight tonnage or income turnover per mile. Such analysis was provided in map form in the twentieth century with the 1963 Beeching Report on the future of British railways (see pages 210–211). That analysis was not comprehensively presented for the nineteenth century, but a mass of material was made available, and notably so for companies reporting to share- and bond-holders, and for state organisations producing reports within government.

Separately, it is difficult to depict (or even sometimes evaluate) improvements (rapid or slow) and periodic upgradings in existing systems, which means that the mapping of transport links, or at least their publication, tend to place an undue weight on new routes. Rail and road improvements in the first half of the twentieth century, a period of consolidation, are generally underrated. This is especially the case because there is an emphasis in mapping on the length and general direction of routes rather than on the time taken to accomplish particular journeys. The latter could vary greatly. As a result, information tended to be produced for the fastest possible timetabled journey.

There is no shortage of ways in which railways could – and can – be mapped, while the data that is used can – and has – been presented very differently. Should the exact route be shown? Or should a route be presented topologically, with the general direction indicated? In terms of total mileage, there is the need to consider how far to include non-public railways, including those in and around mines. In some cases, for example in northern Chile where copper is excavated and transported, these can be more important than passenger lines (see pages 100–101). Linked to this issue comes another of differentiating between freight-only lines and those that are mixed usage because they also carry passengers. As far as considering mileage, there are also the questions of distinguishing between track owned, track operated, and whether only part of the track is used, not least if the line is single or double-tracked. At every stage, in historical terms, there was also the question of what attention to pay to railways that were planned but left incomplete or never even started. They can provide interesting, even dramatic, lines on maps, but frequently were speculations of only slight significance.

There are many other issues involved in deciding how best to map railway provision and thus history. One is how much emphasis to put on stations and how to differentiate between them, for example in terms of their nodality or prominence as a place to change trains, or, separately, their aggregate usage (see pages 250–251). More generally, the system characteristics may be difficult to depict, not least if there is a convention of, say, depicting all stations or lines as equal. So also with the workforce: the early railways in Europe were built by free workmen, but in the Spanish colony of Cuba prisoners were used, both convicts transported from Spain and captured Carlist rebels. How best to depict this is unclear.

In addition, railway maps rarely cover the manufacturing infrastructure, but the manufacture and maintenance of trains became an important aspect of industry, with major workshops founded, for example in England and Scotland at Ashford, Brighton, Cowlairs, Crewe, Derby, Doncaster, Eastleigh, Greenock, Nine Elms London, Swindon and Wolverton. These workshops not only handled repairs but also built locomotives, and indeed produced many different products for the rail industry. Crewe produced its own bulk steel. Founded in 1903, the Glasgow-based North British Locomotive Company was the largest private locomotive-builder in Europe. Such business concerns, also seen in other states, are a crucial element of railway history, but are only very rarely included in the maps, contemporary or subsequent.

RAILWAYS AND MODERNITY

The rail industry that was proposed, charted, sold and recorded in maps became part of the imagination of modernity, and the maps contributed directly to that outcome. They showed lines cutting across regions, countries, sub-continents and, from 1867, continents. The last was the case with the United States of America, although the line across Panama opened in 1855 was also a transcontinental line linking the Atlantic and Pacific oceans (see pages 60–61). That was valuable for the boosterism for that line, but had scant impact in affecting wider attitudes.

Lines and services defined and redefined areas and what they contained. Investment, governmental support, economic opportunity and technological change all helped that process, and interacted to that end. As a result, in the nineteenth century, advances in transport engineering, especially bridge-building and the use of dynamite, nitro-glycerine and gelignite in tunnel construction, helped overcome some of the problems

In 1976 Rudi Meyer was a skilful graphic designer working as a cartographer for SNCF, the French national rail network. He produced a plan of the network that was widely displayed, including within the trains. Meyer's plan contained no geographic references at all, whether coastline, rivers, mountains or frontiers, but managed to convey the shape of France, the range of the network, the hierarchy of lines (communicated through width) and stations (by size of circle), and the linkages into particular axes converging on Paris. The first TGV line, from Paris to Lyon, was marked on the plan with a separate double-line route.

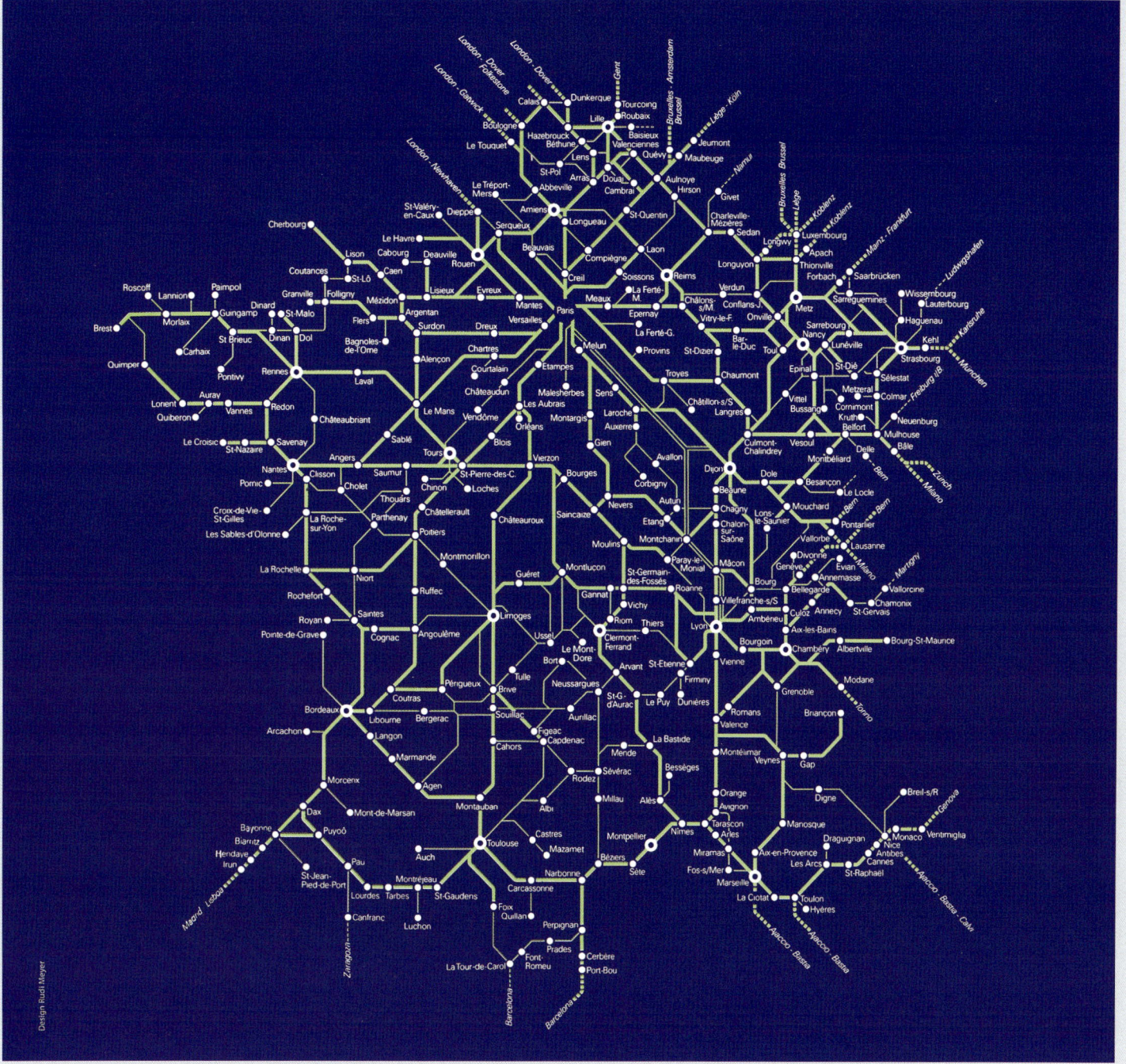

posed by the terrain, notably the Alps (see pages 68–69). These advances literally changed geography, both locally and on a broader scale. The epithets celebrating change were deserved, and certainly when the railways were built and operated. However, many bold plans did not reach that state.

Trains were repeatedly presented as joining and transporting through the overcoming of terrain and distance. Pictures, photographs and films contributed to that outcome, but the rail imagination was overwhelmingly that recorded in maps, and this is that story.

The maps are reproduced in chapters organised primarily on a chronological basis, but within them the arrangement is thematic – principally in geographical terms rather than chronological, which means that the relevant maps for, say, Europe or North America can be found grouped together.

ORIGINS

Rail as a large-scale project essentially started out in order to serve the needs of steam. As a result, rail was the product and means of the Industrial Revolution, a series of changes that rail transport helped speed up, indeed effectuate. This was rail as the moderniser and the bringer of the modern world. Without transport, coal, the fuel that could unlock the ages, was of scant use, but coal with transport was of great value, and thereby the basis for investment and speculation. For a long time the key means of transport was by water because coastal trade, rivers and canals offered the low-cost bulk transport that was not possible on land. Indeed, at the outset of the eighteenth century in Britain, a wagon drawn by four horses pulling 4,000 pounds could rarely cover more than 20 miles, while poorly constructed roads led often to a reliance on light carts or packhorses, which therefore reduced what could be carried.

The eighteenth century saw the development of steam technology, but it was not initially applied to transport. Instead, the change then occurring in Britain was to the existing transport routes consisting of canals and turnpike roads, each of which served as a focus for investment to create new links and improve existing ones. In many respects, railways represented an improvement on roads that was comparable to that of canals over rivers, but it was more than that, and railways also posed particular difficulties. Canal construction was more impressive than river improvement because, like a rail line as opposed to a road, it created a completely new link. As a result, railways, like canals, were also expensive, because large numbers of 'navigators' or navvies had to dig out canals by hand, with tunnels being especially difficult prior to the development of high explosives and mechanical drills. Moreover, railways had the special costs of making the tracks that were absent with the inherited road network.

Railways, like the canals before them, were a response to deficiencies in the existing transport arrangements, to powerful new demands, and to the availability of considerable natural resources. Moreover, for Britain both canals and railways were to provide comparative regional and international advantages. Notably, there were cheaper transport costs and that led to lower production costs, enabling competitive advantages that resulted in a high volume of production that drove further investment.

ENGINES OF CHANGE

Steam technology in the eighteenth century was stationary (it did not form part of a moving vehicle) and expensive, such as the 'two vast engines' seen by John Evelyn in a Cornish tin mine in 1702, 'consisting of several wheels in the nature of pumps to draw the water out the places they intend to search, they are continually going by the force of water'. Major change came with the locomotive engine.

A decade after Evelyn, Thomas Newcomen produced his Atmospheric Engine. In essence this consisted of a beam with a cylinder attached to one end and pump rods to the other. Steam was introduced into the cylinder, then the injection of water condensed the steam, causing the piston to descend under the weight of the atmosphere and thus lift the pump rods at the other end of the beam. The piston was then returned to the top of the stroke by the weight of the pump rods.

Newcomen's engine was progressively improved during the century, not least with bigger and better cylinders and valves, which increased its energy efficiency and regularity. James Watt invented the separate condenser, which greatly increased the fuel efficiency for steam engines, and patented in 1769 the steam jacket round the cylinder. Watt also developed an engine that could transmit power on the up and the down stroke. His first full-size steam engine was installed in 1776, while the new casting techniques and boring machines introduced by John Wilkinson in 1774 and 1781 improved the engine still further.

From that background it was not far to make the leap to locomotives, and this is a reminder of the degree to which the major developments that followed from the 1820s had a long anticipation. The greater understanding of physical laws was vitally important in a manufacturing system that benefitted from enhanced ways of using objects, notably by pushing, lifting or rotating them. Locomotives

'The Effect of a Post-Age. — An Oliver for a Rowland.', a satirical engraving, c.1853, showing a Victoria-era locomotive (Stephenson's *Rocket*) shunting and overturning a George IV-era mail coach as a snail crawls along the telegraph wires.

were the products of a mindset in which innovation was revealed as possible and controllable. A good instance of this was improving efficiency in iron smelting. In 1828 James Neilson patented an increase in furnace fuel efficiency in iron smelting by, counter-intuitively, blowing it with hot rather than cold air. This process also helped in the effective use of black band ironstone and coal.

RAILS AND STEAMPOWER

And so also with the use of rails, albeit at first with that most ancient motive power provided by the horse (see cartouche detail, pages 18–19). The use of horses for railways reflected the animal's more general prominence in transport, industry and agriculture. This was a matter both of overland railways, principally in Britain but also for example in Pennsylvania, and railways within the mines themselves, along which wagons for the excavated coal could be pulled. These wagons were hauled by pit ponies and by human labour, particularly women and children because the men tended to focus on the heavier work of cutting into the coal seams.

Nor was there to be only one type of railway once George Stephenson had worked his technological skill in further increasing locomotive fuel efficiency.

Indeed, travelling from Leeds to Sheffield in 1830, William Cobbett saw: '...the iron furnaces ... the ever blazing mouth of which is kept supplied with coal and coke and iron-stone, from little iron wagons forced up by steam, and brought down again to be refilled.'

The use of the steam locomotive and the development of railways and rail practices were to overcome many of the limitations of earlier transport systems, and not only in terms of motive power. Wagons and carts often provided merchandise with inadequate shelter, and the methods of packing and moving heavy goods on and off carts were primitive. Moreover, due to limitations in transportation methods, droving was the principal way of moving livestock, although it was both slow and the animals used so much of their energy on the move.

The motive force of steam was crucial in bringing about change and the steam locomotive greatly encouraged as well as focused economic development, notably in the internal movement of coal. By raising labour productivity, the use of cheap energy made it easier to pay high wages, and thus sustain demand for all goods and services. The economically revolutionary development of rail and the investment possibilities and engineering problems it raised made the mapping of rail a major task from the outset.

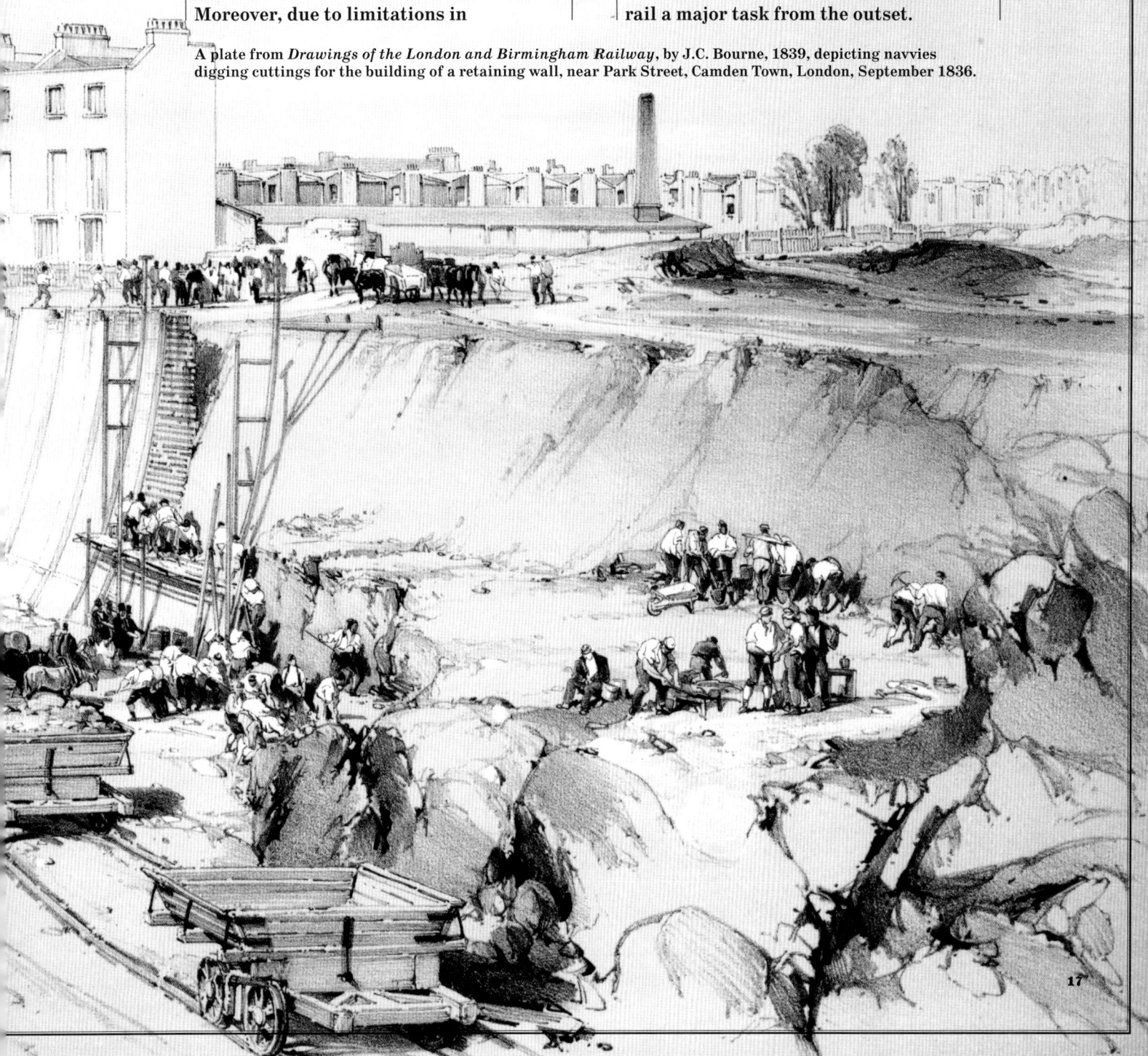

A plate from *Drawings of the London and Birmingham Railway*, by J.C. Bourne, 1839, depicting navvies digging cuttings for the building of a retaining wall, near Park Street, Camden Town, London, September 1836.

IN THE BEGINNING WAS COAL

John Gibson, 'Plan of the Collieries on the Rivers Tyne and Wear also Blyth, Bedlington and Hartley; with the Country 11 Miles round Newcastle', 1787.

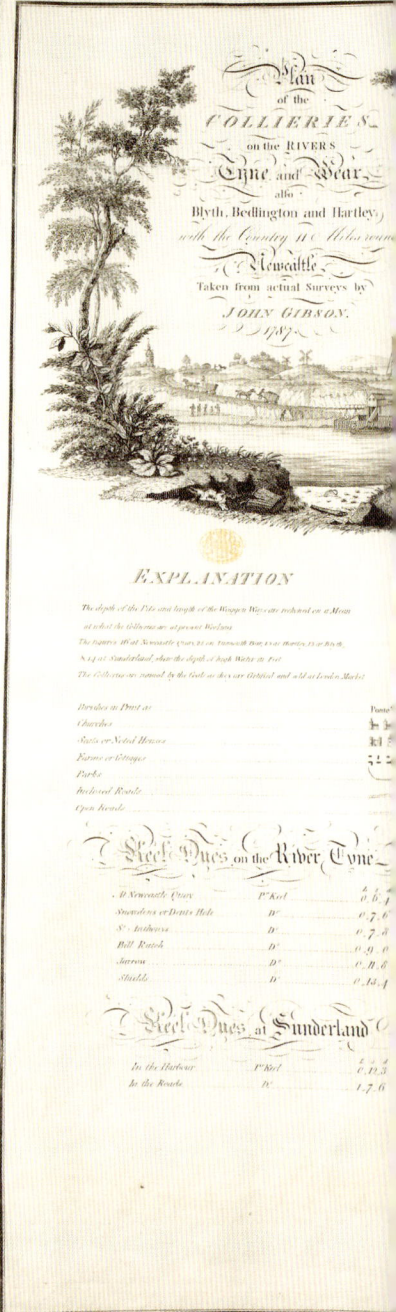

'The great business of the place is the coal trade.... The coal is brought down in wagons along rail roads, and discharged from covered buildings at the edge of the water into the keels or boats that are to convey it.'
Thomas Pennant of Newcastle, 1769

Although tradition is suggested by the dedication to Hugh, 2nd Duke of Northumberland, this copperplate engraving is a map of changes, as very much is the cartouche with its drawing of a full coal wagon moving down the railway to the riverside wharf, the horse acting from behind as a brake, while an empty wagon to the left is pulled up the slope by its horse. The barge on the river is full of coal. The availability of mezzotints, produced from an engraved metal plate, allows for a visually richer cartouche and is an appropriately industrial process with which to depict the region. Glassworks and saltpans on the Tyne, both heavy consumers of coal, are shown. Tables of keel dues on the Tyne and at Sunderland are in the margin.

The duke derived income from his land, which included coal mines as well as farms. Landowners in partnership with investors, who had substantial disposable funds and working capital, played a crucial role in the expensive business of developing coal mines and attendant transport facilities, for example in Northeast England, Ayrshire, Cumberland and Lancashire. Granville, Earl Gower, later 1st Marquess of Stafford (1721–1803), was actively involved in coal, lime and iron ore extraction, and the development of canals and mineral railways, notably in the Staffordshire 'Black Country'.

Harbour facilities were important to the development of railways in the Northeast, and notably the potential choice of river. An Act of Parliament of 1717 had established the River Wear Commissioners in order to develop harbour facilities on the lower Wear, a hazardous anchorage made difficult and dangerous by rocks, sandbanks, the passage of a dangerous bar and exposure to northeasterly gales. After burying, dredging, lighting, pier building and controls over the dumping of ballast, the result was a much-improved harbour entrance and navigable channel that permitted a major growth in trade with the Wear, and thus aided the development of Sunderland. This helps explain the railways developed to connect with the Wear.

The map shows the network of railways along which coal was moved to the rivers Tyne and Wear. This network had expanded with the New Western Way opened in 1739 and the extension of the Wear system to Pelton Fell by 1746. The map and cartouche also show the intensity of the system and the degree to which the invention of the locomotive steam engine was only part of the process. Instead, this development was one variable within an already effective network. The key element was the movement of a bulky product, coal, which was to be joined eventually by other bulky products, notably iron, building materials, grain and livestock.

For bulky freight the improvement of the roads, with the many turnpikes shown on this map, as well as with wheeled transport and better carrying organisation, was of only little help. The map also shows that canals, either as a feeder to the rivers in this region or as a separate system, were absent. Indeed, that lack of canals in the Northeast helped to explain the development of the railway there.

At the same time, railways were seen elsewhere, and sometimes in conjunction with canals. Thus, the 25-mile-long Glamorganshire Canal along the Taff Valley from Merthyr Tydfil to the sea at Cardiff, constructed in 1790–94, was supplemented by 'tramroads' or feeder railways. So also from the East Shropshire coalfield to wharves on the River Severn.

The construction of horse-drawn railways for transporting led to parliamentary discussion, as in the Leeds Coal Way Act of 1758. Developed by Charles Brandling who had a coal mine at Middleton, this wagon-way – the Leeds Coal Way – remained in use supplying Leeds, Ripon and Knaresborough until 1807. This represented a transfer of Northeastern technology, with Richard Humble, the colliery manager, being from Northumbria. Reducing the cost of coal

enabled Brandling to dominate the Leeds coal trade and by the close of the century the average annual output from Middleton was 78,500 tons. There were, however, human costs; the *Leeds Intelligencer* of 29 April 1760 reported: 'Last Saturday, one of the coal-waggons, belonging to Charles Brandling, esq; coming down a hill, nigh Hunslet, overturned upon the driver, and crushed him so terribly, that he died a few hours later.'

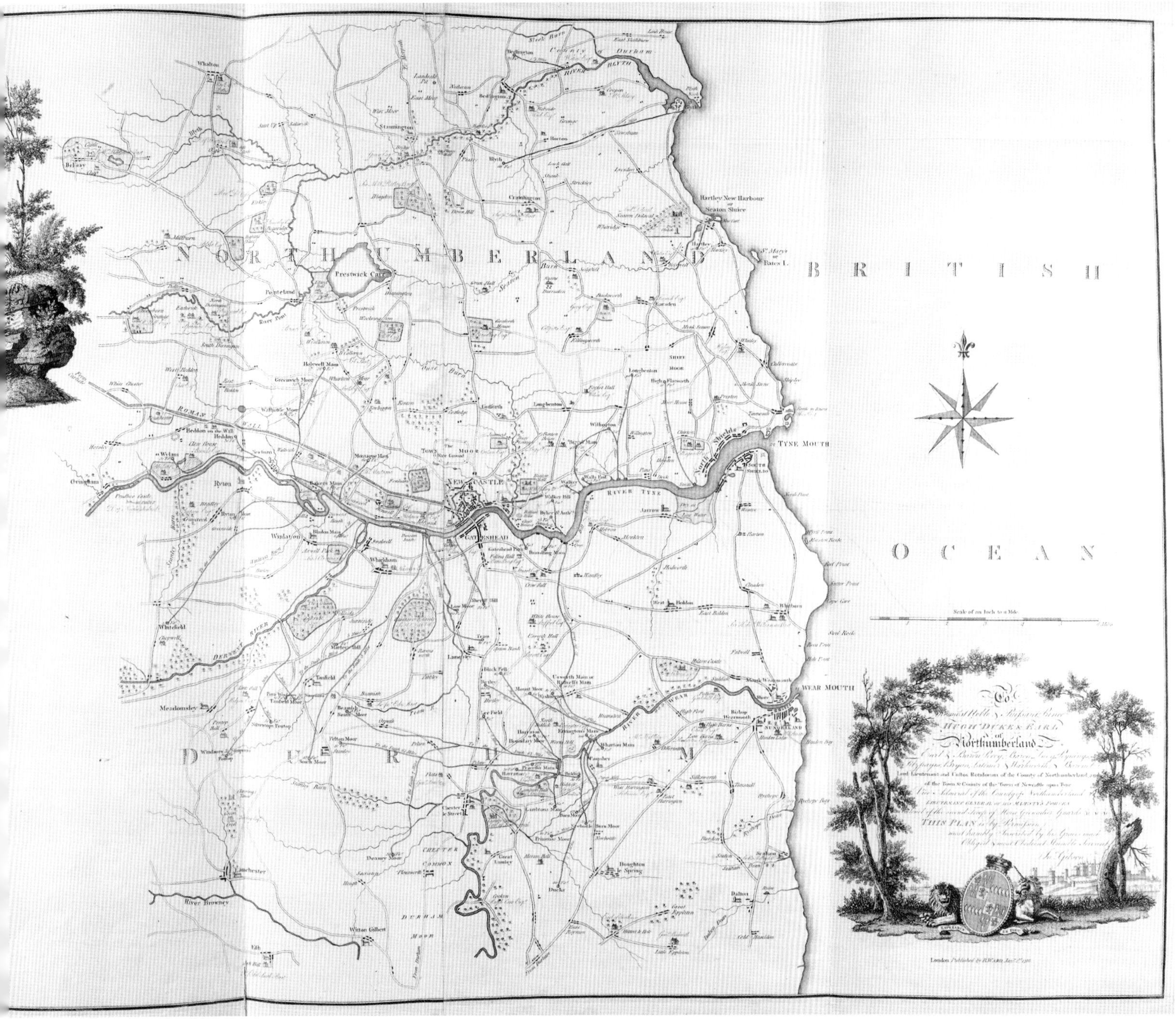

This map by John Gibson drew on the standard format of the county maps of the period, although, covering parts of Durham and Northumberland, it was not itself a county map. Instead, the map shows the extent to which coal and rail could create an economic zone that was very different to the old counties.

1. ORIGINS   19

TOPOGRAPHY ABOVE, MINERALS BELOW

Charles Smith & Son, 'A Geological, Railway & Canal Map of England, Wales and Part of Scotland', 1843.

In 1843 two geological maps of Britain appeared, the one reproduced here, published in London by Charles Smith & Son, and the other commissioned by the Society for the Diffusion of Useful Knowledge, created by the geologist Roderick Impey Murchison and published by Chapman and Hall. The two maps took further John Arrowsmith's comparable 1834 map, the three dramatic maps together reflecting popular interest in geology and also the importance of geological factors for transport systems.

Charles Smith was a publisher and engineer working in London from 1803. He produced many county maps, as well as the *New English Atlas* (1804). Smith's maps were printed from copper plates and coloured by hand with watercolour washes. The colouring was particularly necessary for geological maps, and the specific requirement of this mapping drove forward that particular facet. In contrast, most railway maps did not see a comparable colour palette, other than for the differentiation of the lines of different companies.

Geological factors were important in two major respects. One was an issue discussed in most prospectuses, and in the correspondence, namely the ease of the terrain for railways. This was a matter of elevation, the tackling of which entailed expense and difficulty for construction as well as operating issues for locomotives. The other was working the strata, with some more resistant to being dug out.

These issues focused attention on two major problems with early railway schemes: the difficulty of matching the realities of expenditure to the ambition of proposals, and the more specific technological and engineering challenges that were faced. As a result, detailed surveying and mapping were required to understand the topographical situation and gauge the engineering tasks. There was indeed a contrast between the general geological background depicted in this map, and more local surface geological features that were revealed through survey.

The 1836 prospectus for a London–Dover railway emphasised its nearly level route, which would ease construction and operating costs, and its deep cuttings being in chalk. In contrast, igneous rocks, particularly granite, were far harder to work; while escarpments of any type posed the challenges of gradients and tunnelling. Weak strata and rock structures were not good for railways. The 'geological character of the district' was mentioned in prospectuses when it was seen as 'favourable to engineering lines' – as with that for the railway across the Durham coalfield, the London and Edinburgh Direct and the Darlington and Hawick Junction Railway.

Secondly, geology meant raw materials that could be exploited – notably coal, iron, lead, copper and building materials. Charles Smith's map shows the presence of coal, iron, lead, copper, tin, gypsum and salt, and the key explains the 'mineral character' of the strata. All were seen as being of great benefit because they made freight income likely. Moreover, being offered the opportunity to benefit from mineral rights made it far more likely that landowners would permit building lines across their land. Profit and potential had to be brought into line.

Charles Smith's map brought to the age of railways the initiative and advance in geological mapping provided in William Smith's *A delineation of the Strata of England and Wales, with part of Scotland* (1815). Alongside a geological map, William Smith produced a stratigraphic table, geological sections and county geological maps, collectively intended to form a geological atlas. He marked canals, tunnels, roads, collieries and horse-drawn railways.

Charles Smith shows railways with reference to the length of each line for which an Act of Parliament had been obtained. The inclusion of canals in the map, and the space devoted to them, captured what was to be a competition that lasted for several decades. Canals represented existing transit rights and investment, and did not require powering by coal; they remained profitable and competitive, such that any emphasis on railways can lead to an underplaying of the continued role of canals. Nevertheless, the latter were slow, affected by problems such as leaks, drought and ice, and did not offer income from passenger traffic.

1° 30′ 0 30′ 1° 30′ 2°

56° 30′ 55°

REFERENCE TO THE MINERAL CHARACTER.

	Names of the Formations & average thickness.	Mineral Character	Subdivision
A	Craggy about 50 ft.	Shingle beach with layers of Sand. Shells, fish teeth & bones.	a Norwich Red Coralline Crag
B	Freshwater about 150 ft.	Marls & Limestones, full of freshwater Limnæa & Planorbes, divided into two series by Marls & Estuary Shells.	
C	Lower Tertiary about 400 ft.	Clay above & Clay mixed with Sand below, a great quantity of Septaria & Sea Shells.	c London Clay, cc Plastic Clay, Greensand at the bottom.
D	Greensand about 500 ft.	Sands, pebbly & cherty, slightly calcareous with green grains in the beds the lower layers irony with fossiliferous & calcareous Clay.	d Uppr Greensand dd Gault; ddd Iron Sand.
E	Chalk about 1000 ft.	Carbonate of Lime imperfectly indurated with layers of Flint nodules at regular intervals.	Upper Chalk, Lower Chalk, Chalk Marl & Red Chalk.
F	Upper Oolite about 1000 ft.	Limestone, partly oolitic & partly cretaceous with nodules & ramifications of Chert, Greensand & nodular beds on a thick layer of blue Clay with Lignites & Septaria.	Portland Rock & Kimmeridge Clay.
G	Wealden about 900 ft.	Various coloured Sands & Clay interspersed with Lignites, Conglomerates & Calcareous portions, Limestones & Iron.	g Weald Clay gg Hastings Sand & Purbeck Clay
H	Middle or Coralline Oolite about 600 ft.	Blue & yellow Oolites sometimes pisolitic full of Coral inclosed in a mass of Cherty Sandstones with thick blue Clay.	Upper Calcareous Grit, Coralline Oolite, Lower Calcareous Grit, Oxford Clay & Killoways Rock.
I	Lower Oolite about 400 ft.	White & yellow Oolite, shelly Limestone, variously interspersed with Shale, Sandstone, Clay & Marl.	Near Bath: Cornbrash, coarse Limestone, Forest Marble, coarse shelly Oolite, nodular Sandstones & Clay, Great Oolite, Fuller's earth Marls & Limestone. In Yorkshire: Cornbrash, shelly Sandstones, Shales, Coal, Slate, calcareous shelly Sandstone beds & Oolite
J	Red Sandstone about 900 ft.	Clays, Sandstones & sandy Conglomerates of various colours as red, white, green &c.	Variegated Marls, Gypsum, rocky Salt, red & white Sandstone; JJ Red Marl.
K	Magnesian Limestone about 300 ft.	Limestone with white, yellow, purple & reddish grey powdery Oolite crystalline with strings & nests of Calc spar, red, purple & yellow Sand with red & white Clay.	Grey laminated Limestone, red & white Clay, Gypsum, yellow Magnesian Limestone, Marl slate, red, yellow & purple Sandstone.

	Names of the Formations &c.	Mineral Character	Subdivision
L	Lias about 500 ft.	Blue Clay with Pyrites, Bitumen, Ironstone courses, Septaria with beds of Sandstone, sandy Limestone & blue & white Lias Limesto.	Upper Lias Shale of Yorkshire, Marlstone, Lower Lias Shale, Bath Clay, Lias Limestone, Blue & White Lower Lias Marl.
M	Coal about 3000 ft.	Sandstone, Shale, Clay, Coal, layers of Ironstone nodules, Pyrites Micaceous Sandstone in thick beds, dark Bitumen, laminated & white Clay.	m Millstone Grit & Limestone Shale.
N	Old Red Sandstone varies from 100 to 10,000 ft.	Sandstone, Conglomerate, Clay & concretionary Limestone of various colours.	Quartzose, Conglomerates, coloured Marl, Flagstone & concretionary Limestone.
O	Carboniferous Limestone about 2400 ft.	Limestone, Chert, Sandstone, Coal & Shale with layers of Ironstone nodules	
	Wenlock & Dudley Rks.	Strata of Limestone & Shale.	Concretionary grey & blue Subcrystalline Limestone (Wenlock & Dudley Limestone) Argillaceous Shale, dark grey & liver coloured, with nodules of earthy Limestone (Wenlock & Dudley Shale.)
P	Ludlow Rks. about 2000 ft.	Sandstone, Limestone & Shale.	Laminated Sandstone, Aymestry Limestone, Sand, Shale & Flagstone with concretions of earthy Limestone.
	Horderley & May Hill Rocks about 2500 ft.	Conglomerates, Sandstone and Limestone.	Thinly bedded Limestone, greenish Sandstone, Thickly bedded red, white, green, & purple Freestone conglomeratic Quartzose Grits, & Gritty Limestone
	Builth & Llandeilo Rk. about 1200 ft.	Dark coloured Flags calcareous with some Sandstone.	
Q	Clay Slate from 8000 to 9000 ft.	Argillaceous indurated fissile Rks. Conglomerates & Limestone bands	Dark Flag slate coloured, dark Flag like Limestones, dark & soft Slate.
R	Granite	Crystallized Felspar with variable portions of Mica & Quartz.	Hornblend Rock & Slate, Schorl Rk. & Quartz Rock.
S	Syenite	But little known.	

	Names of the Formations &c.	Mineral Character	Subdivision
T	Greenstone	Felspar & Hornblend of a green, greyish green & black colour.	Porphry, compact Felspar &c.
U	Diallage Rk. Serpentine	Felspar & Diallage; when pure it is Serpentine when crystallized forms Diallage or Gabbro.	
V	Basalt & Trap Dyke	Auguites or Hornblend & Felspar.	
W	Millstone Grit & Limestone Shale.		
X	Cambrian Rocks	Plynlimmon Rocks, Bala Limestone Snowdonian & Skiddaw Rocks.	
Y	Lower Silurian Rocks	Caradoc Sandstone, Llandeilo Flags.	
Z	Mica & Chlorite Slate	But little known.	
♀	Upper Marine from 100 to 200 ft.	Bagshot Sand, London Clay, Plastic Clay & Sand.	
♁	Diluvial & Marsh Land		

NOTE.

Granite & Clay Slate	Produce Tin & Copper as at Tavistock &c.
Clay Slate	Produces Lead, Silver, Cobalt & Manganese as at Aberystwith & sometimes Copper as in the Isle of Anglesey.
Carboniferous Limestone & Millstone Grit	Produce Lead as in Derbyshire & sometimes Copper as at Ecton in Staffordshire & Iron as at Ulverston.
Coal Measures	Produce Iron in great quantities as in South Wales.
Magnesian Limestone	Produces Lead & Zinc.
New Red Sandstone	Produces no metallic ores of importance but contains beds of rock salt & brine springs.
Lias	Produces Alum & Iron Pyrites as at Whitby in Yorkshire.

Nose

Hd

Bill of Portland

London to Liverpool, by Inland Navigation.

	Miles	Miles
London Bridge up River Thames to		
Brentford	13¾	
enter Grand Junction Canal.		
Bull Bridge	6	19¾
Branch to Paddington 14 Miles.		
Uxbridge Wharfs	6	25¾
Rickmansworth Wharf	6½	32¼
Berkhempstead Dº	14	46¼
Bulbourn Junction	6	52¼
Branch to Wendover 6¾ Miles.		
Mersworth Bridge	2	54¼
Branch to Aylesbury 6 Miles.		
Fenny Stratford Wharf	14¼	68½
Cosgrove Junction	11¾	80¼
Branch to Buckingham 10¾ Ms. through Blisworth Tunnel 2980 Yds.		
Blisworth Wharf	9¼	89½
Branch to Northampton 4⅞ M		
Long Buckby Junction	13¼	102¾
Union Canals to Leicester 41 Miles through Tunnel 2072 Yards.		
Braunston Junction	4¼	107
enter Oxford Canal.		
Stretton Wharf	24	131
Longford Junction	11	142
Canal to Coventry 4¼ Miles.		
enter Coventry Canal.		
Marston Bridge	3½	145½
Canal to Ashby de la Zouch 30 Ms.		
Nuneaton	3¾	149¼
Atherstone	4¾	154
Fazeley Junction	10	164
Birmingham & Fazeley Canal to Birmingham 14¾ Miles.		
enter Fazeley Canal.		
Fradley Junction	11	175
Grand Trunk to Riv. Trent 26 Miles.		
enter Grd Trunk, or Trent & Mersey C.		
Rudgley	7¾	182¾
Haywood Junction	5¼	188
Staffordsh & Worcestersh. Canal to the Riv. Severn at Stourport 46½ Ms.		
Stone	9½	197½
Stoke Junction	7¾	205¼
Can. to Newcastle under Line 3½ Ms.		
Etruria Junction	2	207¼
Canals to Caldon and Uttoxeter 30 M. thro' Harecastle Tunl. 2888 Yds.		
Middlewich	18	225¼
through Preston Tunnel 1241 Yds.		
Preston Brook	16¾	242
enter Duke of Bridgewater's Canal		
Runcorn Junction	6	248
enter River Mersey.		
Liverpool	16	264

London to Manchester.

	Miles	Miles
to Preston Brook (Column 2)		242
enter Duke of Bridgewater's Canal.		
Altringham	16	258
Manchester	7	265

London to Birmingham.

	Miles	Miles
to Braunston Junctn. (Colmn. 1)		107
enter Oxford Canal.		
Napton Junction	7	114
enter Warwick & Napton Canal.		
Warwick Junction	14	128
Branch to Warwick ¼ Mile		
enter Warwick & Birmingham Can.		
Stratford Junctn. Canal (near Baderley Wharf)	7½	135½
Canal to Stratford 13 Miles.		
Birmingham	17½	153

London to Worcester.

	Miles	Miles
to Stratford Junctn. (near Baderley Wharf) as above		137½
by Stratford Junctn. & Stratford Can.		
King's Norton Junction	14	151½
Canal to Birmingham 5 Miles		
enter Worcester & Birmingham Can.		
Worcester	24	175½
Junction with Riv. Severn	1	

London to Bath, Bristol &c.

	Miles	Miles
London Bridge up River Thames to		
Kingston Bridge	20½	
Staines Dº	16½	37
Windsor Dº	8½	45½
Maidenhead Dº	6½	52
Great Marlow Dº	7½	59½
Henley Dº	8½	68
Junction of Riv. Kennet	8	76
enter River Kennet.		
Reading	1	77
Newbury	18½	95½
enter Kennet & Avon Canal.		
Hungerford	9	104½
Great Bedwin	5	109½
Devizes	20½	130
Semington Junction	7	137
Wilts and Berks Canal to the River Thames near Abingdon 52 Miles.		
Trowbridge	3¼	140¼
Bradford	2½	142¾
Bath	9¼	152
enter River Avon.		
Bristol	17	169
River Severn	9½	178½
up Riv. Severn to Gloucester 45 Ms.		

London to Gloucester. and Shrewsbury.

	Miles	Miles
to the Junction of the River Kennet (Column		76
Wallingford Bridge	17	93
Culham Dº (near Abi… ngdon)	12½	105½
Branch to Abingdon 1 Mile		
Oxford	8	113½
Oxford Canal to Braunston … 57½ Ms.		
Lechlade	29	142½
Inglesham Bridge	½	143
enter Thames and Severn Ca… nal.		
Cricklade	7¾	150¾
Siddington Junction	5¾	156½
Branch to Cirencester 1½ M… iles.		
Brimscombe Port	12¾	169¼
Stroud Bridge	2¼	171½
Walbridge Junction	¼	171¾
enter Stroud Canal.		
Wharfs on River Sev… ern	8	179¾
enter River Severn.		
Gloucester	10	189¾
Tewksbury	11½	201¼
Worcester	17½	218¾
Canal to Birmingham 31½ M… iles.		
Stourport	13	231¾
Canal to Kidderminster 4 M… iles.		
Bewdley	4	235¾
Bridgenorth	14	249¾
Colebrokedale	9	258¾
Shrewsbury	18	276¾

London to Leicester and Nottingham.

	Miles
to Braunston Junctn. (Colmn. 1)	
enter Grand Union Canal.	
Foxton Junction	23
Union Canal to Market Harborough 5 Miles.	
enter Union Canal.	
Leicester	17
enter Leicester Navigation.	
Junction with River Wreak	6
Navign. to Melton Mowbray 10 Miles.	
Loughborough Junction	8
enter Loughborough Canal & Navig.	
Junction with Trent Navn.	9
Trent Navigation to Grand Trunk Canal at Wilden Ferry 2¾ Miles.	
enter Trent Navigation.	
Nottingham	8

London to Derby.

	Miles
to the Trent Navn. (as above)	
cross Trent Nav. & enter Erewash C.	
Derby Canal Junction	2
enter Derby Canal.	
Derby	8

Note. Although the preceding Rou… necessarily limited to the principa… most direct from London; yet it is… med that most of the chief cross co… cations, will likewise be made out wi… lity by means of the Branches from… aided by the annexed List.

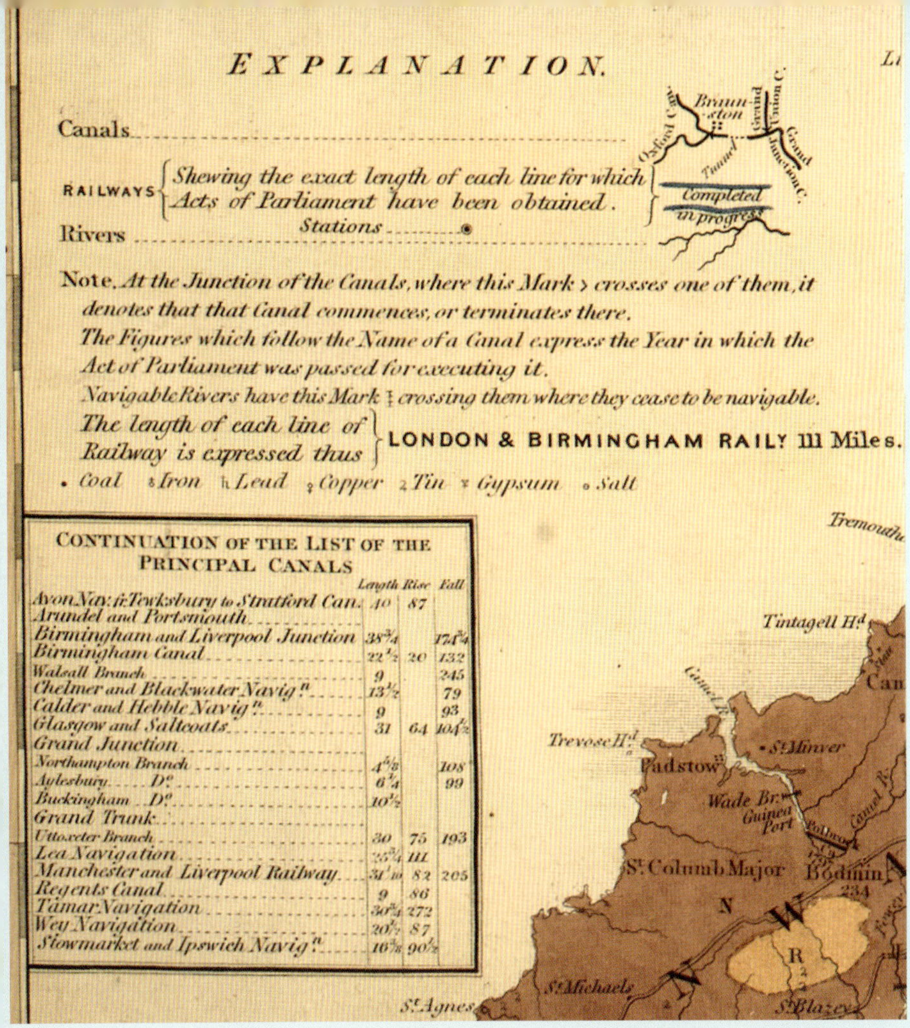

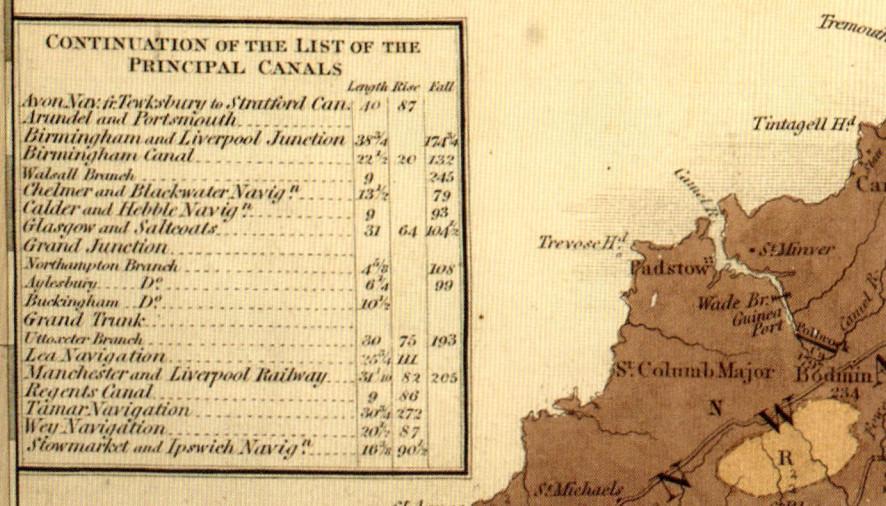

Moreover, canals were not cost-free for maintenance or operation.

Improvements in the transportation of bulk goods cut the price of coal, and thus expanded demand, which was bad for any mining areas with poor transport, such as the Fife coalfield. This development accentuated the significant differences between coal-producing areas as well as within them. The spread of the railway ensured from the 1840s a more reliable supply of coal to London, which was by far the leading market in Britain.

With the growth of the rail network, it ceased to be common to map it against a geological background, as the network had become too extensive. Moreover, in terms of construction, terrain and not geology was the key issue. Linked to this, coal and iron were frequently marked not with deposits but in terms of specific sites such as mines.

THE IMPORTANCE OF THE CIVIL ENGINEER

James Wyld, 'Plan Shewing the Proposed Line of the London and Greenwich Railway', 1832.

James Wyld the Elder (1790–1836), the publisher of this map, was probably the leading cartographic entrepreneur of the age, and a man of more general energy with a great commitment to progress through change. Geographer to the King, William IV (r.1830–37), and a founder of the Royal Geographical Society in 1830, Wyld was a mapper-of-all-works who had taken over William Faden, the leading London mapmakers, in 1823. Wyld did not specialise in railway maps (a developing and profitable world), although his son, James the Younger (1812–87), was to produce *Wyld's Great Western Railway Map* (1839). As a reminder that such activity was not free from difficulties, James the Elder was also involved in legal cases with unsuccessful railway companies and he produced maps of the London system that included planned lines not actually built. That James the Elder was associated with the London and Greenwich Railway indicated the importance of this project.

The map included the existing land use, not least in terms of buildings, roads, waterways and principal buildings, with the proposed line lightly drawn on this background. Indeed, visually, the railway had only a limited prominence; yellow was for roads and blue for waterways, but no distinctive colour was associated with the railway. The inclusion of a scale was useful, while the compass direction explained that the alignment did not put north at the top, a choice presumably taken in order to align London Bridge and Greenwich on a west–east axis. The text printed on the map helped explain the route, needed most notably in Bermondsey, close to the London terminus, where existing land usage was most crowded and complex.

The picture included was also useful in providing an explanation of the viaduct nature of the railway, which reached Deptford in 1836 and Greenwich in 1838. An elevated line avoided the need for level crossings, and the risk both of flooding – which was serious prior to the efforts made to control, indeed canalise, the River Thames – and of people or animals straying onto the line, which was

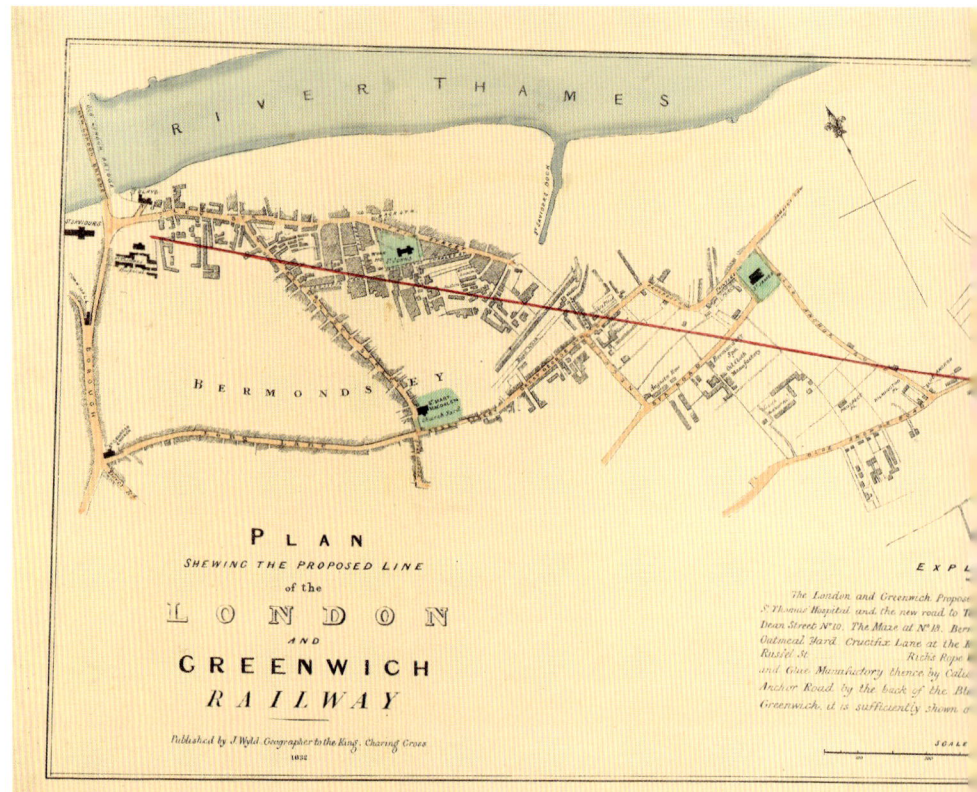

a major issue until rail-observance practices and precautions developed.

The key figure in the project was George Walter, who ran an insurance company that was the base of the new rail company for which he served first as Secretary and then as Resident Director (managing director). The engineer was Colonel George Thomas Landmann, a military veteran of the Peninsular War (in 1808–14, against the French in Spain and Portugal) who had retired from 1824, and then, as an active civil engineer, was prominent in the London and Greenwich Railway and the Preston and Wyre Railway and Harbour Company, among much else. Landmann was typical of the role played in the development of railways by those with military experience, a feature also seen in America.

Hugh McIntosh, the contractor, worked more on canals and docks, but he brought the project to fruition, with the *Gentleman's Magazine* declaring in 1837: 'This great

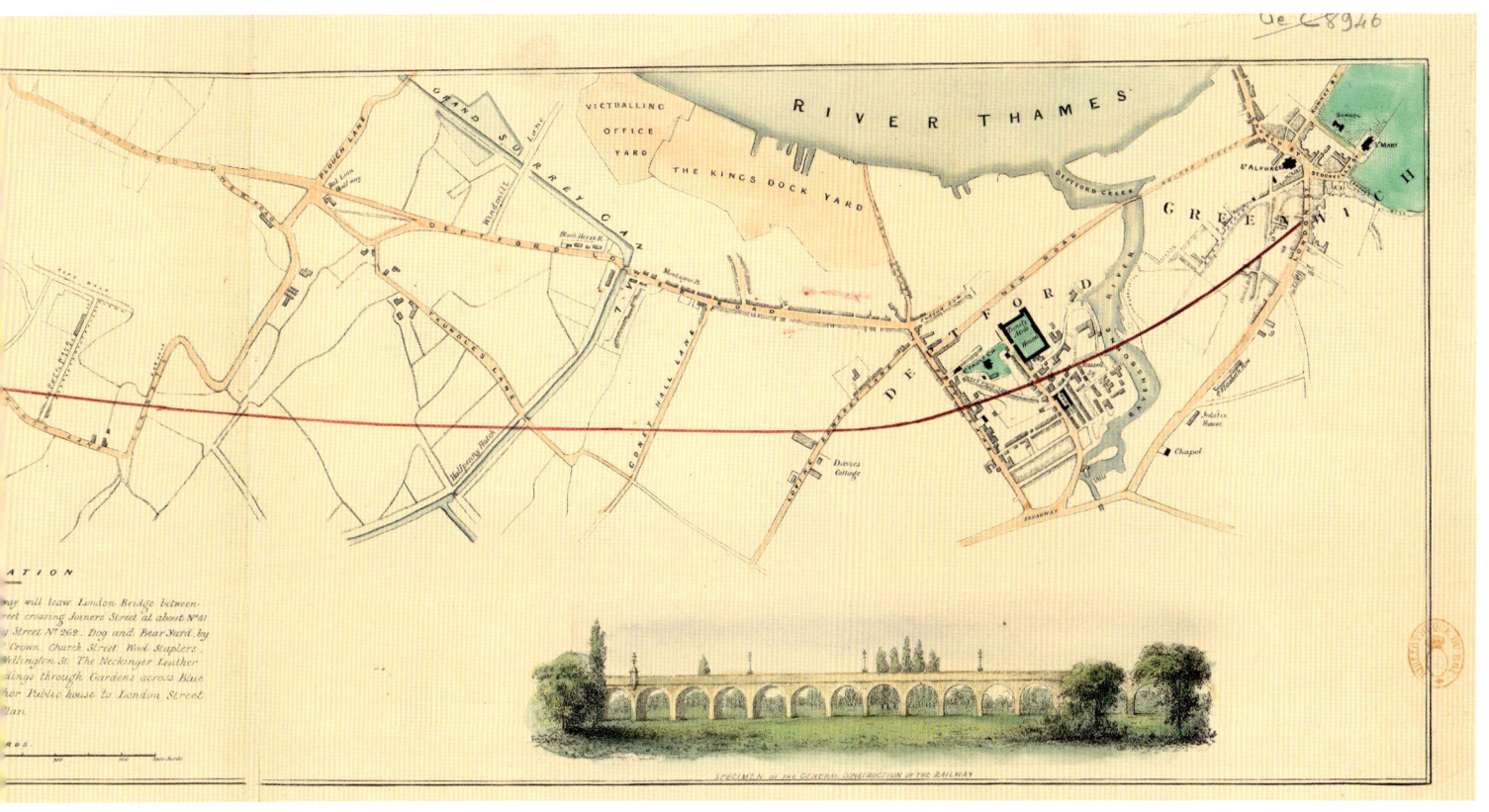

national work reflects the highest honour on the gallant proprietor, Colonel Landmann, no less credit on the contractor, Mr Macintosh, under whose orders no less than 60,000,000 bricks have been laid by human hands.'

The map provides little indication of the difficulties of the task, not least those posed by Deptford Creek. For the train to cross, a balanced bridge was required so that masted ships could still pass up the Creek, but that bridge was labour-intensive to operate and, separately, there were problems with the foundations of the line in this area. There was also experimentation on the line with the locomotives, which saw for the first time horizontal cylinders mounted at the front outside the frame.

The number of passengers rose to over two million in 1844. This indicated the commercial appeal of suburban rail services. However, the line made a loss due to its heavy indebtedness from the original capital expenditure. This was a problem for all rail companies, and more particularly the small ones. The plan had been to press on from Greenwich to Gravesend and also, separately, to Dover, but this was rejected by Parliament in 1836. This rejection limited prospects, and the possibility of raising fresh equity.

As earlier with canals and turnpikes, Parliament proved the major arbitrator of the new system. This helped stabilise the action-reaction processes of policy formation and also shape both lobbying and government responses. This made recourse to Parliament more predictable and attractive for rail companies. More generally, public policy was important in offering a safe context for the time horizons required by investment in significant infrastructure. In turn, that policy, which helped sustain low interest rates, hinged greatly not just on the law but on its enforcement via the courts and parliamentary scrutiny and observance. Maintenance of contract extended widely, which encouraged investment in railways.

1. ORIGINS 25

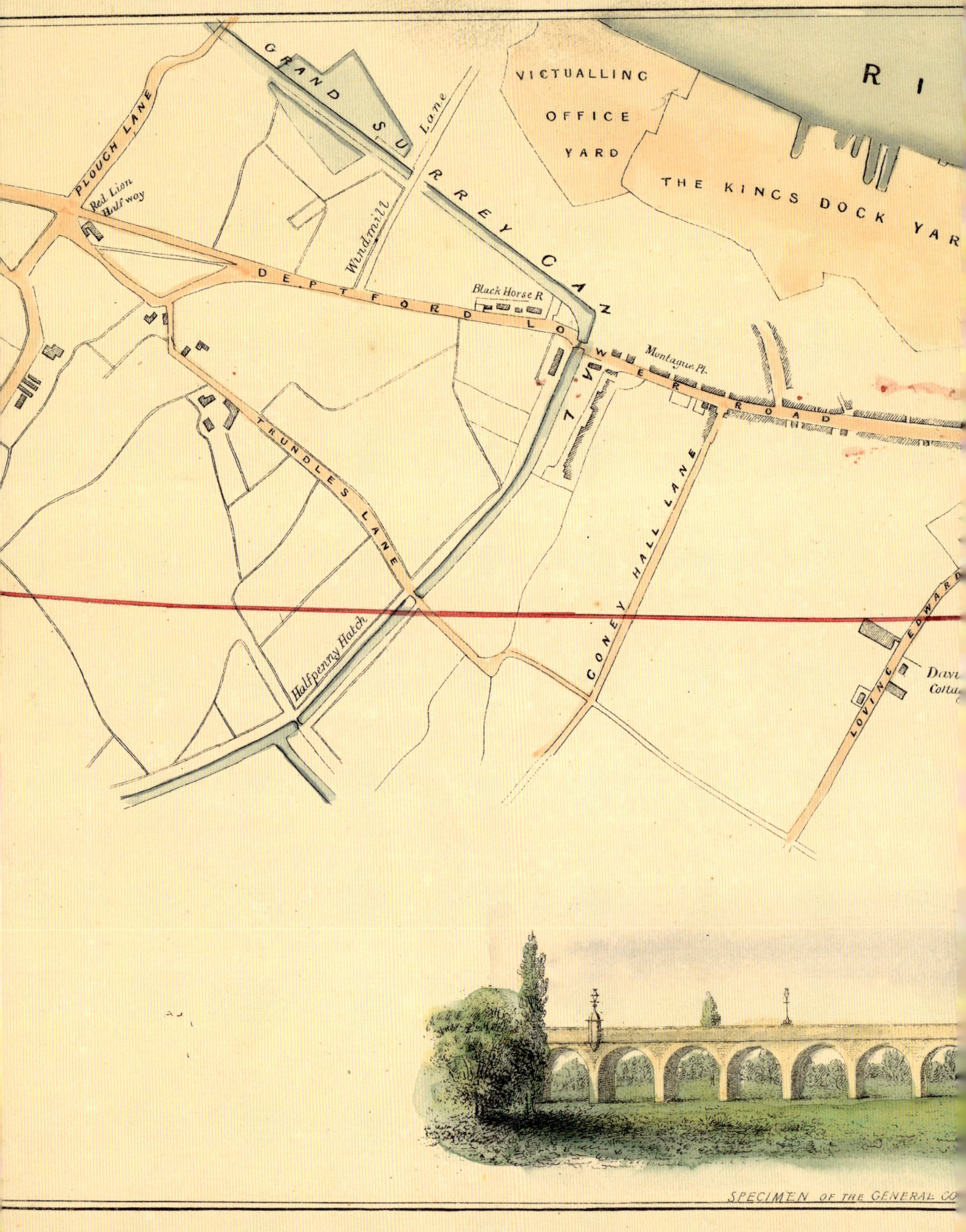

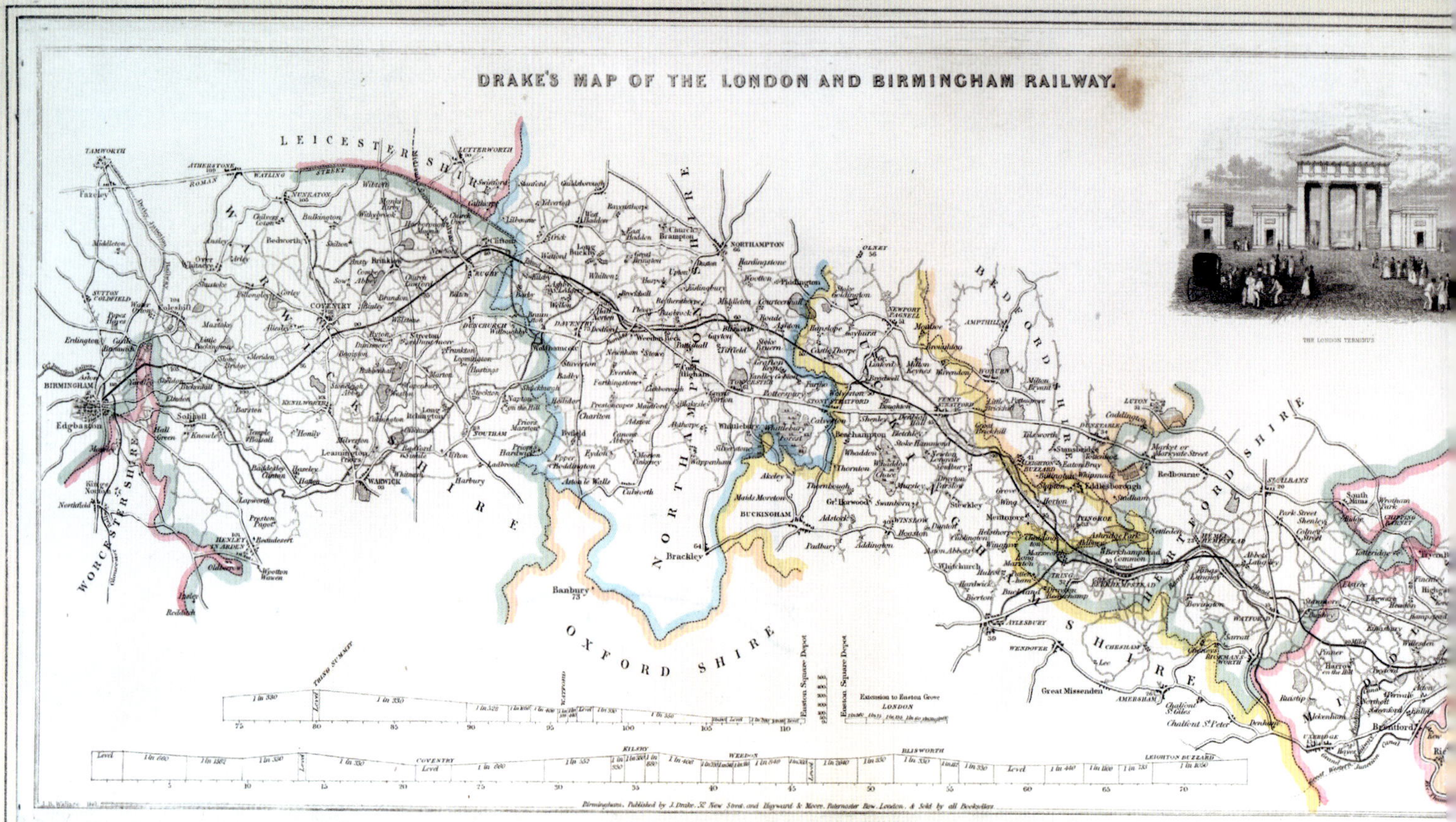
DRAKE'S MAP OF THE LONDON AND BIRMINGHAM RAILWAY.
THE LONDON TERMINUS
LEICESTERSHIRE
WARWICKSHIRE
NORTHAMPTONSHIRE
BEDFORDSHIRE
HERTFORDSHIRE
OXFORD SHIRE
WORCESTERSHIRE
BIRMINGHAM
COVENTRY
NORTHAMPTON
BUCKINGHAM
Euston Square Depot
Extension to Euston Grove LONDON

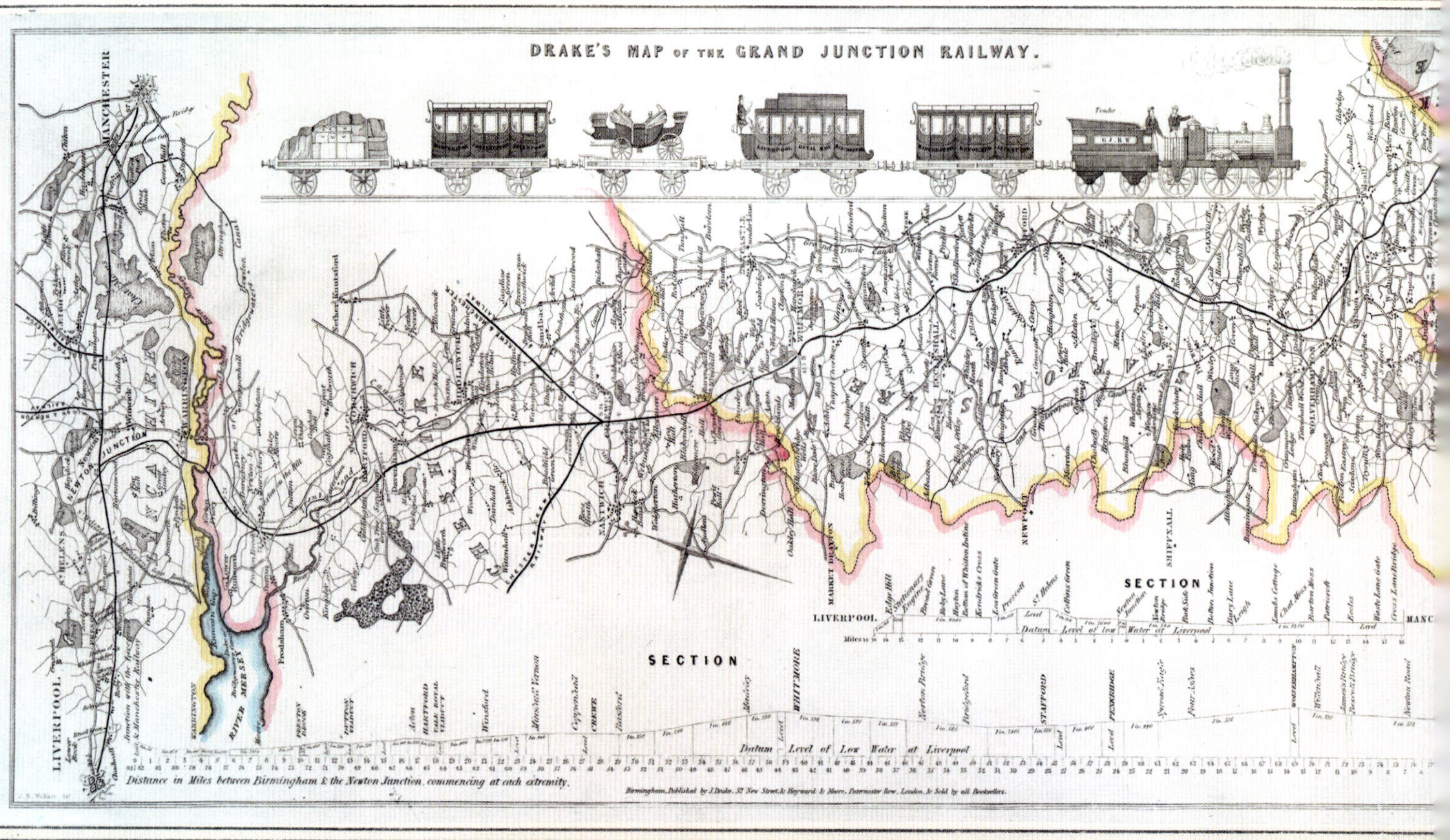
DRAKE'S MAP OF THE GRAND JUNCTION RAILWAY.
MANCHESTER
LIVERPOOL
SECTION
Datum Level of Low Water at Liverpool
Distance in Miles between Birmingham & the Newton Junction, commencing at each extremity.

A CORE LINE

J.B. Wallace, 'Drake's Map of the Grand Junction Railway', 1839.

Opened in 1837, the Grand Junction Railway (GJR) ran from Birmingham via Crewe and Warrington to join the already established Liverpool and Manchester Railway at Newton Junction. This route meant a longer journey from Birmingham to either Manchester or Liverpool, but it was less expensive to build to that junction than to provide two direct routes. The northern end of the route to Newton entailed taking over (purchasing in 1835) the Warrington and Newton Railway, a branch from the Liverpool and Manchester Railway, which had opened in 1831. The GJR established its main railway works at Crewe, where it built its first locomotive in 1843.

The map includes both a pictogram depicting a train, the locomotive *Wild Fire*, with carriages and wagons, and a mass of information. John Stagg of Birmingham constructed *Wild Fire* in 1839, with the prototype built in 1837.

The map appeared as one sheet of a double-sheet that included 'Drake's Map of the London and Birmingham Railway', and thus the route from London's Euston to Liverpool and Manchester. Indeed, the southern sheet carries a view of 'The London Terminus', which was the arch at Euston Station.

The compass indicator (on the Birmingham–Lancashire section) indicates that the map is configured to the northwest, which was the basic alignment of the line. The scale is five miles to the inch horizontal. The map includes diagrams showing a level section of the railway (the line in height), as well as distances. County boundaries are outlined in colour, and they would have helped readers to clarify the location.

James Drake, the publisher, was based in New Street, Birmingham, and the map was drawn by J.B. Wallace and engraved by J. Archer. The map was sold with a sheet bearing timetables, distance tables and details of fares. A First Class journey from Birmingham to Liverpool cost 21 shillings and took four and a half hours.

Drake had written and published *Drake's Road Book of the Grand Junction Railway* in 1838. The second edition of this was illustrated with numerous engravings and included 'The Visitor's Guide to Birmingham, Liverpool and Manchester', as well as the revised version of the map in the first edition, the revision that appeared separately in 1839. The interest in the railway was also seen in the 1839 illustrated book *The London and Birmingham Railway* by Thomas Roscoe, a prolific writer.

The line was a prime area of population growth. London was the major city and the leading recipient of migrants within the British Isles. Yet, industrial and commercial towns in the Midlands and North became in the nineteenth century relatively far more prominent, notably Birmingham, Liverpool and Manchester, although also Leeds, Newcastle and Sheffield. In his *A Dictionary Geographical, Statistical, and Historical, of the Various Countries, Places, and Principal Natural Objects in the World* (1849), J.R. McCulloch referred to Manchester as 'situated close to an almost inexhaustible coalfield' and as 'the first manufacturing town of the world'. Whereas in 1801, Liverpool's population stood at 82,000, within fifty years this had grown to 376,000.

The depiction of railways in the map was one of a number of linear transport means, the others being rivers and roads. Indeed, the early mapping of rail drew on the need to show new turnpike roads and canals, and to relate them to existing settlements. The mapping of roads as a result of the turnpikes was a crucial background to that of rail, while the detailed knowledge of local topography shown in canal construction was applied to the railways. Surveying was a key skill for canal construction, with terrain, slope profile, drainage and soil type each highly significant at the local level, and therefore affecting more general issues of the viability of particular routes. Just as there was continuity in mapmaking, so the same individuals acted as surveyors for these different transport types.

The Grand Junction route was one of the core lines in England. Highly profitable, it became (through a merger) the basis of the London and North Western Railway (LNWR), formed in 1846. This fundamental character very much remains the case. Thus, the current plans for a high-speed line, HS2, in Britain addresses the London to Birmingham route, although the Manchester extension has been ditched.

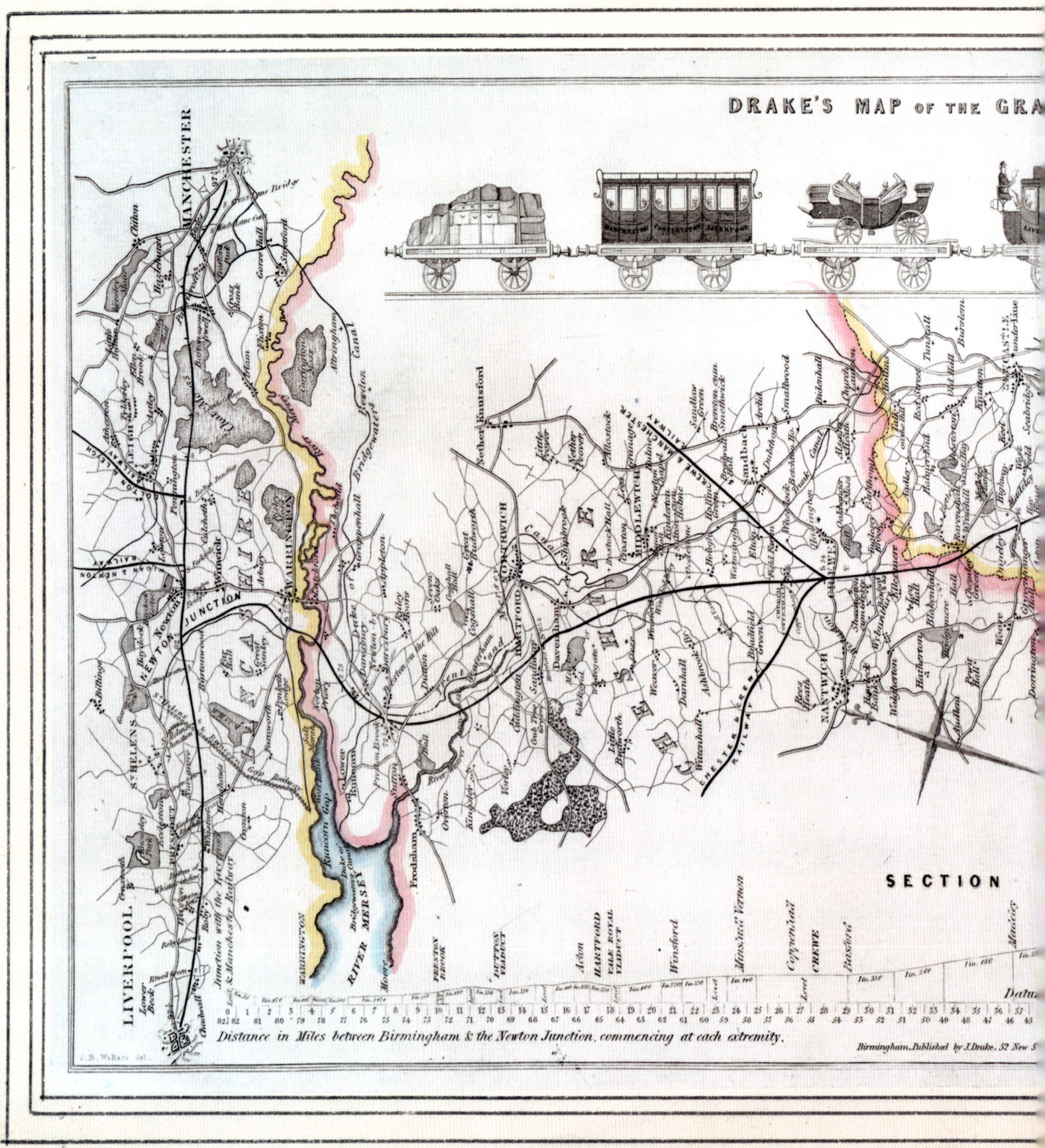

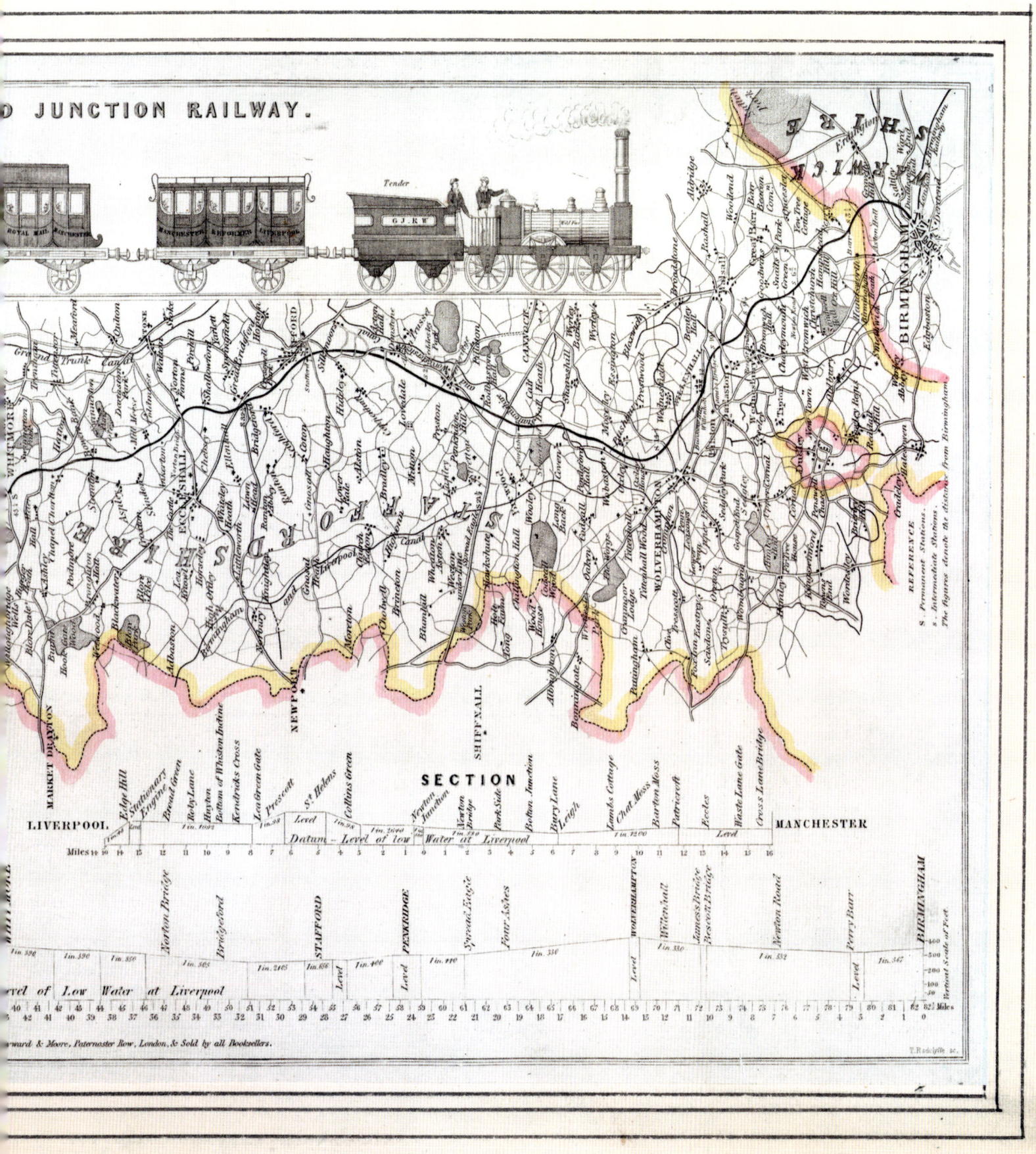

1. ORIGINS 31

A SYSTEM EMERGES
Charles Fowler, 'Fowler's Map of the Railways in Great Britain', 1841.

This map presents a system in the process of development, with the lines overpainted by hand in red already constructed and those in plain black not yet laid. The map was published in *Fowler's Railway Traveller's Guide*, which appeared in Leeds in 1841. Charles Fowler, a Yorkshire surveyor and engineer, was also responsible for railway surveys – for example for the Scarborough–York line, the plan of which was produced in 1841 and is marked on this map. Shown overpainted in blue, this line was to terminate at a junction at York with the York and North Midland Railway (Y&NMR), which is shown in red. In 1819, Fowler published a plan of Leeds and in 1831 one of Leeds and its environs. He was later described as an architect and valuer as well as a surveyor.

Fowler was not to have the success of the Manchester-based George Bradshaw, a noted engraver and printer, largely of maps, who in 1839 produced the world's first compilation of railway timetables, a title that continued until 1961 when the last edition of the British guide was published. However, Fowler's map was impressive in showing the extent to which the early lines left much of the country uncovered while concentrating on links between key ports.

The railway focus on the coalfields is readily apparent. The railway 'system' in Britain followed the pattern seen earlier with canals and turnpikes, which stood in contrast to the situation elsewhere in Europe. In Britain, the co-operation of parliamentary authorisation and individual entrepreneurial bodies reflected the absence of a centrally directed national policy, let alone a body such as a transport ministry. The possibility of creating companies was thus a permissive national policy in Britain, and not a prescriptive one. Rather than following some master plan, the rail system came in large part to reflect the ambitions and dynamism of individual companies (as earlier with particular turnpike and canal trusts), and what they hoped would be the ability of specific routes to produce revenue, either directly or as part of a network.

Thus, the potential role of lines in intra-regional communication was seen as significant. These connections helped to constitute a national system. However, just as with canals and turnpikes, this was due to perceptions of commercial opportunity in defining necessary and profitable links, and not to national planning, whether or not directed for economic purposes.

Parliament oversaw the system. This oversight involved the interaction of local, regional and national interests and views; and all in terms of a sense of opportunity, public activism, financial commitments, specific economic interests and contention over location. The complex links between regulation, oversight, arbitration and representation at the national and local levels were encouraged by legal debate and political strife.

The 'mania' of speculation in 1844 was to bring ambitious expansion, a process eased by the encouragement of company formation with the Joint Stock Companies Act of 1844. This replaced company formation only by royal charter or Private Act of Parliament. Moreover, whereas canals had generally only had investment in the form of large shareholdings and from those close to the user and operator of the canals, railways and mines attracted small investors. Developing and sustaining their interest was in part done by boosterism, both in company reports and in newspaper coverage. Because expansion of the system reflected competition for actual and anticipated markets, duplication was an aspect of this, as happened with Epsom, which became a London commuter settlement. The London, Brighton and South Coast Railway came to Epsom in 1847, the Epsom Downs Branch line in 1865 and the London and Southwestern Railway (LSWR) in 1859.

The density of stations and intensity of services also increased. Thus, Ealing Haven (later Broadway), opening in 1838, was the first halt on the Great Western Railway (GWR) line west from Paddington, the next stops westwards being Hanwell and then Southall, with trains stopping at each six times per day in both directions. Over time, the frequency of service increased and more stations were built on this line: Acton in

1868 and later West Ealing. There was also a branch line from Southall to Brentford, which opened in 1859 and remained until 1942.

Meanwhile, the stock market had developed and by 1840 the capital of British stock companies was about £210 million. In turn, the attraction to shareholders was greatly increased by the Limited Liability Act of 1855, which increased the willingness of the public to hold rail shares and thus the possibilities for capital formation for new projects. Limited liability also proved of benefit for the private engineering companies producing hardware and components (for example, lights) for railways, including locomotives.

The early start of rail in Britain left it with specific geographical characteristics. The first locomotives had relatively little power compared with what was to come, and they could not haul trains up steep gradients. As a result, British rail construction went for what by later standards were particularly flat lines. That encouraged the avoidance of terrain features, and therefore the need for a grasp of the geographical situation that was provided by accurate maps.

FRANCE ENTERS THE COMPETITION
'Railway from Paris to Marseille, the part Between Avignon and Marseille', 1842.

This map, cartographer unknown, from 1842 was one of several to accompany the various proposals discussed in a government evaluation of the options for an important section of the rail system, that for linking the interior of France to the Mediterranean, and thus bringing to fruition the Rhone route southwards. This proposed line would provide a link from Paris to the Mediterranean, and thus for a north–south route through France. Taking a railway on to reach the Mediterranean also meant avoiding the need for the cost and disruption of trans-shipment onto river, canal or road at Avignon. That had been a major cause of the development of road links to Marseille. The railway thus maximised the transport potential of the route. This had even greater significance because Marseille was the port for France's developing presence in Algeria, where an expeditionary force had seized the port of Algiers in 1830, beginning a lengthy process of conquest.

In part seeking to catch up with Britain, the French government of Louis-Philippe, which emulated Britain as well as competing with it, authorised nine major lines in 1842 and a shared public-private agreement was reached. The government was to provide the land and prepare the road-bed, while the companies built and ran the lines.

Whereas the actual choice of route between Lyon and Avignon posed few problems arising from the hilly topography and tributary valleys at either side of the Rhone Valley, the situation thereafter was different. The route between Avignon and Marseille had to tackle low-lying, inundated terrain; the very extensive delta of the Rhone and its swampy environs; and also hilly terrain. There was also the need how best to establish links with existing towns, notably Arles and Aix, existing waterways, notably the Canal d'Arles, and roads. Three alternatives were distinguished by colour, and there was both a scale and the inclusion of pertinent material. The numbers printed alongside towns indicated their population.

In the event, in 1842, the decision was taken for the route via Arles. A key supporter in the National Assembly was Alphonse de

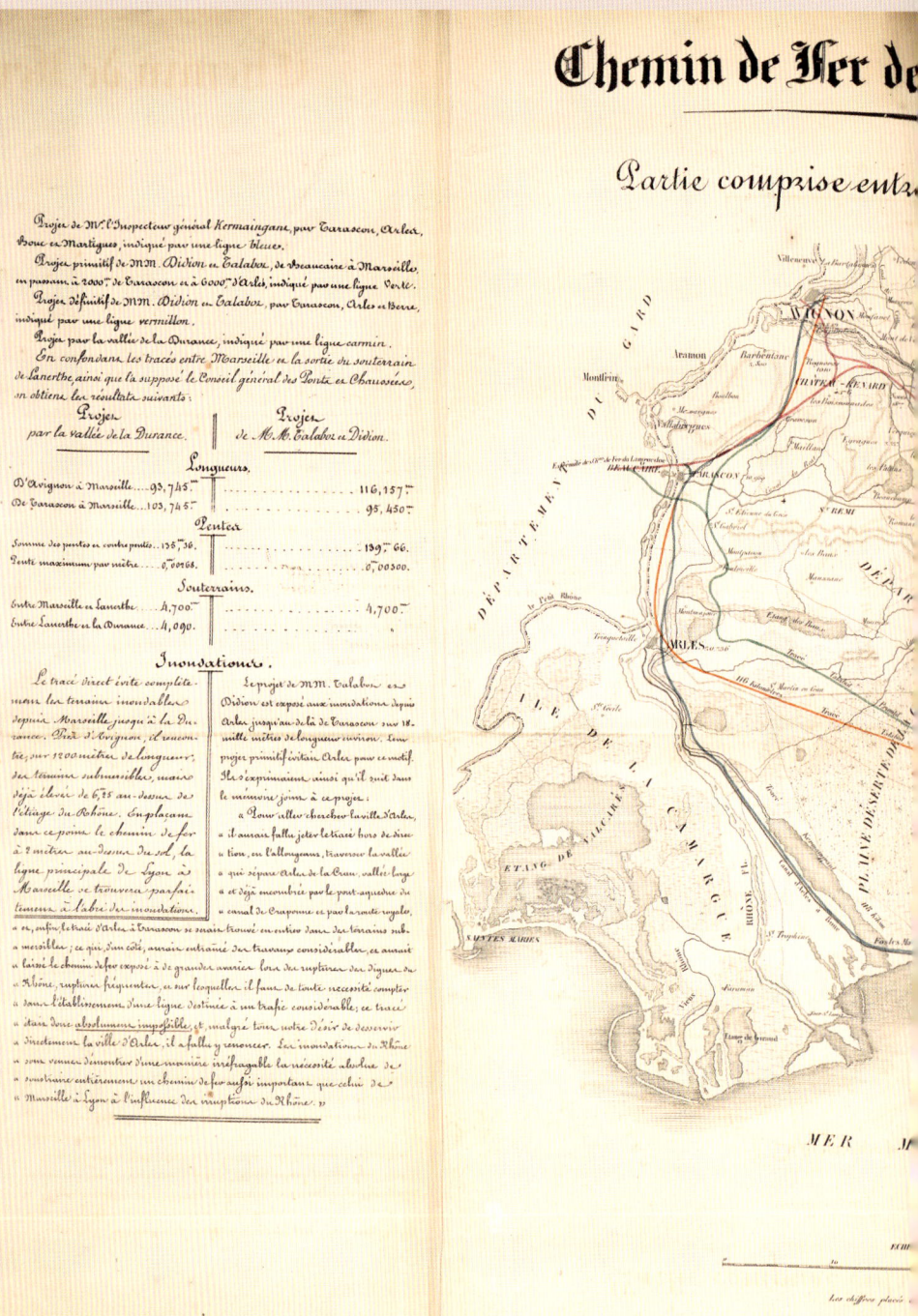

Lamartine, who was to be rewarded in 1848 by gaining the National Assembly constituency of Bouches-du-Rhône. The line (marked in red) went from Avignon across the River Durance via Tarascon (from where there was a branch to Nîmes, further west), and then to Arles, where the station was opened in 1848. It then turned southeast to St Chamas (the location of a nearby ordnance factory), passing along the northern shore of the Étang de Berre, a brackish lagoon, to reach Marseille-Saint-Charles station, which was opened in 1848 on the site of a former cemetery.

The first section of the line built was opened in 1847 and it led from Rognonas south of Avignon, to Pas-de-Lanciers near Marseille, with the extension to Marseille itself following in 1848. As today, building in and close to towns was more difficult due to more fragmented landownership as well as more intense land use. In 1849, Avignon was connected with Rognonas. The entire line to Paris was completed in 1856.

Louis-Napoleon Bonaparte, president from 1848 to 1852 and emperor as Napoleon III from 1852 to 1870, was a keen supporter of the scheme. His uncle, Napoleon I, while exiled on St Helena, had told Lieutenant Colonel Wilks, who was governor there: 'Your [British] coal gives you an advantage we cannot possess in France.' Napoleon III worked hard at countering this disadvantage.

As an important backing for the line from Avignon to Marseille, the Parc des Ateliers in Arles was developed from 1844 as a construction and repair site for locomotives belonging to the Avignon-Marseille rail company, which in 1852 became the Chemin de fer de Lyon à la Méditerranée.

The idea for a system radiating from Paris, which was essentially political in character, vied with others focused on economic needs, for example linking Boulogne to the textile and steel industries of Champagne, and thence to the Saône-Rhône axis, crossing, in Champagne, with an axis between Alsace and the lower Seine.

SEEKING SECURITY AGAINST PRUSSIA
Innsbruck Redlich, 'The Railway from Innsbruck to Munich', 1859.

Finished in November 1858, the line between Innsbruck, the capital of the Austrian duchy of Tyrol, and Munich, the capital of the kingdom of Bavaria, was celebrated in this route map, which is surrounded by illustrations of prominent sights en route, as well as aspects of railway construction, including the tunnel at Rattenberg and the bridge over the Alpine-fed River Isar. The map explains the route that was followed, one that took advantage of the valley from Rosenheim to Innsbruck. More direct routes from Munich south to Innsbruck would have entailed crossing very direct Alpine terrain and that would have posed major difficulties for locomotives in pulling loads uphill.

As elsewhere in Germany (see pages 132–133), power politics was a key element, not so much in the details of the route, but rather in the motivation and general alignment. There was historic tension between Austria and Bavaria. This included conflict in the 1700s and 1740s, Bavarian concern about Austrian expansionism in the 1780s, and, as a result of Bavaria's alliance with Napoleon I, the Bavarian occupation of Tyrol in 1805–14 – an occupation that had led to a bitter rebellion. However, the situation changed in the 1850s, with joint concern about Prussian ambitions. This led to the decision to link the Bavarian and Austrian networks. The Bavarian line between Munich and Rosenheim was completed in 1857, while the Lower Inn Railway in Austria was built in 1853–58. Under the treaty of 1851, Bavaria agreed to build railways from Munich to Salzburg and from Rosenheim to Kufstein; while Austria agreed to build a line from Salzburg to Bruck an der Mur and a line from Kufstein to Innsbruck.

A different range of material was offered in the map (see below) of the Saxon-Bavarian Railway Company that ran from Leipzig to Hof, the first section of which was opened in 1842. Construction was approved in 1840 and started in 1841, with the line reaching Altenburg in 1842. The drawings show the two dramatic viaducts required to cross the Göltzsch and Elster valleys, one 598 metres tall and the other 280 metres. These proved difficult and expensive to build, resulted in financial difficulties and led to Saxony buying the private company in 1847. That section was opened in 1848 and the last section in 1851. Again, like Bavaria and Austria, this was a railway outside the Prussian system.

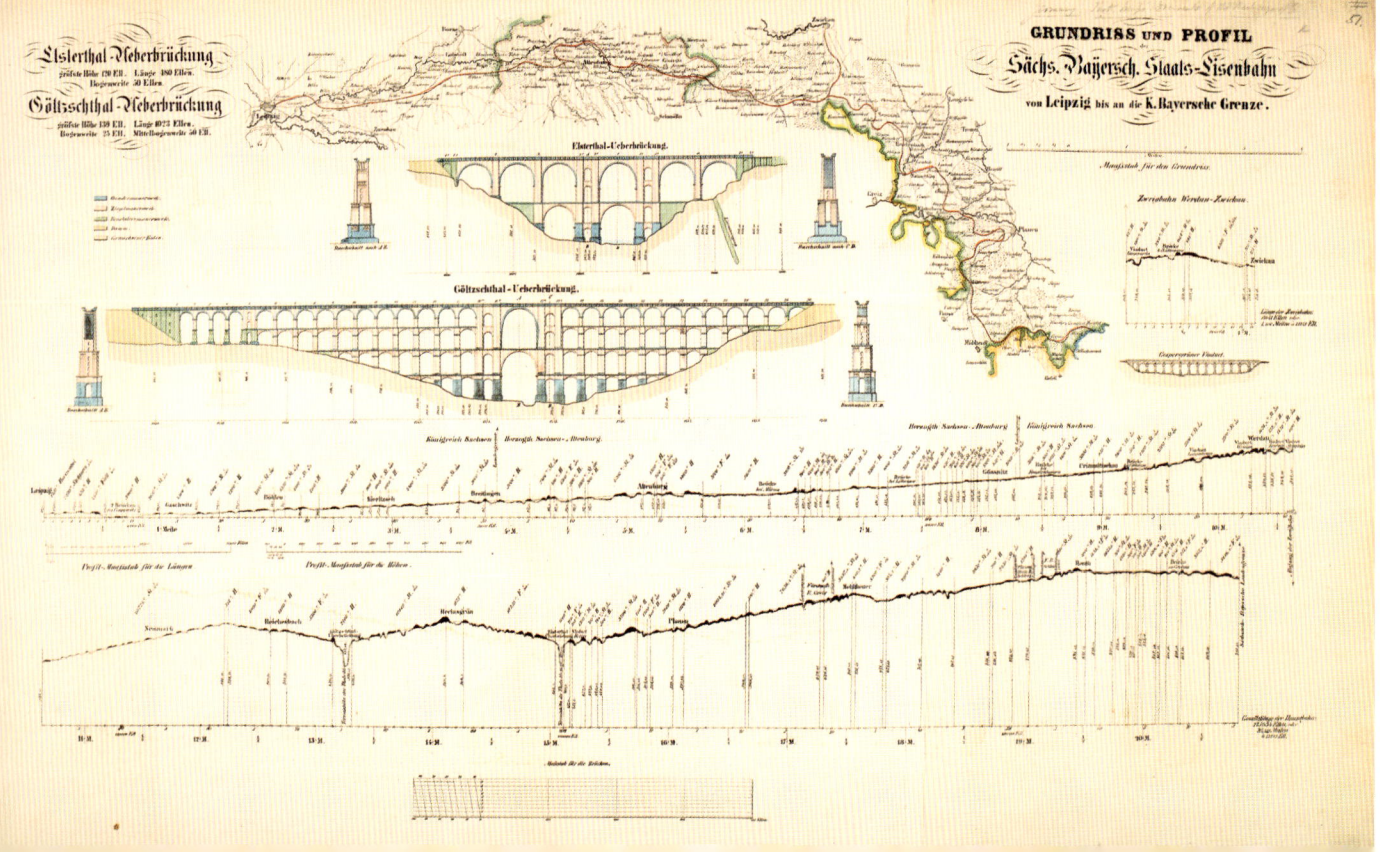

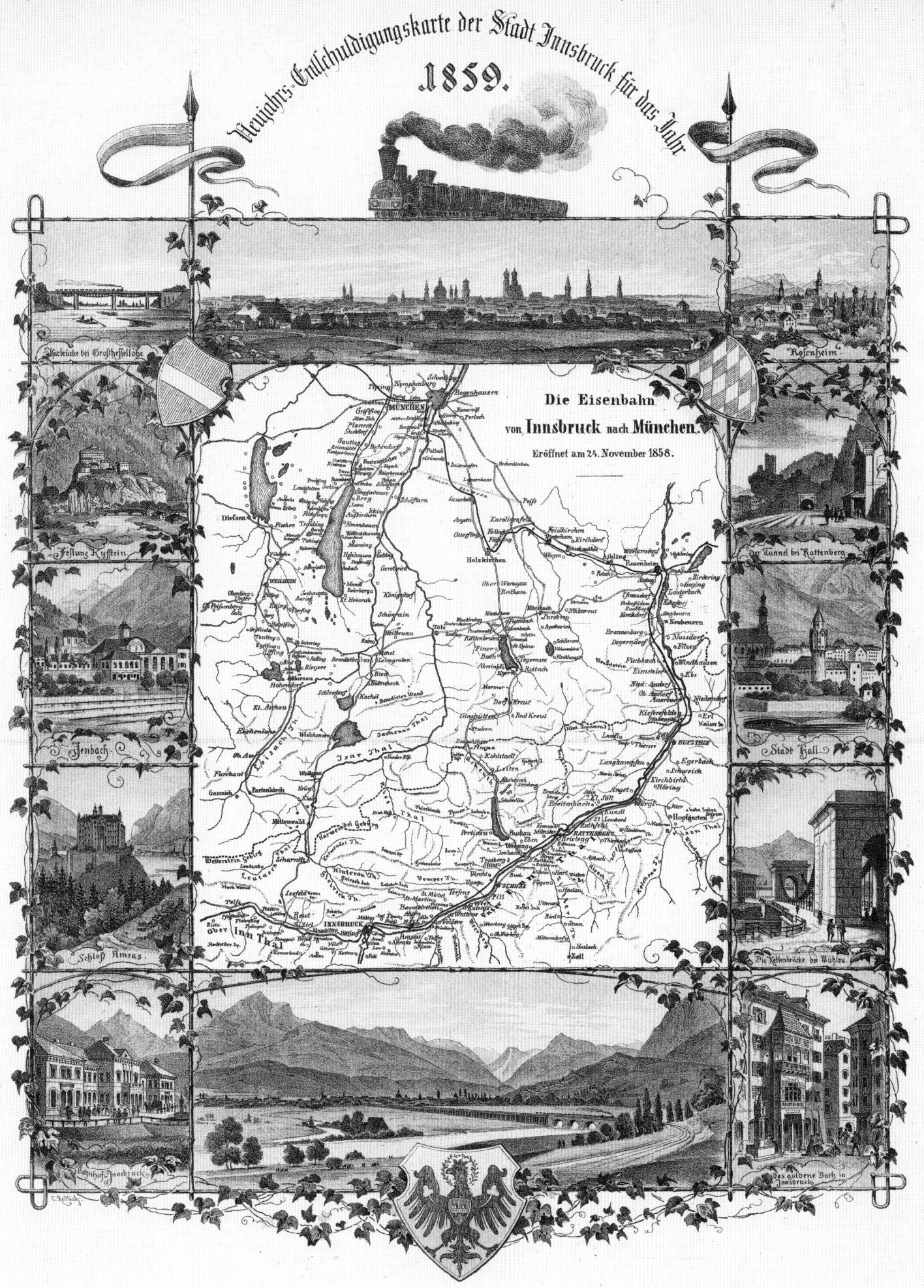

THE ORIGINS OF AMERICAN RAILROADS

Henry Schenck Tanner, 'Map of the Canals & Rail Roads of the United States', 1830.

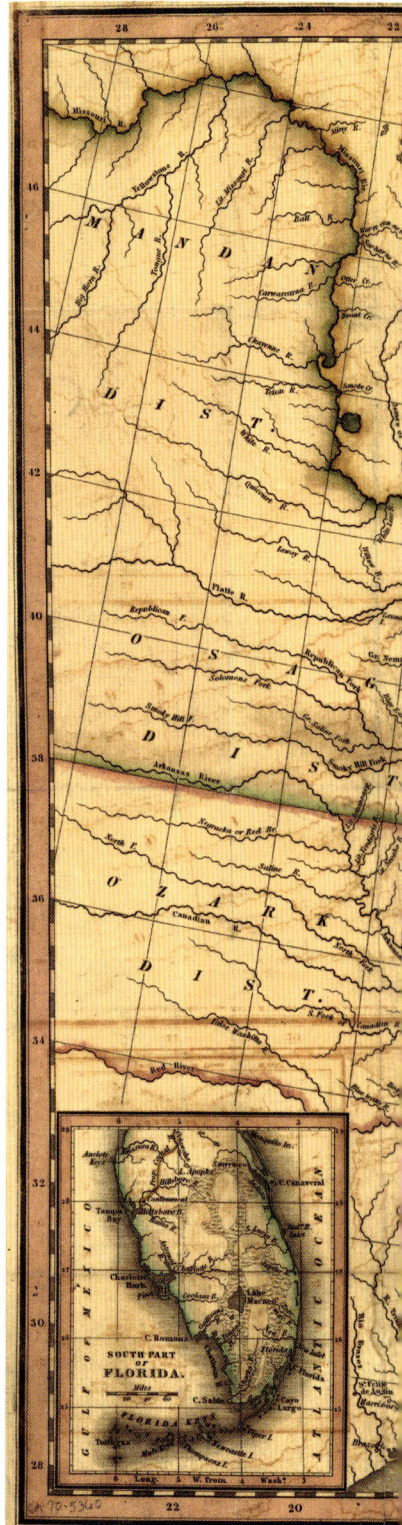

The earliest American map showing railways dates from 1830. The working lines are in blue, while proposed lines are in green; canals are shown in red and yellow. New York-based Tanner (1786–1858) was a prolific cartographer and he also produced that year a 'Travellers Pocket Map of Virginia' with the same information. However, his map, like so many produced for a system that was in a process of rapid development, and where reliable information was limited, was more ambitious than accurate. In particular, Tanner anticipated success, showing the Columbia–Philadelphia and Harpers Ferry–Winchester lines as operational, although they did not begin running trains until 1832 and 1836 respectively, and the Winchester and Potomac Railroad (W&P) was not surveyed until 1831–32. Moreover, the Baltimore and Ohio Railroad (B&O) did not reach Harpers Ferry until 1834. It is instructive that the four sections that were included in this map were all of canals, and not of railways, because that showed the focus of attention.

The trajectory of early American railways was similar to that in Britain, with horse-drawn railways coming first in the shape of the three-quarter-mile-long Leiper Railroad in Philadelphia in 1810–29, and a three-mile-long Granite Railroad in Quincy, Massachusetts. The use of locomotive steam engines then followed the news, example and export of British success.

Incorporated in 1827, the B&O opened its first 13-mile stretch in 1830, and started regular scheduled passenger services. Initially a Maryland railway, this continued through Virginia from 1837, reaching the Ohio River at Moundsville, Virginia, in 1852. The building of this 380-mile-long line across the Appalachian Mountains was a major and costly achievement for the American engineers, Irish navvies and Anglo-American financiers, not least in the absence of any comparable experience. Baltimore lacked a valley route into the interior, and could not take a northern route because Pennsylvania did not wish to encourage competition for Philadelphia. The Ohio River Valley in turn linked up to the Mississippi river system, providing an instance of the linkage between rail and steamship.

Boston's centrality to the developing system in Massachusetts had not been inevitable. In 1824 Worcester, Massachusetts, was a major centre of manufacturing and transporting goods to Boston. However, Irish labourers were brought over to build the Blackstone Canal from Worcester to Providence, Rhode Island. Almost as soon as it was finished, Massachusetts embarked on building a railway from Boston to Worcester with the same Irish labourers. This, the Boston and Worcester Railroad, effectively put the canal out of business. The next leg was called the Western Division Railroad from Worcester to the New York line, and once that connected to Albany the railway was renamed the Boston and Albany Railroad (B&A).

In 1816 the exiled Napoleon, having complained about the advantage Britain derived from its coal, added:

> 'But the high price of all articles of prime necessity is a great disadvantage in the export of your manufactures ... your manufacturers are emigrating fast to America.... In a century or perhaps half a century more: it will give a new character to the affairs of the world. It has thriven upon our follies.'

In 1830 this still seemed a distant prospect, but by the 1850s there was already a clear sense of change in process.

1. ORIGINS 39

A TRANSCONTINENTAL RAILROAD
Office of PRR Surveys, 'Map of Routes for a Pacific Railroad', 1855.

In 1853 Jefferson Davis, Secretary of War from 1853 to 1857, established the Pacific Railroad Surveys (1853–55) to study possible routes to link the Atlantic and Pacific oceans. The five surveys led to a 12-volume compendium, *Reports of Explorations and Surveys, to ascertain the most practicable and economical route for a railroad from the Mississippi River to the Pacific Ocean*. The reports were accompanied by many detailed maps completed by government surveyors, as well as this general 'Map of Routes for a Pacific Railroad', stretching from Canada to the United States' newly established southern border with Mexico. The Gadsden Purchase of land from Mexico in 1853 was a treaty that provided the United States with thousands of square miles of land that enabled the planning of a line west from El Paso. From 1861 to 1865 Davis was President of the Confederate States.

The general map drew on the five surveys and their particular routes. Thus, for example, after Lieutenant John Gunnison was killed with seven of his men by Ute at Sevier Lake, Lieutenant Edward Beckwith took charge of the Central Survey. Having surveyed passages along the Weber and Timpanogos canyons he proposed three routes in his zone, which essentially covered Utah. The northernmost crossed the Green River near Black's Fort and continued past Fort Bridger, which is now in Wyoming, and along the Weber River to Ogden, where it turned south. Ogden had been founded in 1846 as Fort Buenaventura. The middle route left the Wasatch Mountains via Timpanogos Canyon. The southernmost route ran westwards from the Oquirrh Mountains. The northernmost route, which was advocated by Beckwith in his report, was chosen and it was roughly along this path that the transcontinental railway was eventually built and Ogden became a major rail junction.

The map reproduced here is significant as evidence of what is so often lacking in the reproduction of historic maps, namely the role of what in effect were drafts. Relief is shown by hachures and spot heights and is incomplete; prepared by a young lieutenant, Gouverneur Kemble Warren, a note at bottom left explains: 'This map is a hurried compilation of all the authentic surveys and is designed to exhibit the relations of the different routes to each other.'

The focus on San Francisco in the aftermath of the California Gold Rush is apparent, as is the range of possible originating points for the transcontinental railway from Duluth on Lake Superior south, via Chicago, to New Orleans. The surveys were published by the War Department from 1855 to 1860.

Two of the five surveys were southern routes, the southernmost to San Diego, providing the route where in 1881 the Southern Pacific Railroad completed the second transcontinental railway. Earlier, support for this line was in part in pursuit of Southern plans to secure a future that strengthened the South by producing a rail link both to the Southwest and to the Pacific, and thus a key economic advantage. Mapping was a crucial part of the surveys. Indeed, in his preliminary report to Congress of February 1854, Davis noted:

> 'A map is projected to embrace all the reliable surveys and observations that have been made west of the Mississippi, and is now in a condition to receive the trace of each survey as it comes in; so that all the surveys that have been made, or that are making, can be presented in one connected view.'

The surveys were carried out by officers from the Corps of Topographical Engineers, a branch of the US Army established in 1838 when the Topographical Bureau expanded and was organised as a fully fledged staff corps. Military surveyors frequently

clashed with the local populations of Native Americans. Indeed, the exercise of force was a powerful factor in this mapping. Thus, in 1849 the Topographical Engineer Lieutenant James Simpson produced a map to show Lieutenant Colonel John M. Washington's punitive expedition of that year against the Navajo. Simpson was able to explore the Canyon de Chelly only after Washington had defeated them. Simpson later became Chief Engineer of the Interior Department and advised on the route for a railway line from Texas to California, and oversaw the construction of the transcontinental railroad. In 1851 there was an expedition that sought information about the region, not least to open up a possible rail route to California. The expedition had several skirmishes, especially with Mohave and Yuma in the Colorado Valley, although it failed to find an acceptable rail route.

1. ORIGINS 41

SPREADING THE NEW AGE 1860–85

'The latest is best ... not to believe in the nineteenth century, one might as well disbelieve that a child grows into a man ... without that Faith in Time what anchor have we in any secular speculation.' This remark, made in 1857 by the painter William Bell Scott (1811–90), was exemplified in his commanding painting *In the Nineteenth Century the Northumbrians Show the World What Can be Done with Iron and Coal* (1861), which was set in the vibrant industrialisation of Newcastle and sought to capture 'everything of the common labour, life and applied science of the day'. Scott depicted workers at Robert Stephenson's engineering works, one of the largest manufacturers of railway engines in the world (although the company went bankrupt in 1898), as well as the steam of modern communications and telegraph wires (see page 44).

Rail spread to new countries and in the meantime the scale of rail rapidly increased, both in terms of the intensification of existing systems, as in Britain, and the spread of lines to parts of countries not hitherto reached, as in Spain. In Britain, which was the originator of rail, the volume of freight carried rose from about 38 million tons in 1850 to 513 million tons in 1912 (see pages 72–73). The transport system facilitated an economic expansion that lessened the problems of unprecedented population growth. As with many changes in railway history, this in part was a matter of technological improvements, although there was no equivalent to the container 'revolution' since the 1950s. More important was investment in rolling stock, the greater capacity of a larger system, the pulling power of more powerful engines, organisational change and entrepreneurial activity. Thus, the ability to send goods at night, which (in part) was a matter of developments in signalling, meant 24-hour activity. So also with the development of production systems for a range of goods, such as signalling equipment and coal scuttles.

The spread of the rail system introduced new routes not used by roads or canals, but in Britain the situation was broadly similar to that of the canals in the late eighteenth century, with a

American Progress, a chromolithograph *c.*1873, after a John Gast painting of 1872, shows trains among the forces streaming westwards across the prairie, technology justifying the continental expansion of the United States.

concentration initially on the coalfields. Thus, in the Cumbrian village of Tindale, where coal had been mined and lime produced from the early eighteenth century, a zinc smelter was opened in 1845. The nearby presence of coal and lead mines, a water supply and a railway all proved helpful.

RAIL – THE SOCIAL DISRUPTOR

Railways in Britain and elsewhere represented a form of social compact, with the change it represented assisting all interests, although unequally so. The propertied middle class played a major role, but in co-operation with both the social élite, who were compensated for providing the land to enable railway construction, and those who had less of a propertied stake in society. The last group gained paid employment, but the terms of that could be a troubled process. The rail industry insisted on firm working conditions, which could be both onerous and a challenge to labour customs. A free-market society was not one of equal freedoms.

43

In the Nineteenth Century the Northumbrians Show the World What Can be Done with Iron and Coal, painted by William Bell Scott, 1861. Workmen forge a locomotive wheel, while visible in the background is Stephenson's High Level Bridge and the River Tyne.

However, entrepreneurialism is no respecter of rank and the era of rail, like the Industrial Revolution itself, helped to narrow the social divide by facilitating social mobility, enabling talented individuals to rise from humble origins. The birthplace of George Stephenson (1781–1848) was a stone tenement in the Tyne Valley at Wylam shared by four families, with the Stephensons having two rooms. His father was a coalminer, and George himself started work at the age of ten. George's son, Robert (1803–59), a great engineer, took forward his father's role, with both men crucial in the development of British railways. Robert was also a key bridge-builder, notably the High Level Bridge at Newcastle (1848), the Britannia Bridge across the Menai Strait to Anglesey (1850), the Royal Border Bridge at Berwick (1850) and the Victoria Bridge at Montreal (1860). He dined with Queen Victoria and was buried in Westminster Abbey.

Yet for most rail workers there were no equivalent means to rise and progress. This was particularly so for construction, which was the largest use of labour for rail in this period. Most of this work was non-specialist, in that casual workers moved into and out of rail, yet arduous and frequently dangerous, points that are easily forgotten when looking at maps.

Station staff were numerous and large numbers of people worked in the freight movements at marshalling yards. Again, opportunities for upward mobility were limited. Instead, it was largely engineers who could benefit the most.

QUESTIONS OF FINANCE AND CONTROL

The increase of freight volume and therefore profit was significant for the financing of railways because capital markets were still poorly developed. The precarious nature of many banking firms and the vulnerability and short-term nature of most credit created difficulties for the financing of the railways, more particularly of their construction rather than operation. There were also frequent bankruptcies. Thus, in 1875 the two Quebec railways, the North Shore Railway, which joined Quebec City and Montreal, and the Montreal Colonisation Railway, which prospected a line west to Ottawa, went bankrupt. Given the various interests bound up in rail, not least that of Montreal within Quebec, the provincial government felt obliged to take over, uniting the two as the Quebec, Montreal, Ottawa and Occidental Railway (QMO&OR).

That intervention might appear to meet modern norms for legitimating state control, but in this case there was simply a transfer of cost. Moreover, as a result of the burden the provincial government was weakened so that by 1885 the railway had been added, for cost reasons, to the Canadian Pacific (CP) system. As with other politics of rail, there were also earlier consequences that it was not possible to present in map form. For example, the emphasis on railways along the north shore of the St. Lawrence River led to complaints from the south shore, while, separately, Quebec complained about the dominance of Ontario (see pages 56–57).

The spread of rail continued to be slow in some areas, notably Africa, where the first railway, Alexandria–Cairo (1854), was followed by Durban–The Point (1860), but by relatively little else.

THE MULTI-FACETED DEEP SOUTH

G.W. and C.B. Colton, 'Map of Alabama, Describing the line of the South & North Alabama Railroad. And its Connections.', [undated].

Produced by the leading New York map publishing house established by Joseph Hutchins Colton, this map very much presents the railway in terms of economic potential. This is shown by dividing Alabama into five regions from north to south: the 'Stock & Grain District', the 'Mineral Region', the 'Mixed Agricultural & Manufacturing Region', the 'Cotton District' and the 'Timber Region', which were readily distinguished by colour. Iron furnaces and operational coal mines were also shown, as were the Shelby Iron Works, the coal fields, the iron ore deposits, the canals and the roads. The railways were classified as 'In Operation', 'Graded' or 'Projected'. The map is easy to read.

The Alabama State Legislature chartered the railway in 1854, and work started in 1859 on the section linking the Cahaba River to the Alabama and Tennessee River Railroad at Calera, which was known as Lime Kiln Station. Providing the fertiliser needed for the often poor soils of the South, lime kilns were powered by coal. There was also grading of the line, preparatory to the laying of the rails, on to Elyton, which was later to be called Birmingham. There was more work during the American Civil War, with slaves used for the purpose. The railway delivered wood (for charcoal) and coal to the Shelby Iron Works, which had been built in 1844, with a rolling mill following in 1860 that was able to produce finished bar iron. A larger furnace stack was built in 1863. The iron was used for constructing the Confederate warships at Selma, including the *Huntsville*, *Tennessee* and *Tuscaloosa*, as well as for Confederate railway rolling stock. Destroyed in March 1865 by raiding Union troops, the furnace was rebuilt by 1869 in order to provide iron, a key requirement for railway rolling stock.

Work on the railways in the South had been limited during the Civil War, in large part due to a shortage of resources of all types. Work on this line resumed in 1869, after the Civil War, with the railway from Calera to Montgomery finished by the end of 1870, but progress was slower further north towards Decatur, the crossing point on the Tennessee River. This crossing and line were designed to help provide an outlet for the South to Midwestern markets. Indeed, the line was part of the integration of the South into the American economy during Reconstruction, a period of Southern history when Northern interests were to the fore, as was the attempt to remould the South (railways being seen, like education, as a way to serve all and to modernise). Refinancing to meet debts, however, proved necessary in 1871 and this led the Louisville and Nashville Railroad to take effective control. The following year, the first trip from Montgomery to Louisville was completed.

The map is important both for its emphasis on resources and because it also showed the South & North Alabama Railroad in relationship to other lines. The latter factor was a measure not only of connectivity but also of the distinctive advantages of particular lines. As a classic instance of the 'silences' of maps, this one provides nothing of the crucial, contested present of the railways, namely Republican-led Reconstruction and the degree of Democrat-led resistance it encountered. In 1865 federal troops in Alabama began to oversee Emancipation, which was resisted by white supremacist groups, notably the secretive Ku Klux Klan and the more overt White League. In 1874 violent intimidation of black Republican voters by white supremacists led to the overthrowing of Alabama Reconstruction. The building of railways served the strategy of Reconstruction, with the state government, notably the governors, pushing a bold policy of railway subsidy, which led to thousands of new jobs in construction. However, that helped lead to corruption and high taxes, which further discredited Reconstruction among unsympathetic whites. The state government under Reconstruction, especially Governor William Smith (1868–70), had to finance millions of dollars of railway bonds, which proved a heavy burden. The Democratic Party era that began in Alabama in 1874 was unwilling to take on such financing, the general economic crash of 1873 having resulted in hostility to the risk associated with public debt.

BOOSTING THE LAKES

Sheble, Smith & Co., 'Map Showing the Lake Superior and Mississippi Rail Road', 1869.

Published in 1869 by the Philadelphia-based map publishers Sheble, Smith & Co., which published in the following year a map of the New Jersey railways, this map emphasises the significance of economic resources. It is an instance of the cartographic boosterism that was so important in American railway cartography. Thus, the line from Duluth on Lake Superior to St. Paul was located in part with reference to other railways, both finished and unfinished (as coded in the key). There was also a demonstration, using colour, of 'the extent of its land subsidies' (designed to support the railway), and then overprinting to show economic potential, with 'Rich Wheat Lands', 'Rich Prairie Wheat Lands', 'Heavy Hardwood Forests' and 'Vast Pine Forests' all marked, and in large letters, on the map. 'Rich' and 'Vast' were clear value-added terms, part of the boosterism of the period. The presence of iron, gold, silver and, repeatedly, copper were noted in small capital letters. Indeed, iron production was to grow rapidly once the railway was opened. There was also the placing in the map of a summer isothermal line for 70°F, which helped establish the growing season.

Begun in 1863, the line was finished in 1870. It ran in large part along the valley of the Saint Louis River, which was just as well because that provided a flat and largely drained route, although there was still the need to cut the way through difficult forest and valley bottoms meant more tributaries and poorer drainage.

Duluth had only been developed as a town in the 1850s, and key elements then were the road to St. Paul and the Sault Ste. Marie Canal, which together suggested that, with the addition of the railways, Duluth would have access to both Atlantic and Pacific and be able to pass Chicago, which did not seem to have the same extent of potential access to the Pacific because it was further away. Indeed, in 1869–70, in population terms, Duluth was the most rapidly growing city in the country, although that was from a small base.

This railway line was hit hard in the so-called Panic of 1873 because Jay Cooke, its Philadelphia-based financier, had become badly overextended. In large part this overextension was a consequence of his focus on the development of the Northern Pacific Railway as a means to construct a new transcontinental line.

The Lake Superior and Mississippi Railroad (LS&M) was reorganised in 1877 as the St. Paul and Duluth Railroad and in turn was purchased in 1900 by the Northern Pacific. Aside from the line between St. Paul and Duluth, there were branches to Minneapolis, Taylors Falls, Kettle River, Cloquet, Grantsburg and Superior (a city in Wisconsin). This map is more colourful than many of the booster maps of the period, but obviously devotes no attention to liabilities and difficulties, not least the cost of extracting the minerals.

AN AMERICAN DISJUNCTION

Ensign, Bridgman & Fanning, 'The New Military and Rail-Road Map of the United States, showing the Depots and Stations', 1865.

The American Civil War (1861–65) was a classic instance of the need to reframe a rail system for very different purposes. Although the violent emergence in 1861 of two rival blocs that were each homogenous was better for either rail system than a civil war of two sides across the entire networked country, neither system had a coherence that matched these blocs. This was particularly because far from the slave states and the linked plantation economies that comprised the South, four states in the North – the border states of Maryland, Delaware, Kentucky and Missouri – also permitted slavery, as did the western portion of Virginia, which was to become the state of West Virginia. The Civil War did see extensive mapping, notably by the Union, but rail was not its focus.

The existing rail mileage in the Confederacy, formed in February 1861, was far shorter than in the Union, and was mostly designed to link ports to hinterlands – as with Jacksonville and the Florida, Atlantic and Gulf Central Railroad – rather than to provide an overarching network, which had not been possible due to insufficient capital. Moreover, major problems were caused by different gauges, which are not apparent in the maps, although what does emerge clearly is the weakness of the South's rail infrastructure, and notably in comparison with the far greater density of railways in Ohio and Pennsylvania.

An aspect of this made clear in the map (shown here is a 'Detail of the Southern States' from Ensign, Bridgman & Fanning's 1865 map) is the density and significance of junctions and major bridging points. There were fewer junctions in the South than in the North. There were rail-steamer junctions across the Mississippi in the South, rather than bridges: the first bridge across the Mississippi had not been finished until 1856 and was in the North, between Rock Island, Illinois, and Davenport, Iowa. It had taken three years to build during peacetime, which showed the issues posed by adding to rail infrastructure during the Civil War. That at St. Louis took from 1867 until 1874.

The significance of rail junctions such as Macon in central Georgia emerges from this map. Train tracks ran thence to Atlanta, Columbus and Savannah, and the importance of that interconnecting rail hub explained the two (unsuccessful) Union cavalry attacks in 1864.

The limited nature of the Southern rail system prior to the Civil War emerges clearly in the 1860 map by A.M. Gentry on the line of the Texas and New Orleans Railroad, with a clear distinction between what was actually built (to Houston), partially finished (to Columbus), under contract (to San Antonio) and 'chartered but not under construction' as far as Guaymas on the Gulf of California.

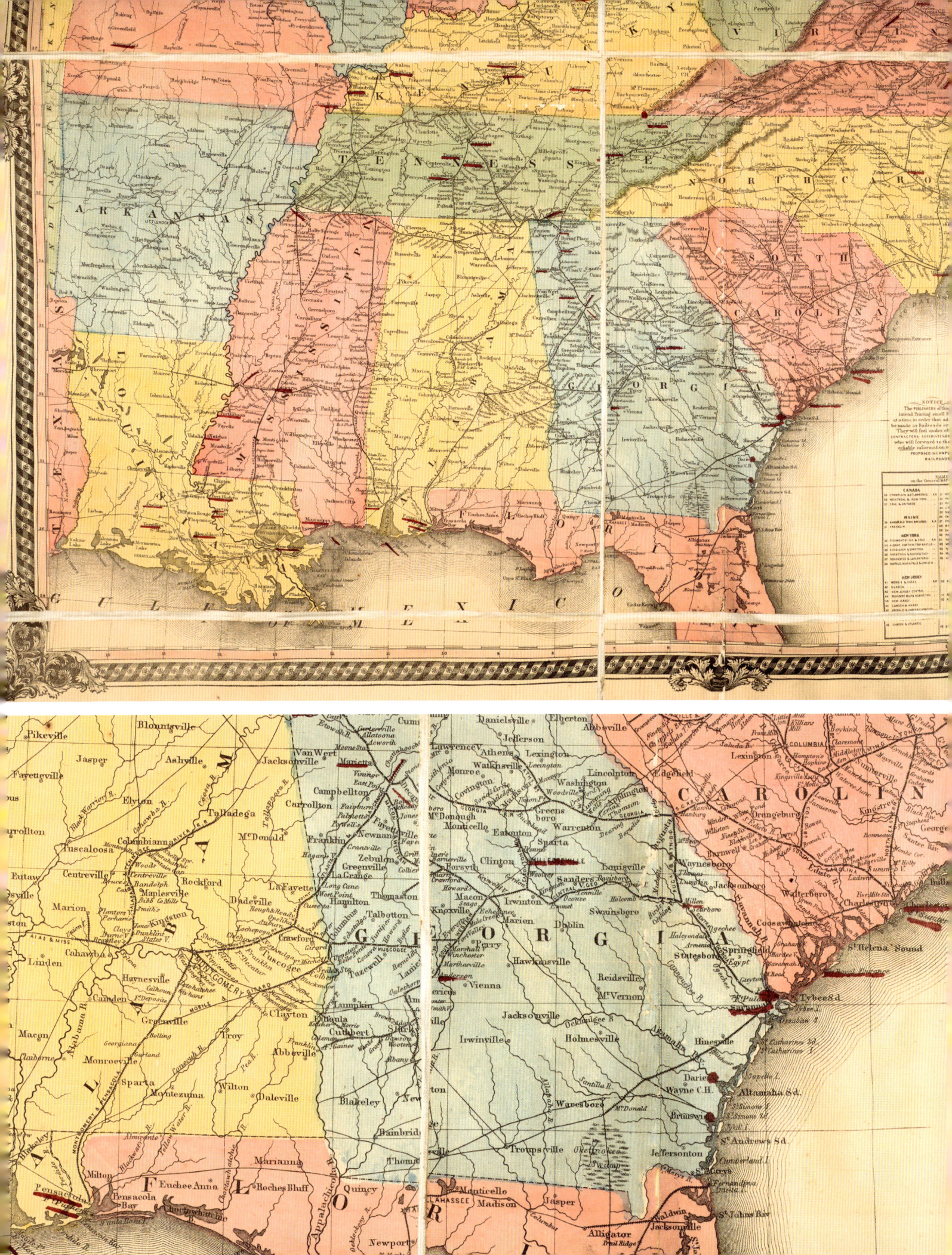

THE WAR AROUND WASHINGTON, DC.
Louis Prang, 'Birds Eye View of the Seat of War', 1862.

Published by Louis Prang (1824–1909), a talented German-born Boston printer who had moved into lithography in 1856, and who after the war was to play a major role in the development of American Christmas cards, this bird's-eye view reflects Prang's skill of being able to create profitable images from the war. Aside from maps that were like pictures, he also produced illustrations of the battles themselves. This map – or rather picture of the battlespace – is both instructive visually and draws the viewer into the map in a fashion that ordinary projections and perspectives fail to do.

The map displays the region around Chesapeake Bay and the Potomac River, the central space in the war in the East from its beginning in 1861 to its close in 1865. Prang includes Washington, recognisable by his drawing of the Capitol building, Baltimore, Harpers Ferry, Richmond, Manassas Junction (Bull Run), Fort Monroe and Norfolk Harbour, the last a direct reference to the prominence of the naval dimension.

Railway lines are shown as part of the campaign space. This is an accurate account of the tactical, operational and strategic significance of rail. The area covered by the map had already seen serious Confederate raids on the Baltimore and Ohio (B&O) line in 1861, which led to an end to services, obliging the Union to use lines further north. That the Union had these lines was an aspect of its railway capacity. Such strategic depth is an important aspect of the military significance of rail, because it reduced the vulnerabilities arising from the location of the front line, with the resulting exposure to opposing forces, and also had there only been a single line.

In 1862, due to the density of their forces in the area and the deficiencies of the Centerville Road, not least the amount of fodder eaten by the ox teams, the Confederacy turned to rail and constructed a spur off the Orange and Alexandria Railroad (O&A) at Manassas Junction towards Centerville: the Centerville Military Railroad, five and a half miles long, intended to support the Confederate defences. The iron for the track came from storage and from raids on the B&O lines, notably the Great Train Raid of May 1861, which included the seizure of 14 locomotives at Martinsburg, their disassembly and movement by horse-drawn teams to Virginia. The speed of the railroad's construction ensured that no ballast was used while the spacing of the ties was twice as usual. The withdrawal of the Confederate troops to the Rappahannock River in March 1862 led to the speedy abandonment of the line.

The visual placing of rail in the map is enhanced by the trains shown on railways, for example near Lawrence (top, centre). The importance of particular river rail crossings also emerges from the lack of other crossings – for example, on the Rappahannock River below Fredericksburg, or for many miles further upstream. This enhanced the significance of Fredericksburg as a target and a major battle was to be fought there. A scale of sorts that counters the impact of the perspective is provided by the mileages printed at the bottom of the map.

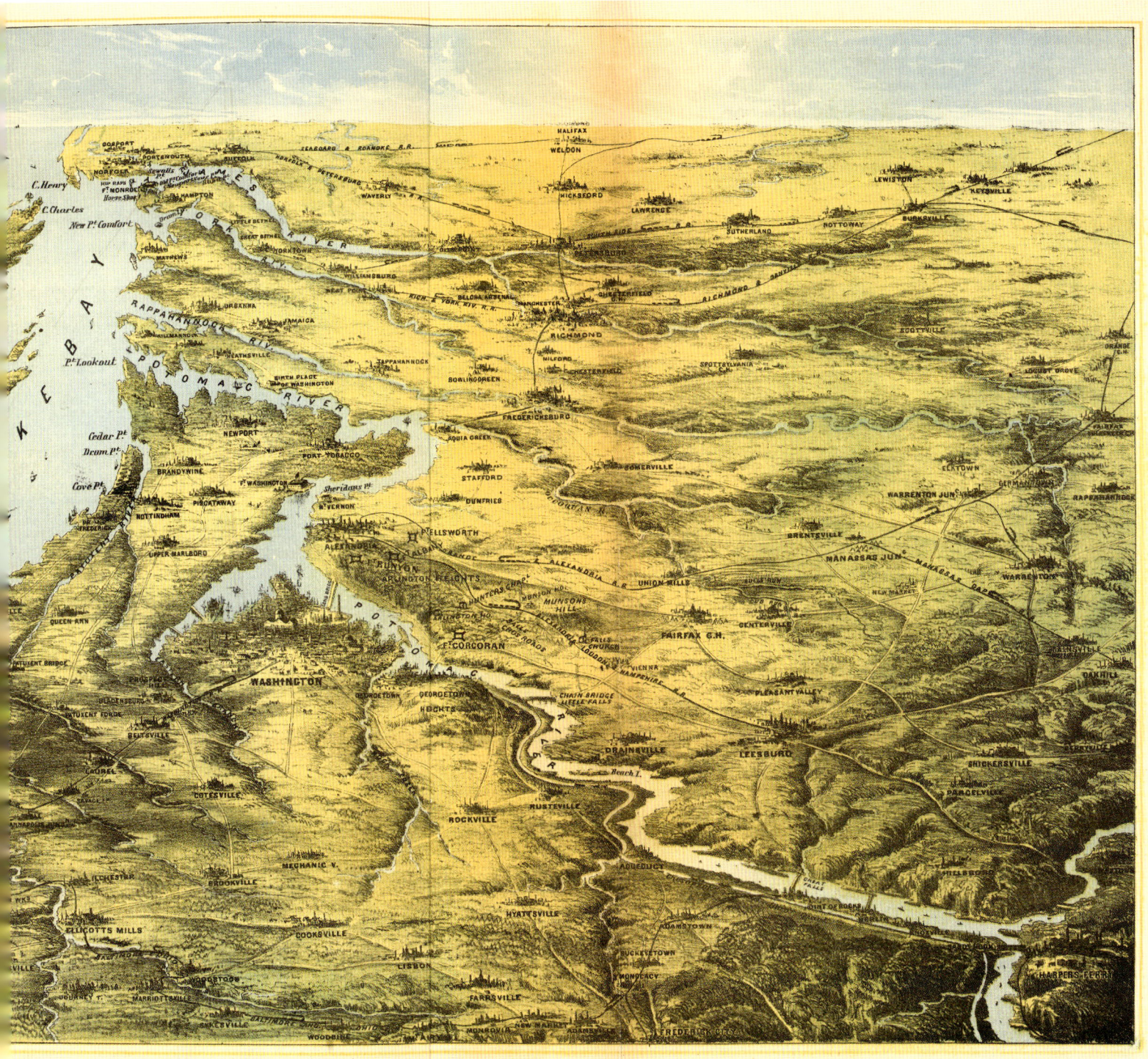

CREATING CANADA

Perry & Spaulding, 'Watson's New Township Map & Rail Road Map of the Dominion of Canada: With Plans of the Cities of Montreal, Toronto, Quebec and Ottawa', 1873.

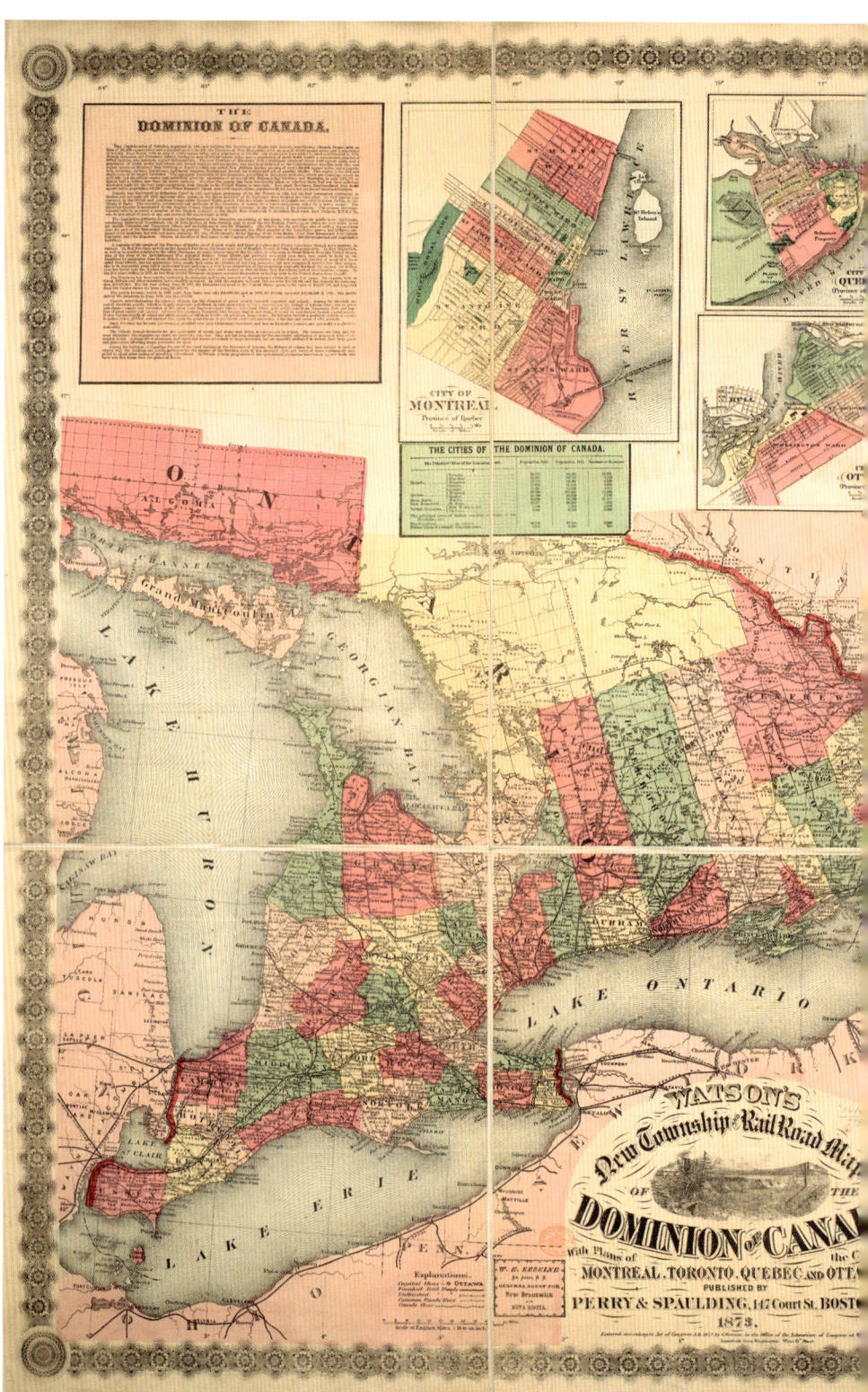

Showing railways finished and unfinished, including on the American side of the border, this map was in part a clear appeal to investment and settlement. The text of the map noted: 'More than 3,000 miles of railway are already in operation.... the Canadian Pacific ... the means to be provided by grants of land ... on each side... the price to be paid for the annexation of British Columbia to the Confederation.'

As with many remarks about railways, none of these were value-free. As in the United States, Argentina, Australia and more generally, discussion of land grants and settlement made little or no reference to the indigenous populations, although the situation was better in New Zealand where the Maori were stronger. Nor were all outsiders treated equally. In particular, and most clearly in North America and Australia, it was settlement by Europeans rather than by East Asians that was desired, notably not by Chinese.

This map includes the information: 'The official tables of returns of railway traffic show a steady increase... in 1871 over CAN$14 million.' It was necessary to emphasise such figures in order to attract and hold governmental and investor support (both shareholder and bondholder), not least because tax revenue was limited for the former and there were attractive alternatives for the latter.

In Canada, like the United States, railway was to become a narrative about transcontinental railways, but this was not the early history of rail transportation in the country. In eastern Canada, there were opportunities for river, lake and canal travel that were lacking further west. Waterways provided a route from the Atlantic to Lake Superior, whereas in the west there was no comparable route into the interior from the Pacific and no valley through the Rockies to offer an equivalent.

Despite this contrast, railways were important in eastern Canada. That this was so owed much to the extent to which the waterways, including the lakes, and notably their shores, were frozen for the long winters. In addition, the waterways between

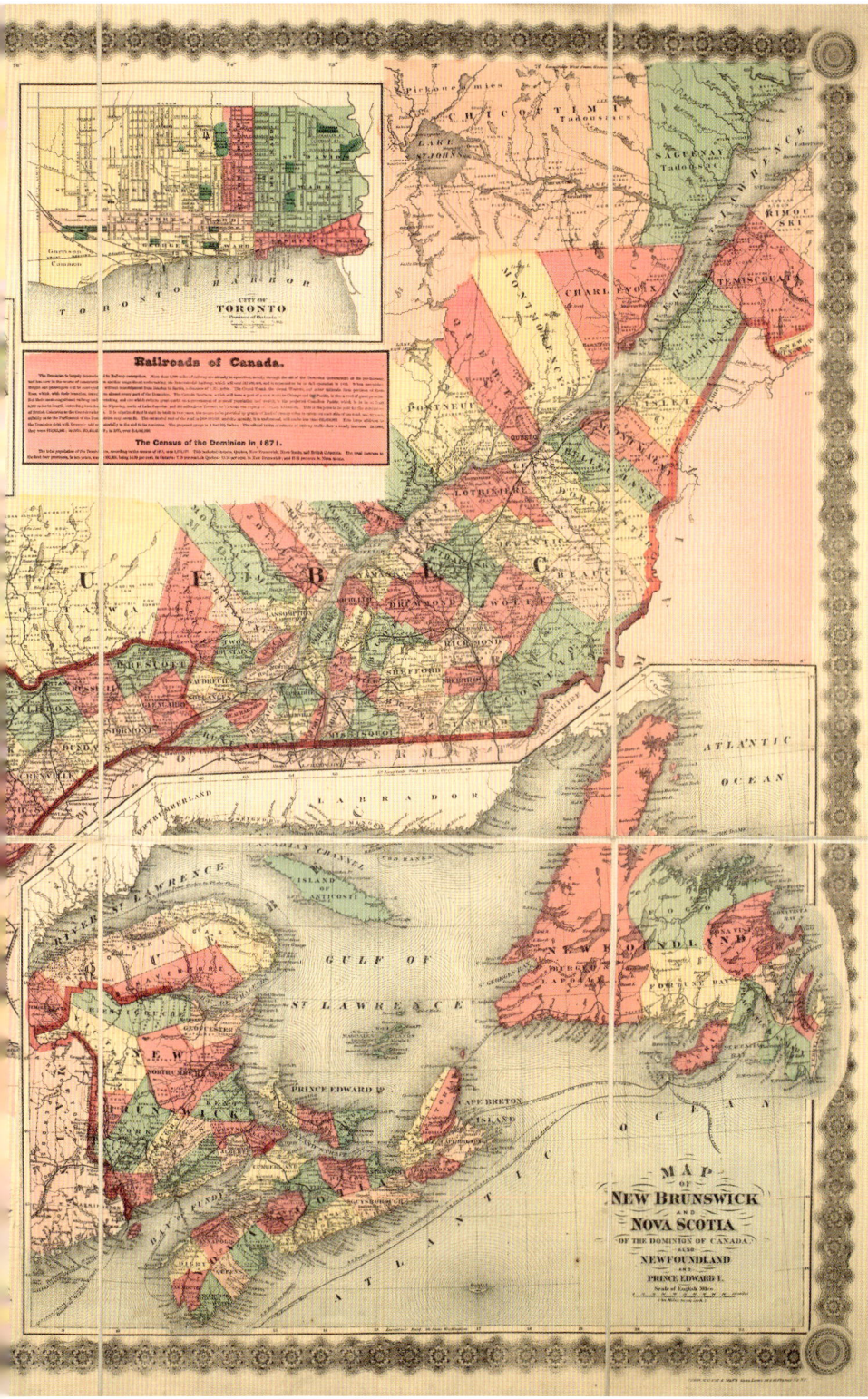

the lakes and along the St. Lawrence River included rapids, waterfalls and shoals. These difficulties meant that canals and railways were put forward as solutions, initially to provide and/or bypass portages but eventually of a longer and more significant character. The surveying of potential routes for railways also involved security considerations. Thus, the potential routes for a railway between Halifax and the St. Lawrence River, surveyed in 1864–65, were shown on a map, which helped to clarify the decision to choose a route as far from American territory as possible.

In addition, the railways were seen as a way to open up the interior, and the goal of doing so drew on American examples, Canadian entrepreneurship and British finance, with Canadian political interests playing a crucial intercessionary role. A key instance was the Grand Trunk Railway from Sarnia to Montreal, which was completed in 1860, and the Ontario, Simcoe and Huron Railway, completed from Toronto to Collingwood on Georgian Bay in 1855. Further west, Toronto was better located for such rail construction.

Railways played a major role in Canadian politics from the 1840s to the 1930s, with the Grand Trunk dominating the political scene in the 1850s and 1860s. There was a close relationship between railway building, railway finance, Canadian Confederation and nationalism. Indeed, the national dream of Canada was closely linked to rail and, more particularly, the idea of a transcontinental railway, although in practice many Canadian interests preferred transport links to the American seaboard, and thus to warmer waters and American markets.

Railway financing, as referred to in the text of this map, was crucial politically because railway debt was an important factor in Canadian Confederation, not simply for building the transcontinental railway but also to bring recalcitrant smaller colonies onside. The Confederation project, a means to form a national instead of a provincial identity, was also linked to the British financing model of Canadian railways, which ensured that private corporations played the key role in developing the country. British financiers also wanted to protect their existing stake in Canadian railways, notably the bondholders in unprofitable Canadian railways. Less positively, the strong focus on financing rail, while valuable, was also potentially stultifying in terms of Canada's national economy.

SETTLING THE NORTHWEST

C.S. Lott, 'Map of part of Manitoba and the Canadian Northwest Territories: Alberta, Assiniboia and Saskatchewan...', 1890.

This lithographed map was produced in Winnipeg, Manitoba, in 1890 on behalf of the Land Commissioner. The remainder of the title of the map reads: '...showing system of Land Survey and the Lines of the Canadian Pacific Railway Company.' The transcontinental railway had been built in 1881–85 and in the map the completed lines are marked in solid black, the incomplete ones in interrupted black lines. Notably, other companies' lines are also shown. First Nations reserves are marked 'I.R.' for 'Indian Reserve'. Terrain is marked by hachures not contour lines.

The map was produced in a difficult period for the expansion of rail, because the world economy was not growing at the anticipated rate, which increased the competitive pressures for Canada, not least with the United States where the growing season was longer. Early frosts and drought hit hard, the latter being a persistent problem on the Prairies, and one not shown on the map. British settlers preferred other destinations, including the United States and Australasia, which helped explain the interest in placing the land seen in this map, as well as the bankruptcy of the Manitoba and North Western Railway (M&NW) in 1894, a company that might have thrived if more people had immigrated. Not indicated in this map are the steep gradients involved in crossing the Little Saskatchewan and Assiniboine valleys, which placed a limit on the length of any of the M&NW's trains and therefore increased operating costs. Earlier, gradients had been one of the factors in the choice of a route across the Rockies with the Yellowhead Pass rejected in favour of the Kicking Horse one, but access to cultivatable land was more significant. The construction of the M&NW began in 1881 after it had received a land grant of 1.4 million acres, and in 1889 it reached Yorktown, shown in the map. Had the company met its deadlines the land grant could have reached 2.5 million acres.

The first branch line in Alberta ran from Dunmore to Lethbridge, marked on the map. The line was that of the North Western Coal and Navigation Company and it was used to move coal from Lethbridge to sell to the Canadian Pacific Railway for its trains.

Edmonton and Calgary were connected in 1891, initially by the Calgary and Edmonton Railway, which was later purchased by the Canadian Pacific Railway. The proposed line shown south from Calgary to Fort McLeod was designed to open up agricultural land to

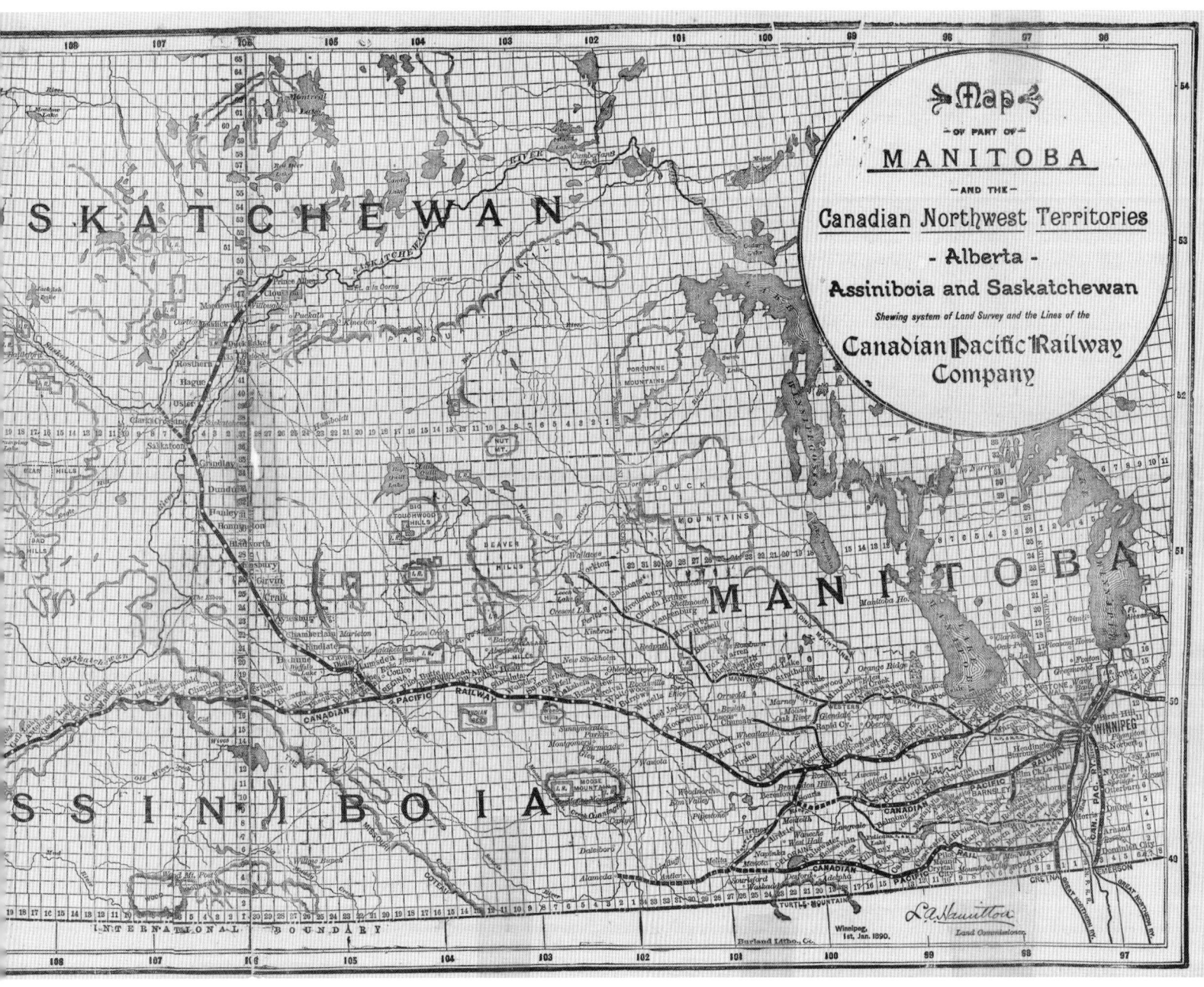

the west and south of Fort McLeod, possibly eventually to British Columbia. Part of the rail-bed was completed, but not the line as a whole. This map is very much that of a rail system that is expanding, but uncertainly so.

There was to be another pulse of transcontinental railway-building in Canada in the early twentieth century, as both the Grand Trunk Pacific and the Canadian Northern sought to compete with the Canadian Pacific. The inevitable result was bankruptcy and nationalisation by 1917. Railway branch lines were completed by the end of the 1920s, when the Great Depression brought Canada's railway age to an abrupt close.

LINKING ATLANTIC AND PACIFIC
W.J.L. Wharton, 'Isthmus of Panama, Showing the Proposed Panama Canal and the Railway', 1885.

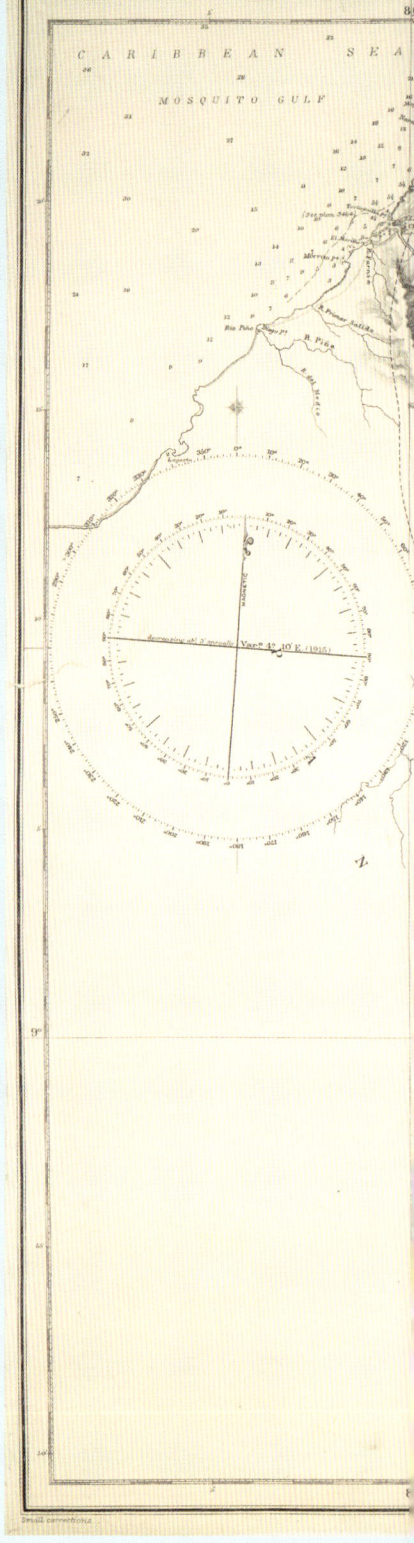

The California Gold Rush, which began in 1849 and grabbed the imagination of the age, led to extensive interest in getting to the Pacific. A satirical print of that year by the New York lithographer Nathaniel Currier included prospectors heading to California by 'Rocket Line' and 'Air Line', both powered by steam. Rail appeared far less risky.

The California Gold Rush spurred on the settlement of America's West Coast. There are no precise figures, but it is estimated that over 40,000 people arrived in California by sea in 1849–50, and over 80,000 came overland. The journey from Independence, Missouri, to the Pacific coast possibly took four to five months but could not be done when the Rockies were blocked by snow. It was a difficult if not hazardous route. The sea route from the East Coast via Cape Horn took four to six months, but that over the isthmus of Panama took fewer than three and could take closer to two. Malaria and yellow fever, however, made it risky. This led to American investment in an 'inter-oceanic', indeed transcontinental, railway across Central America in order to provide an easier and safer alternative to mule trains across a disease-infested landscape.

Authorised by treaty with Colombia (which ruled Panama) in 1846, the Panama Railroad Company was incorporated in New York in 1849 and work began in Panama in 1850. Having cost US$8 million and the lives of thousands of workers (some from China, others former slaves from the West Indies), mainly due to disease, the 47.6-mile (76.6-kilometre) route from Colón to Balboa was opened for coast-to-coast travel in 1855. The Admiralty map makes clear the difficulty of the construction, not least coastal swamps and having to dig across the Continental Divide, but it could not show disease, alligators, mosquitoes, sandflies, heavy rainfall and the wood rotting in the heat and humidity. The key solution to crossing the coastal swamps was to use pile-driven timbers, which could be done by employing a steam-driven pile driver, but most of the rail-bed had to be prepared by hand-cutting.

Separately, the railway tycoon Cornelius Vanderbilt (1794–1877) tried, with his Accessory Transit Company, to link the Atlantic and Pacific by developing a route across Nicaragua. Vanderbilt's plan reflected the extent to which water routes were still seen as an alternative to rail, an equation affected by the absence of coal in Central America, although steamships were also necessary. He backed first a canal and then the use of Lake Nicaragua and the San Juan River with a road as part of the route. Steamships to and from Nicaragua and on the waterways were part of the equation.

There was also a plan for a transit route across the Isthmus of Tehuantepec in Mexico. Delayed by controversy, permission to build was granted to Louisiana investors later in the decade. Both Nicaragua and Mexico were closer to America than Panama, but both were more politically unstable, and Nicaragua was also volcanic.

The rail link through Panama was important to the development of America's Pacific coast, not least of San Francisco, which was joined to the terminus of the rail route by regular steamship services. As so often, steam by land and sea proved symbiotic. However, the Panama Railroad was to be adversely affected by the transcontinental line through the United States, which faced a very different environmental challenge.

The line through the United States took the bulk of the passenger traffic, but Panama became more significant anew for freight with the building of the Panama Canal in 1904–14. In this construction, narrow-gauge railways played a major role, notably in the movement of supplies as well as of excavators. Thus, the important Culebra Cut, which sliced through the Continental Divide, contained a system of railways to remove the spoil, with about 160 trainloads of material removed daily. The Americans were able to remove about 76 million cubic metres of spoil from the cut. That was a typical instance of the significant role played by railways that only had a temporary presence. Such railways tend to be readily forgotten and poorly mapped.

As a near-contemporary parallel to the cross-isthmus railway, but in a very different

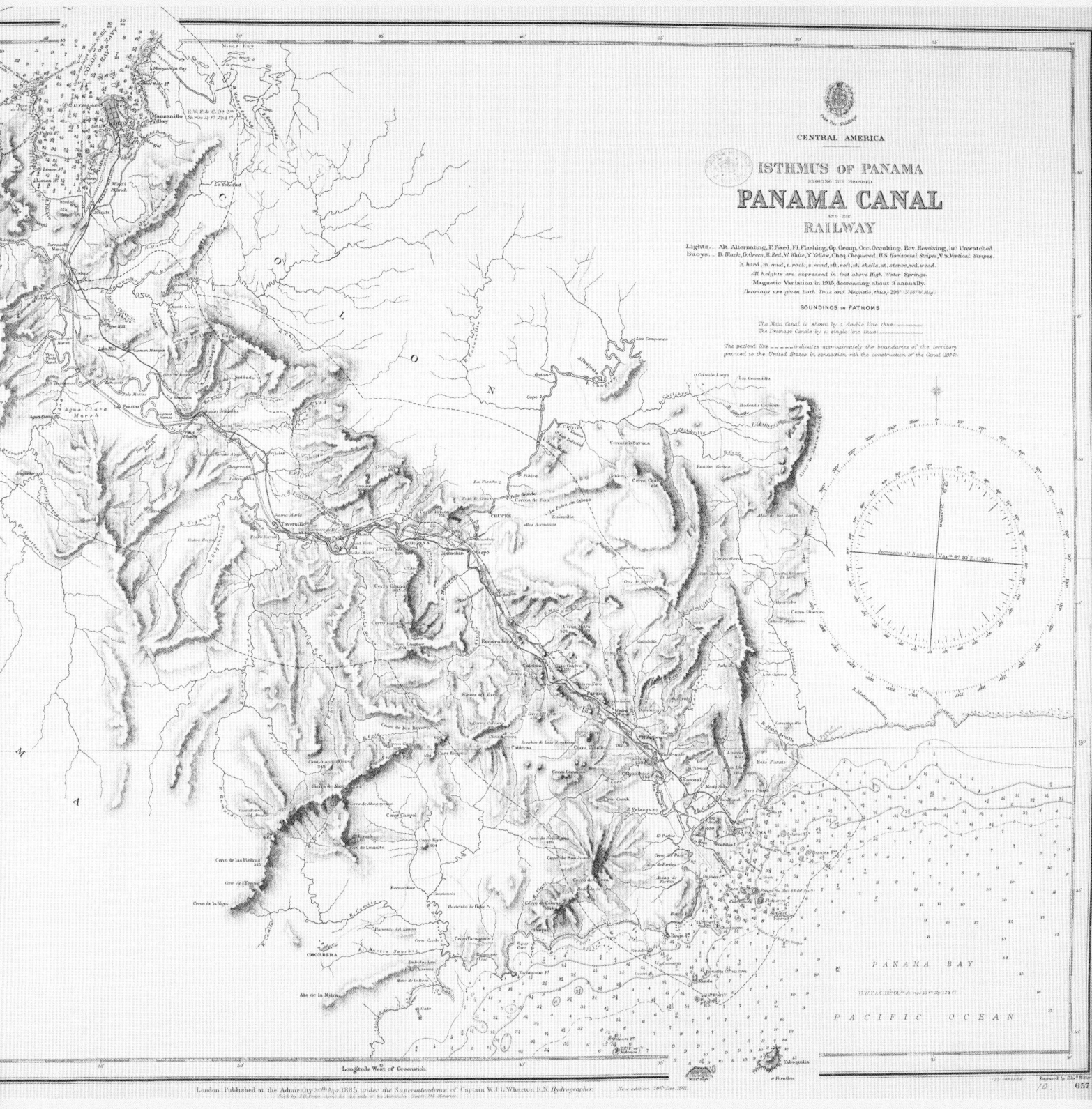

physical setting, the completion of the Alexandria to Suez railway in 1858 made travel easier between Britain and its most significant colony, India. It was followed in 1879 by the Suez Canal, which was easier to construct than its Panama counterpart because it could be at sea level (rather than having to construct locks) and the climate was less deadly.

LINES FOR LIVESTOCK

Whitehead, Morris & Lowe, 'Uruguay Central and Hygueritas Railway of Montevideo', [undated].

Published by Whitehead, Morris & Lowe, a London firm of lithographic printers, as well as engravers, steam printers, bookbinders and export stationers. Based in Fenchurch Street and Tower Hill, the firm also produced banknotes and bonds. This financial background helps explains the provenance of this map, which was produced as part of a process of engaging investor interest for a company – the Central Uruguay Railway (CUR) – that was registered in London in 1876. The CUR, in operation from 1878, took over the railway of the same name founded in Uruguay to work a concession granted by the government in 1866, under which construction had begun in 1867. The first proposed line, from Montevideo north to Durazno, was authorised with the government guaranteeing 7 per cent profits and offering a large part of the initial capital. This promise was not kept, which was not surprising because this was a period of strife for Uruguay. The difficulties of financing expansion domestically led to foreign investment, a process that saw the rearranging and renaming of companies, and an eventual measure of consolidation.

The first section of the railway, from Montevideo, the headquarters and national capital, to Santa Lucia, is marked in black on the map. The line was to be extended northward to Durazno and then to the Rio Negro, which was reached in 1886. The railway's projected western extension was marked in red, and a concession it had been granted was shown to the port of Colonia, where there was a ferry to Buenos Aires.

The railway had arrived relatively late in Uruguay with scant activity till the mid-1870s and by 1884 there was still less than 300 miles by 1884.

The map clearly classifies the range of rail projects and the state of completion. It provides no hints about economic possibilities, but that was a matter generally left to the boosterism of accompanying text, a common theme of which was linkage to neighbouring states.

In practice, the relations with these neighbouring countries and domestic politics were matters of strife. Indeed, there was what in effect was civil conflict from 1832 to 1904, with gaps (notably from 1876) but no permanent stability. Peace with Brazil in 1851 gave Brazil an exclusive right of navigation in the Laguna Merin and the Rio Yaguaron, the natural border between the two states, an agreement that restricted the range of potential exporting zones from the hinterland.

In 1864–65 a fresh civil war saw Brazil and Argentina intervene, which then widened out into the Paraguayan war when Uruguay joined the two powers against Paraguay in 1865. Paraguay's defeat in 1870 was followed in 1872 by a power-sharing agreement between the rival Uruguayan sides, only for there to be a military coup and a failed revolution in 1875.

This was a troubled background for investment although there was a massive increase in livestock rearing and export, of both cattle and sheep. This growth encouraged finance for the development of railways. By 1900 there was over 1,000 miles of track in use, and £11 million of the UK's £36 million foreign investment in Uruguay was in railways. Uruguay produced a massive amount of meat, which ensured that foreign trade doubled in value between 1870 and 1900, a period of low inflation. Distances were far shorter than in Argentina or Brazil, and cattle could be driven to railheads, but Uruguay's rail network was still impressive.

The range of beef products included the meat extract that from 1899 was trademarked as Oxo, as well as the corned (from the grains of rock salt known as 'corns') beef trademarked as Fray Bentos in 1881. It was produced in Fray Bentos, a town on the Uruguay River founded in 1859 where the meat-processing factory was established in 1863.

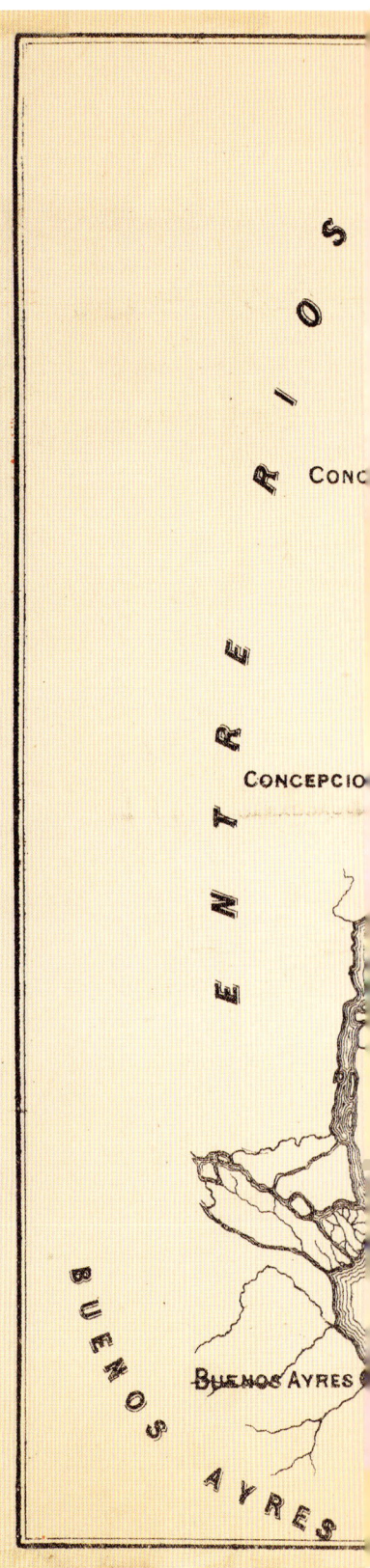

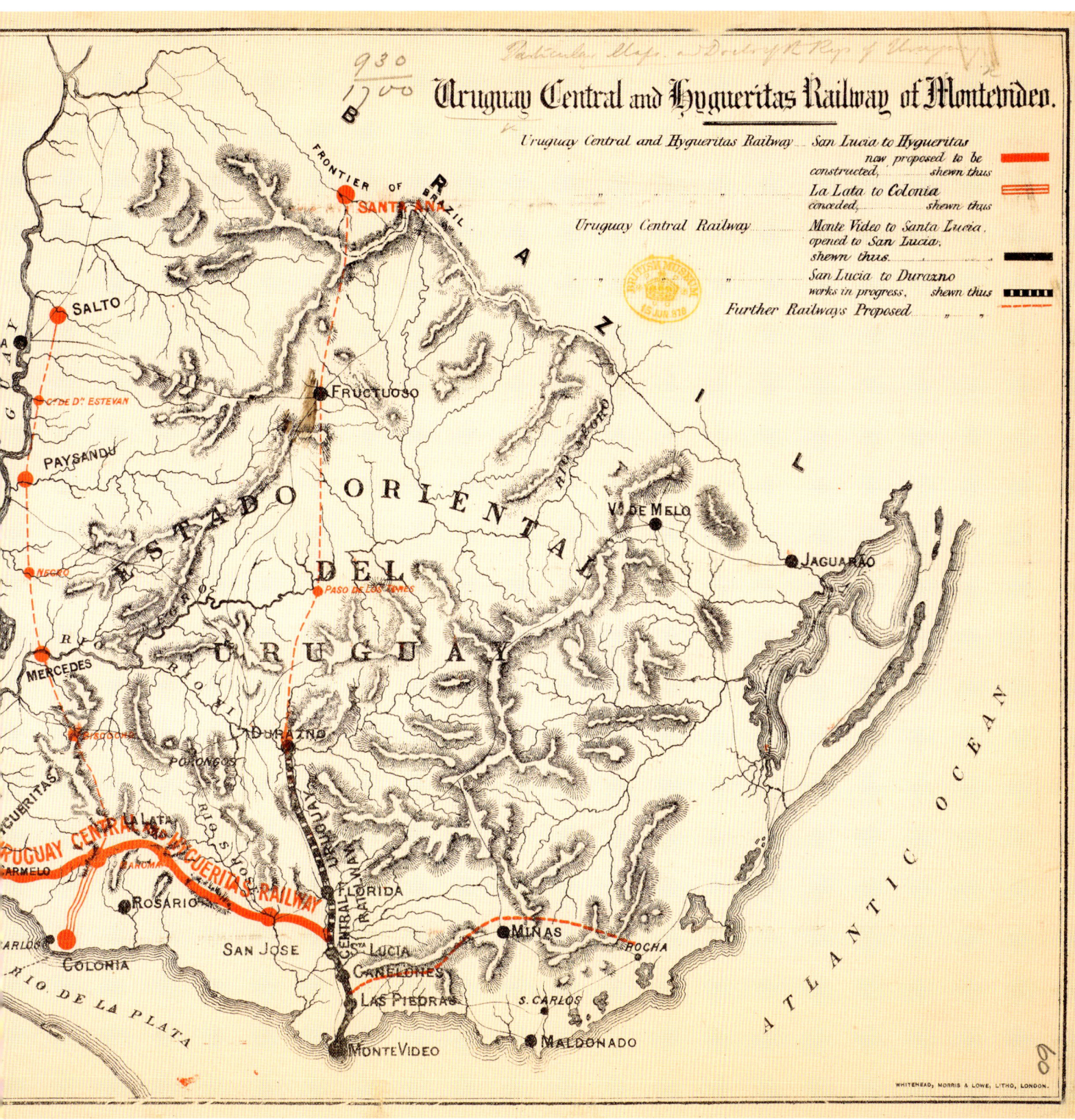

OPENING THE PAMPAS

F.I. Rickard, 'Plan of the City and Suburbs of Buenos Ayres Shewing the Projected Line of Tramway of the Buenos Ayres Street Railway Company Limited Marked Red', 1870.

These three maps show the different scales of railway development in Argentina, which grew rapidly in the late nineteenth century, with British investment leading to a high rate of economic development as well as rapidly rising per capita income. These elements were to be ignored in the 1930s when an anti-liberal nationalism blamed Britain's economic role, including in railways, for the country's poor economic performance, but that is not an appropriate way in which to look at the period.

The first railway had opened in 1857 in what is now the Greater Buenos Aires area. Finance was raised by British entrepreneurs such as Thomas Brassey (1805–70). He played a major role in the development of Argentina's railways, particularly for moving agricultural products to Buenos Aires, and thus in serving the Atlantic trading system.

The original companies served River Plate interests as well as London finance: there was a mutuality at play and one seen across the range of companies. The scale of funds available locally was significant, as was the willingness of the federal government to provide supporting land grants, so that in 1888–89 alone Argentina's federal congress was asked to sanction guaranteed concessions with a capital value of £5.5 million. Government profit guarantees were part of the system.

The guarantees underlined the extent to which railways that were not nationalised could still be part of a public-private partnership. Thus, looked at differently, nationalisation was not necessarily a key stage of development, but rather sometimes a response to particular circumstances, political and/or financial. The limited liquidity of states in the nineteenth century was matched by that of companies, thus helping produce a querulous mutual-reliance. That led to uncertainty on both sides, and this pushed up the difficulty of major schemes. So also did the need to pay for locomotives and rolling stock. Some Argentinian nationalists complained about the British influence.

This 1870 map (right), significantly in English, provides a plan of the city and suburbs of Buenos Aires showing the projected tramway lines marked in red, clearly designed to link up the rail stations: the rail lines are in black. The tramways, initially horse-drawn, began operating in 1863 and Buenos Aires was to have the highest tramway-to-population ratio in the world. Although the map suggests a network, initially only the Retiro terminal for the Northern Railway was served; the Constitucíon terminal for the Great Southern Railway followed in 1866. The city got its first electric tram in 1897.

The second map here, from 1885 (following pages, left), shows the railway that was planned, by British engineers, to go from Buenos Aires to the major river port of Rosario. Mariano Billinghurst (1810–92), the son of a British immigrant and an Argentinian mother, was the concessionary and a railway pioneer, typical of what is best expressed as Anglo-Argentinian. Intended by the government partly in order to ensure competition for the Central Argentine line, the Buenos Aires–Rosario Railway reached Rosario in 1885 and services began the following year. Construction was difficult due to swamps and rivers that had to be crossed, requiring bridges and viaducts. In turn, in 1910 a French company opened a competing line between Rosario and Puerto Belgrano, near the city of Bahia Blanca.

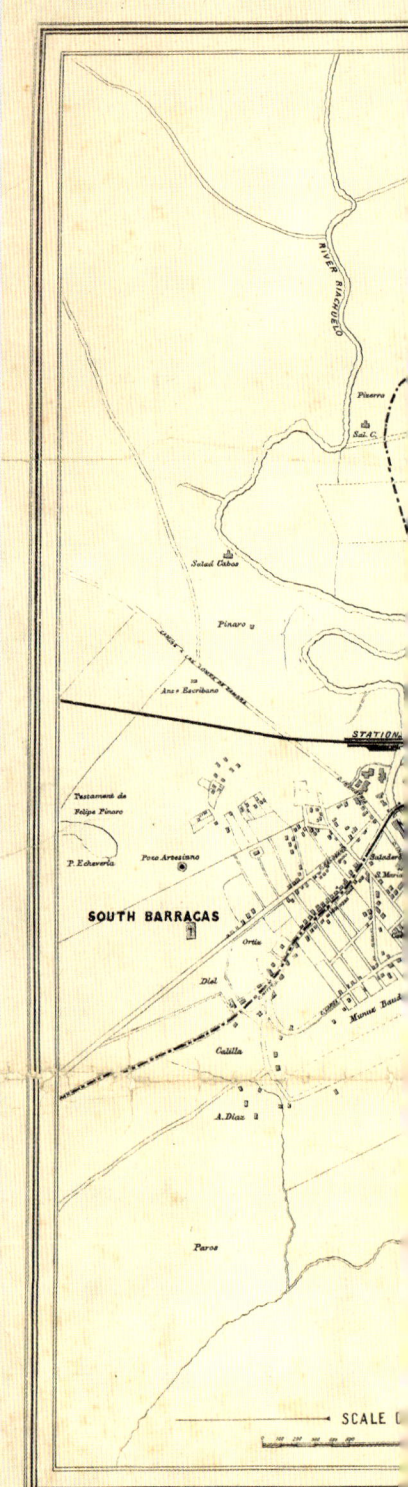

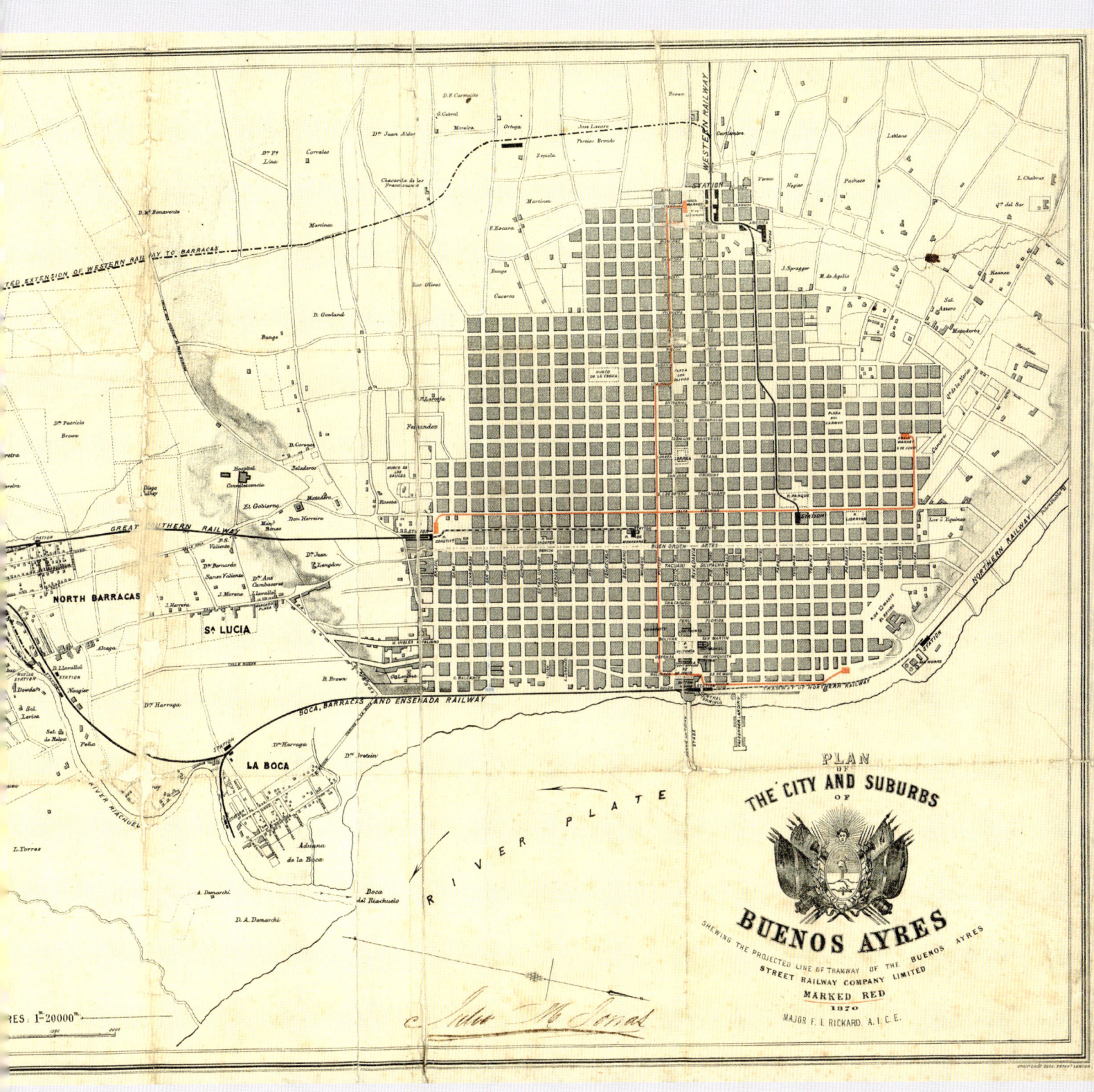

By 1889 (see map, right) there was an extensive rail network on the Pampas and in 1900 Argentina had the seventh-largest rail system in the world. What in 1870 had been only 458 miles of railway in operation, was 2,797 miles by 1880 and 8,771 miles by 1895, thanks in large part to British investment. Capital investment increased from £8.1 million in 1875 to £96.2 million in 1895, passenger numbers increased from 2.6 million to 14.6 million and freight from 0.7 million tonnes to 9.7 tonnes. Part of the Argentinian rail system was used to move beef to ports for export to Europe. This was linked to other technological changes in the late nineteenth century, including large steamships, refrigerated ship holds from 1877 and barbed wire for the management of cattle. The first modern meatpacking plant was built near Buenos Aires in 1883. Serving foreign markets attracted investment, not least in railways, by providing a clear revenue stream. In 1913, when Argentina took 41 per cent of British investment in Latin America, 34 per cent of investment was in railways (compared to 38 per cent in government bonds). This represented a foreign debt that posed a challenge to overall solvency. (Separately, it was British railway-builders who helped introduce football to South America, notably to Argentina and Uruguay.)

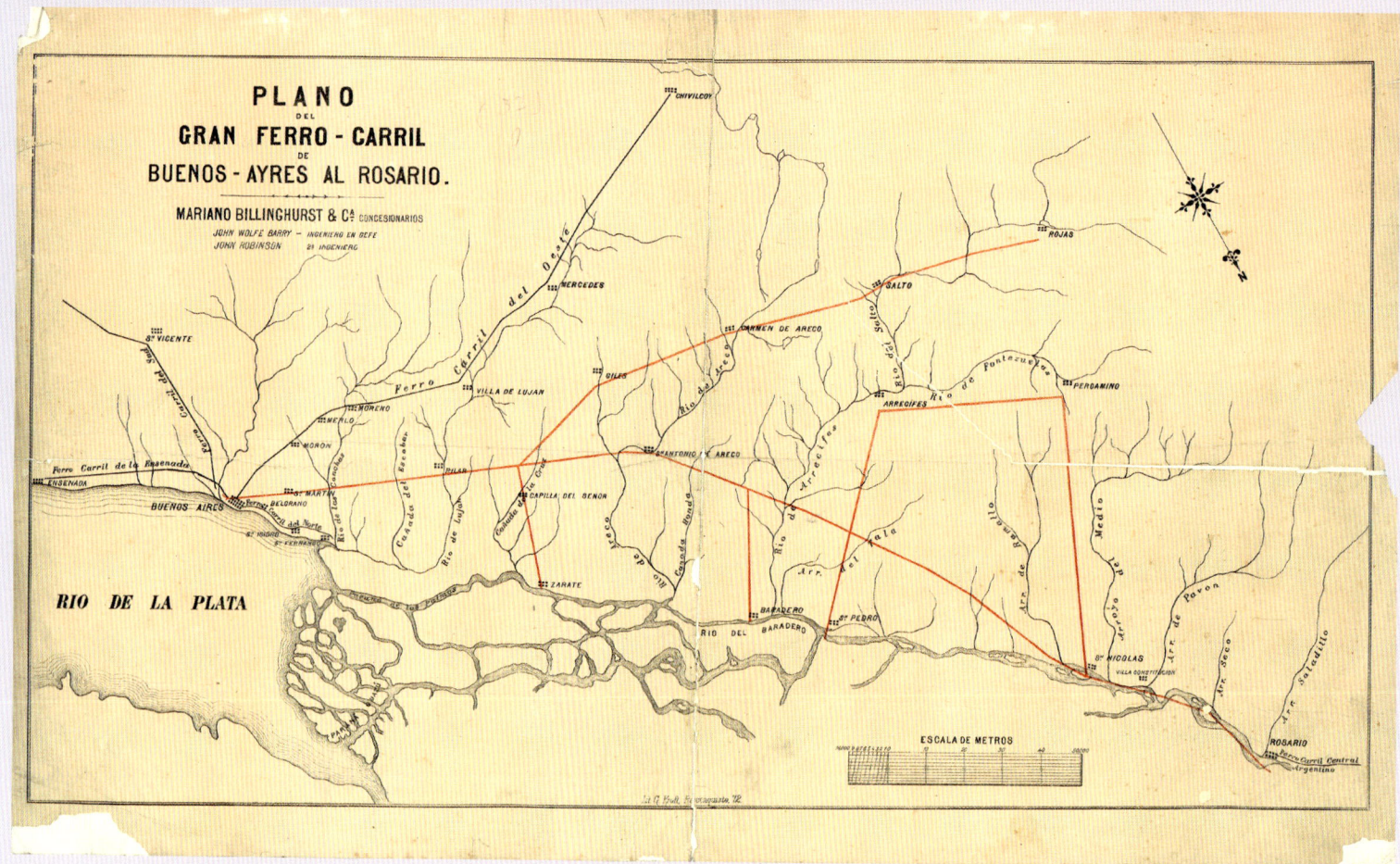

UNDER THE ALPS
Martin Wanner, 'The St Gotthard Railway', 1880.

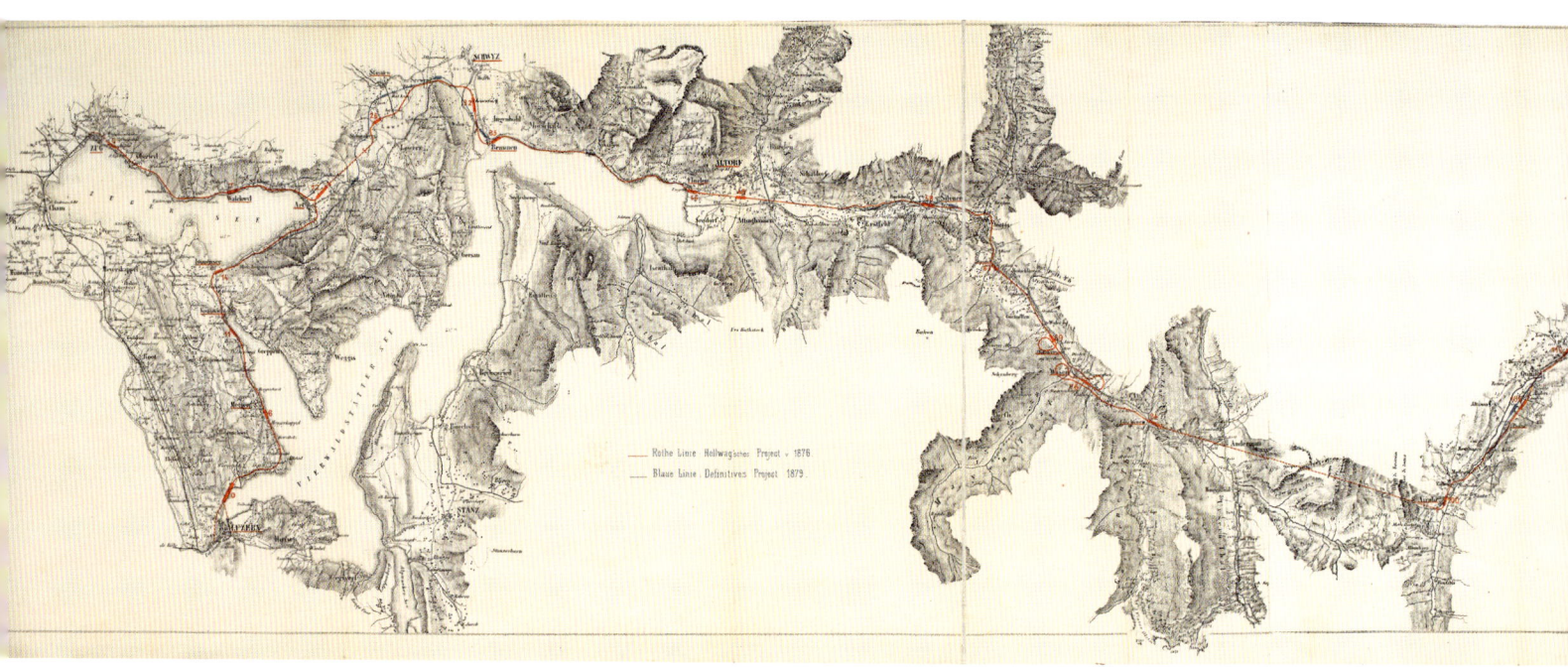

The combination of distance, route and detail poses problems for the mapmaker, which can result in long, thin maps (such as the 1897 fold-out map by Carl Spitteler). Published in Berne in 1880 by Martin Wanner, this map provides greater detail of an otherwise difficult route by using separate maps.

Crossing the Alps was a goal actively pursued from the mid-nineteenth century. This reflected not only the wish for links within countries but also the strategic competition between them, not least that of France and Austria within northern Italy, a competition that led to a major war in 1859. Major tunnel systems were built at the Semmering (1848–54), Mont Cenis (1857–71) and Gotthard (1872–82) passes. Each was an impressive engineering achievement. The mile-long Semmering Tunnel was designed to link Vienna to its Adriatic port at Trieste and thus to help in the consolidation of Austrian control in northern Italy. The Mont Cenis pass, in turn, was intended both to help strengthen Piedmont, a state that still ruled Savoy, and also to provide improved links with France.

The Mont Cenis Tunnel, otherwise known as the Fréjus Rail Tunnel, was at the time (at 13.7 kilometres/8.5 miles) twice the length of any tunnel previously constructed thanks to dynamite and pneumatic drills, powered by compressed air, having enabled construction to proceed quicker than envisaged. The drills were adopted after being demonstrated to the builders in 1855.

The Gotthard line was intended to help Switzerland, not only because all the prior north–south rail lines were outside its boundaries, but also to link Italy with its new ally Germany. The two had co-operated against Austria in 1866. German, Italian and Swiss public and private investors all contributed to the line, which was built and operated by the Gotthard Railway Company. When opened it included the world's longest rail tunnel, built using dynamite, mechanised tunnelling machines, compressed air and water jets. Many workers died in accidents and four were killed when a strike was suppressed by the police in 1875. The Swiss added fortresses to protect the tunnel in the event of

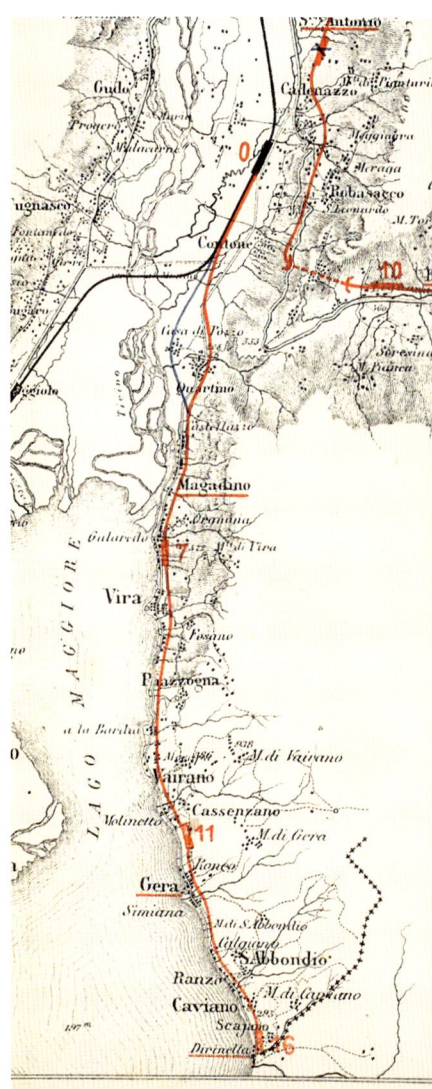

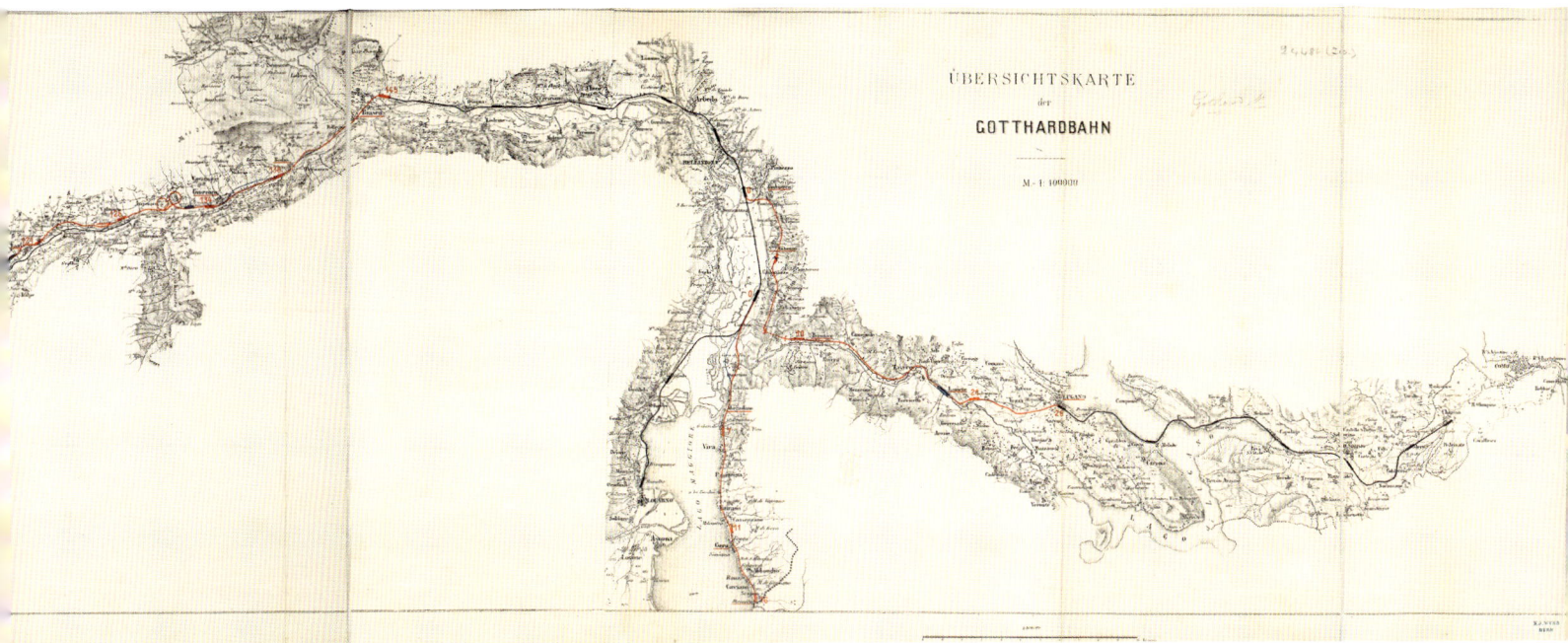

invasion. Similar precautions were to be taken with other tunnels.

High explosive had transformed the challenge posed by mountains, and from the 1880s mechanical drills also became important. The Simplon Tunnel followed in 1906 and the Lötschberg in 1913. Working conditions were harsh and dangerous. Accidents killed 67 people during the construction of the Simplon Tunnel, and troops and vigilantes were used to intimidate anyone protesting the conditions. The Simplon trains were operated with electricity not steam, because the latter posed major problems in underground systems, including those in cities.

Ironically, 1882 saw an intimation of major change. 'The Coming Force – Mr Punch's Dream,' John Tenniel's *Punch* cartoon of 6 December 1882, satirised the enthusiasts for electricity, capturing their sense of excitement and anticipation of new frontiers as an earlier coal-based world of fog, chimney sweeps and coalmen succumbs.

2. SPREADING THE NEW AGE 1860-85 69

MILITARY MOBILISATION AND RAIL

British War Office, 'Military Map of Great Britain: Distribution of Troops 1st May 1883', 1883.

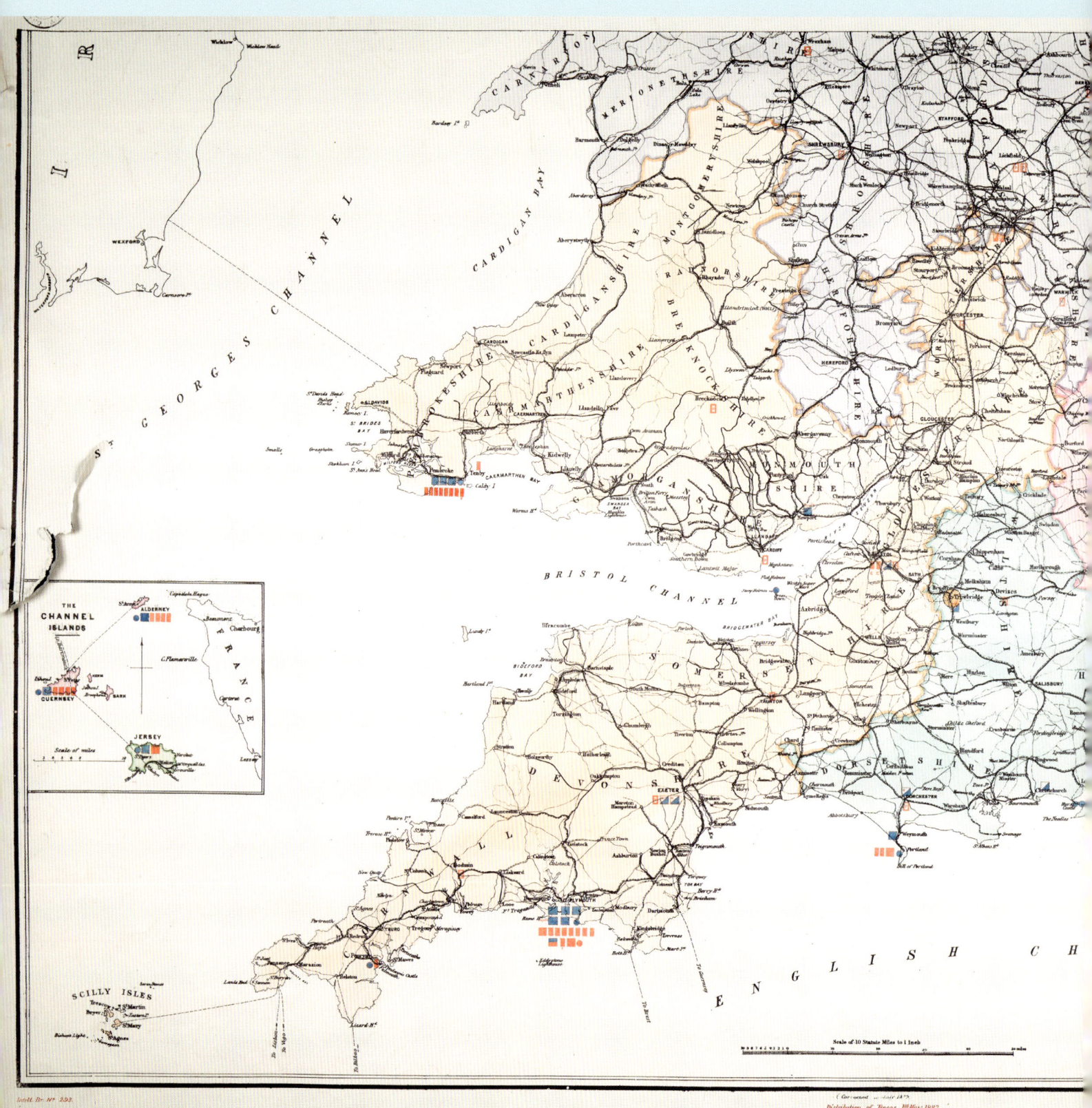

This base map was lithographed at the Intelligence Department of the War Office in October 1881 and as a result it offers the rail system as the background, overprinted with the distribution of troops as at 1 May 1883. There was no accompanying explanation for the map but it underlines the significance of the rail–military interaction in that period. Military Intelligence was tasked with mobilisation plans. As a reminder that rail mapping was often not brought to fruition in any activity, there were no annual manoeuvres between 1873 and 1898. There was a partial mobilisation for the Egyptian campaign in 1882, but there was no crisis in 1883 when the map was produced. The mobilisation plan for home defence, such as it was, drawn up by Robert Hume in 1875, was not updated by serious planning until 1886, when Henry Brackenburg drew up the new one. This map in part reflects the location of regimental depots following the Localisation Scheme introduced in 1872, which had given every cavalry and infantry regiment and artillery brigade a fixed depot and recruiting area.

The defensive focus was on the south coast – and more particularly the protection of the major naval dockyards against French attack, with the Channel Islands (shown as an inset, left) and Dover also seen as exposed positions. Rail served to provide the links for defence. There was no substitute for rapid movement and the military had major requirements for bulk movement, notably for artillery and cavalry. Maps were important for enabling a systematic process of effective and rapid decision-making and for the implementation of strategic plans in terms of timed operational decisions and intended tactical actions.

Somewhat differently, a generation earlier a British military railway had appeared in Captain Frederic Brine's *Map of Sebastopol and Surrounding Country* (1857). Based on surveys by the Royal Engineers, the map showed the positions during the siege of Sevastopol in the Crimean War (1854–56). The 19-mile-long 'Balaklava Railway' was laid in 1855 inland from the anchorage of Balaklava, first worked by horse, and was taken up only in 1856.

2. SPREADING THE NEW AGE 1860–85 71

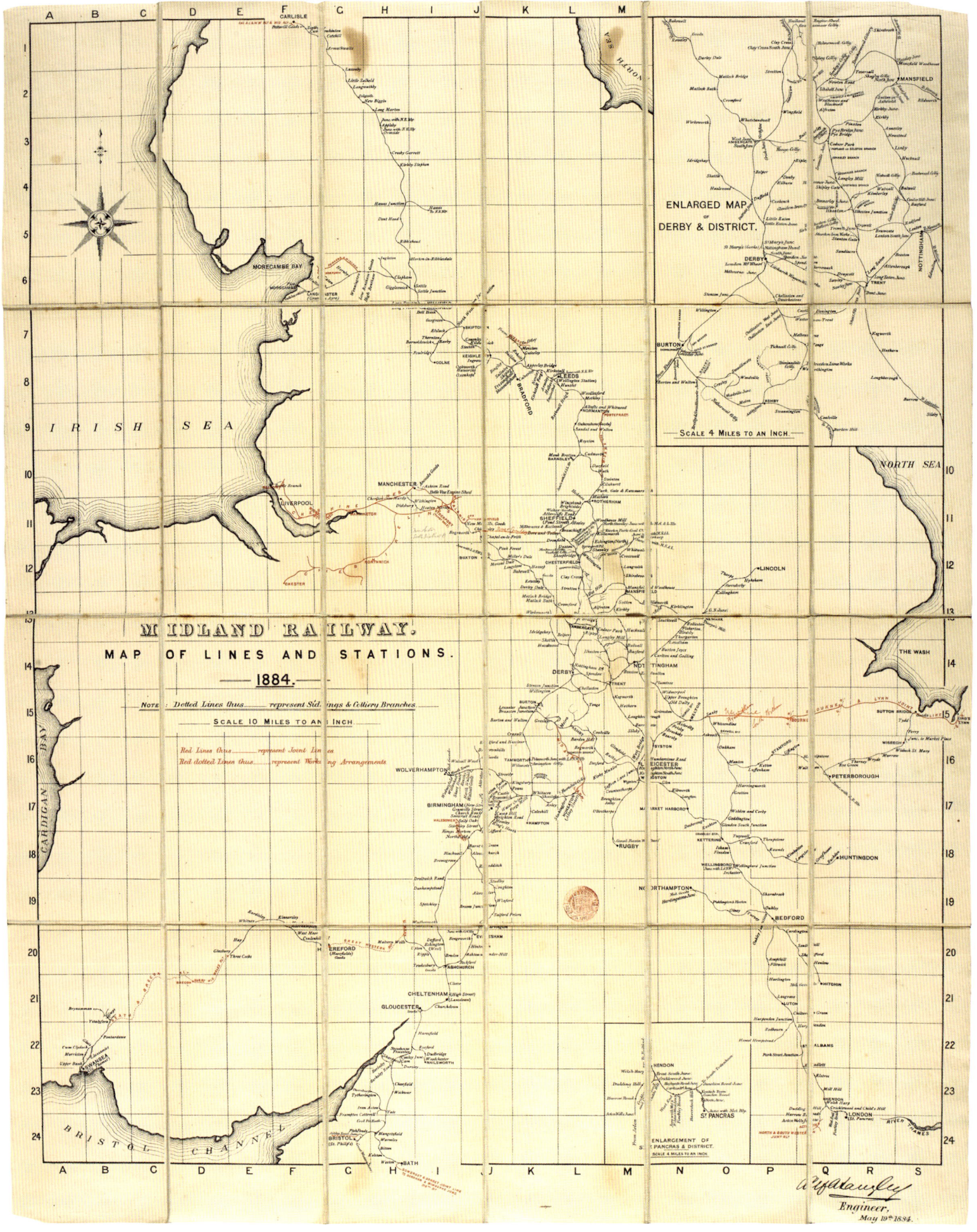

A LAND CRISS-CROSSED WITH RAIL LINES
Alf A. Langley, 'Midland Railway. Map of Lines and Stations', 1884.

Established in 1844 and based in Derby, the Midland Railway was one of the largest in the country. The company originated as a merger of the Midland Counties Railway, the North Midland Railway and the Birmingham and Derby Junction Railway. The Birmingham and Gloucester Railway was added in 1846. Derby, as the map shows, was extremely well connected in terms of regional and inter-regional routes: it was important that it was north of Birmingham. Moreover, the Derbyshire coalfields provided a significant source of income, a position consolidated in 1845–47 as the company absorbed the Sheffield and Rotherham Railway, the Erewash Valley Line and the Mansfield and Pinxton Railway. Other early acquisitions included the Leicester and Swannington Railway and the Leeds and Bradford Railway. There was acute competition with the other major railways, not least over services to Birmingham and Nottingham. Through Midland services to London were introduced in 1858 over Great Northern Railway (GNR) lines, followed by the building of a Midland line from Bedford, with St Pancras station and its hitherto unprecedented single-span overall roof opened in 1868 on the site of a slum called Agar Town.

The map offers guidance to a range of lines: red lines, such as those to King's Lynn, Liverpool and Chester, represent Joint Lines; red dotted lines, such as that to Hereford, Working Arrangements; and on the inset map faint lines are for colliery branches, such as to Ticknell Colliery, east of Burton. The annotation on the map by the engineer Alf A. Langley very much reflected the sense of a map in progress. There was no allowance for terrain, whether mountains or rivers.

Meanwhile, across Britain rail was transforming the local economy. With railways creating ready access to nearby iron ore mines and to coke supplies from Durham, four major iron and steel works were established at Workington, Cumbria, in 1862–74, and the population there rose from 6,467 in 1861 to 23,749 in 1891.

Furthermore, companies and towns that wished to stay at the leading edge of economic development had to become, and remain, transport hubs. Thus, in Britain the biscuit industry showed the significance of rail. In Carlisle, Jonathan Dodgson Carr adapted a printing machine to cut biscuits, which replaced cutting by hand. Helped by the city's position as a major rail junction, both local and regional, and as a leading station on one of the two routes to Scotland, he was able to sell his products, notably Carr's Table Water Biscuits, throughout the country. Similarly, another biscuit-maker in another key rail station was Huntley & Palmers, supplier of sponge fingers to the nation, based in Reading.

More generally, the economies of production of scale combined with rapid, inexpensive, high-volume transportation, entrepreneurialism, capital ability, company formation and advertising through national media made national products both possible and formulaic, as with the drinking of Burton-on-Trent ale in London, a process that was served by warehousing built under the platforms at St Pancras.

SCANDINAVIAN RAIL

Printing Office of the General Staff, 'Maps of the Extent of Sweden's, Norway's and Denmark's Railways', 1863 and 1880.

Published in Stockholm, these two maps indicate not only the rapid increase in the Scandinavian railways but also the distribution of activity. Broken lines identify railways under construction, while the colour captures the gauge used. From 1864 to 1905, Norway was under the Swedish crown. The Denmark–Germany border depicted is that after German territorial gains in the war of 1864, and the 1880 map also shows both the expansion of railways in northern Germany and the linkage to the Danish system.

The maps show internal waterways including rivers, which indeed could be a formidable problem, notably west of Stockholm. No attention in comparison was devoted to mountains, even though they were a major factor, and notably so on the Sweden–Norway frontier. No railways crossed this in 1863, but by 1880 there were two lines in the south, where Christiania, the capital of Norway, was then the name for Oslo.

In addition, in a far more ambitious project, there was an attempt from Sweden to extend the line inland from Sundsvall to the Norwegian frontier, whence there was a line to Trondheim, the Meråker Line, which was built from 1875 to 1879, although not officially opened until 1882. The line from Sundsvall became the Central Line. The line faced the problems of landslide. As the map indicates, there was also an attempt to join that line to the line northwards from Stockholm and Upsala (Uppsala).

The development of Danish railways owed much originally to a British civil engineering partnership, Peto and Betts. It went bankrupt in 1866 and the Danish government took control in 1867, subsequently acquiring other railway line companies, including that from Grenaa to Randers and Aarhus marked on the map.

The first Norwegian lines reflected the country's economic role as a producer of raw materials, in this case timber, with the horse-powered Damtjern–Storflåtan railway opening in 1805 and the Gjøsbubanen following in 1825 and switching to a stationary steam engine.

The 68-kilometre-long trunk line between Eidsvold and Christiana (Oslo) opened in 1854, connecting with steamboats on Lake Mjosa. Steamship lines were the major competition for the railways in Scandinavia, not least because they provided more flexibility. Robert Stephenson was the line's chief engineer, and British shareholders provided half of the capital. The financial success of this line encouraged confidence among investors about developing further Norwegian lines.

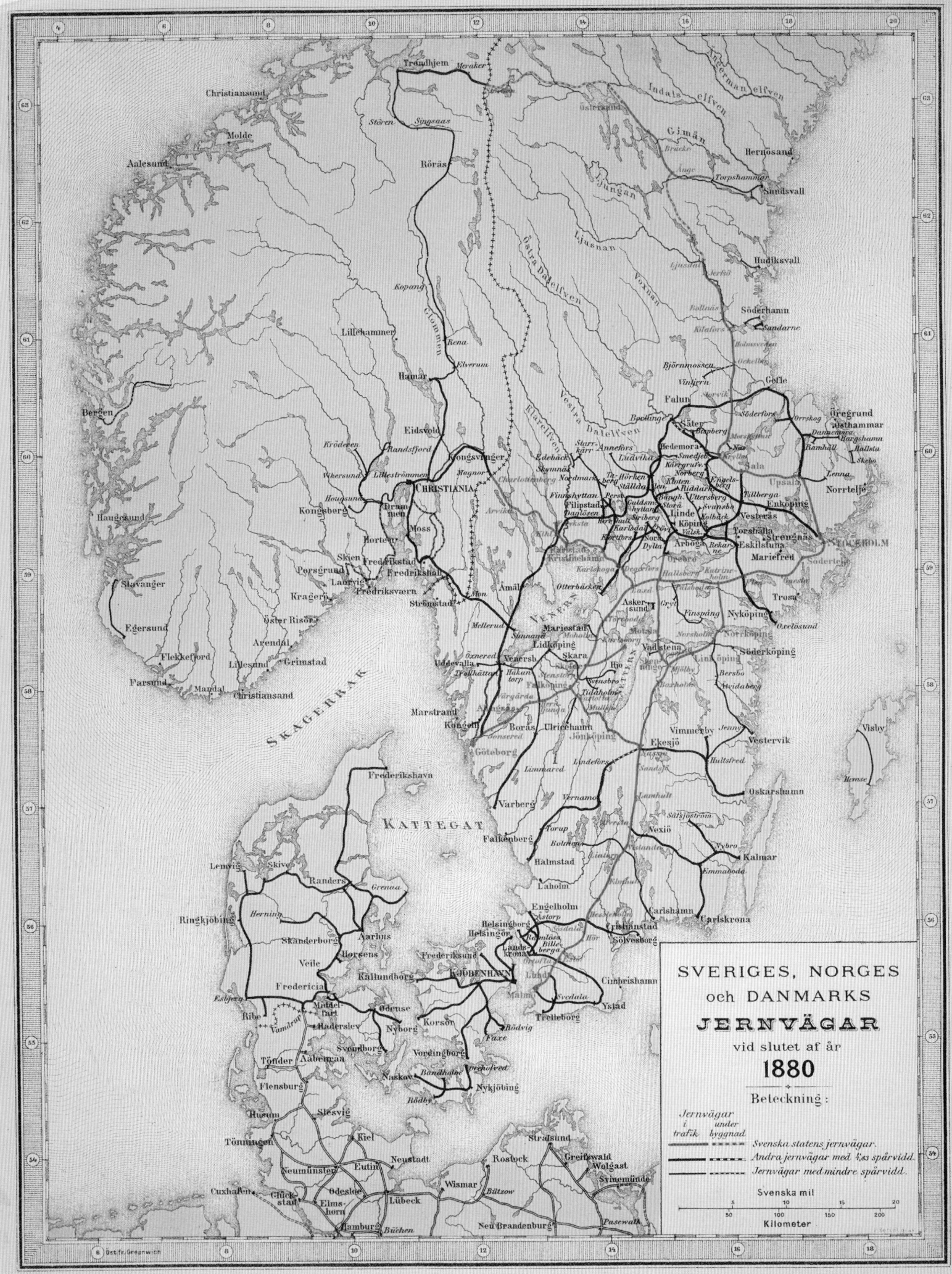

LINES ACROSS THE LOW COUNTRIES
J. Smulders & Co., 'Railway Map of the Kingdom of the Netherlands, 1877.

With the ready examples of Britain and Belgium to follow, the Dutch faced difficulties without similar coal or metallurgical industries. The lack of much coal (most was imported from Germany) meant the economy in the Netherlands did not see the degree of industrialisation experienced in Britain and Belgium, or indeed in northern France and western Germany. This was serious for the railways, not only because the locomotive fuel, coal, had to be imported, but also because the limited industrialisation affected demands for rail services, especially of freight. Indeed, the existing provision of waterways was able to meet most needs, and extended to regular services for passengers. Furthermore, limited industrialisation ensured restricted profits across the economy, which constrained investment and indeed liquidity.

In addition, while flat terrain was generally regarded as desirable, the flatness of the terrain in the Netherlands was made difficult for rail construction by the waterlogged character of much of it, especially in the most economically advanced province, Holland, and by the need to bridge many rivers.

The first Dutch train ran in 1839, covering the 16 kilometres from Amsterdam to Haarlem. King William I supported the scheme, in part to compete with newly independent Belgium. This independence ensured that Antwerp, which prior to the French Revolution had been shut to international commerce, was now a major rival to Dutch overseas trade.

Subsequent expansion saw the Amsterdam to Haarlem line reach Rotterdam in 1847, thus providing a key link to the major port, while another line reached from Amsterdam to Utrecht (1843), Arnhem (1845) and the German frontier (1856). By 1860, only 325 kilometres of railway had been built, but that year it was decided to construct an extensive rail network to serve all areas of the country. The model was the French 1842 plan (see pages 34–35). The government thereafter built most of the lines, but some were constructed by private railway companies. Even remote corners were reached, such as Franeker in 1904. Electrification followed from 1908. Waterborne trade remained significant, but the railways were now the leading sector.

The First World War hit the profitability of Dutch railways, in large part due to a British blockade aimed at stopping German trade. The Dutch were also affected by a shortage of coal. In 1938 two private companies were merged to create the Dutch state-owned company after the remaining shares in the companies were purchased.

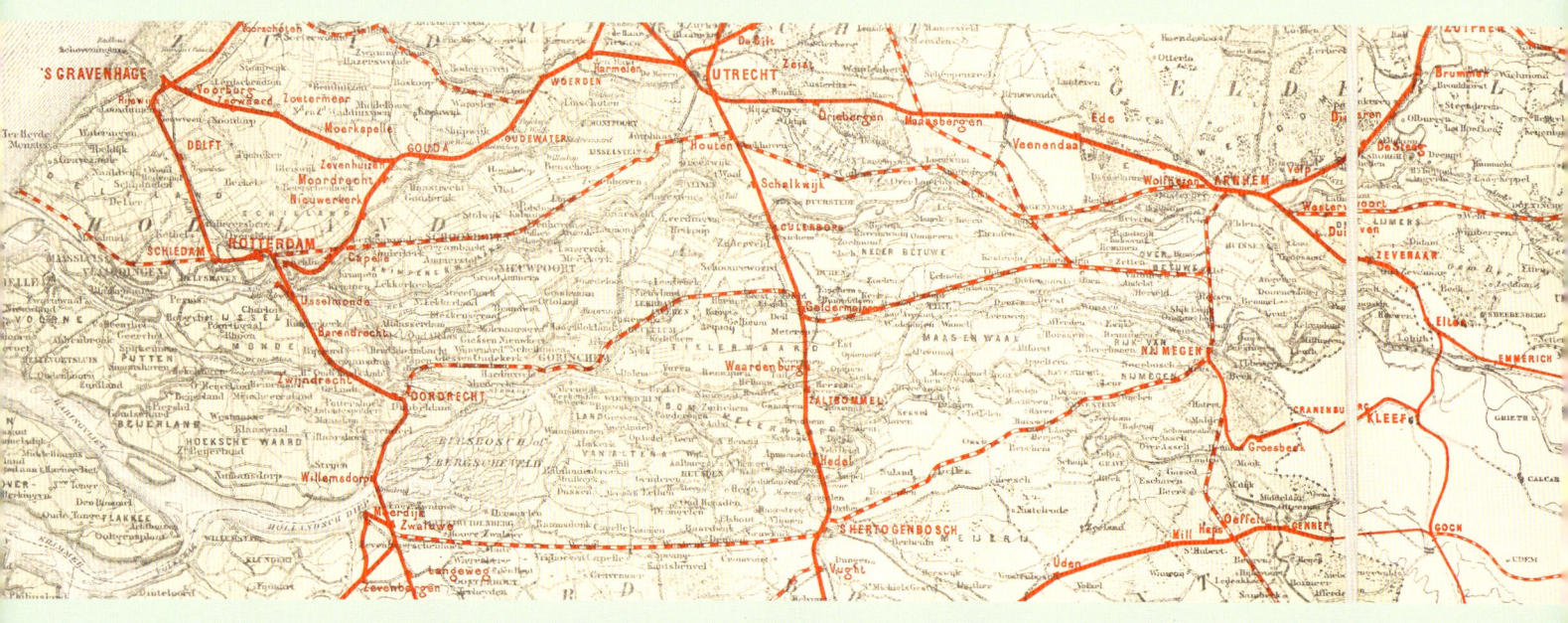

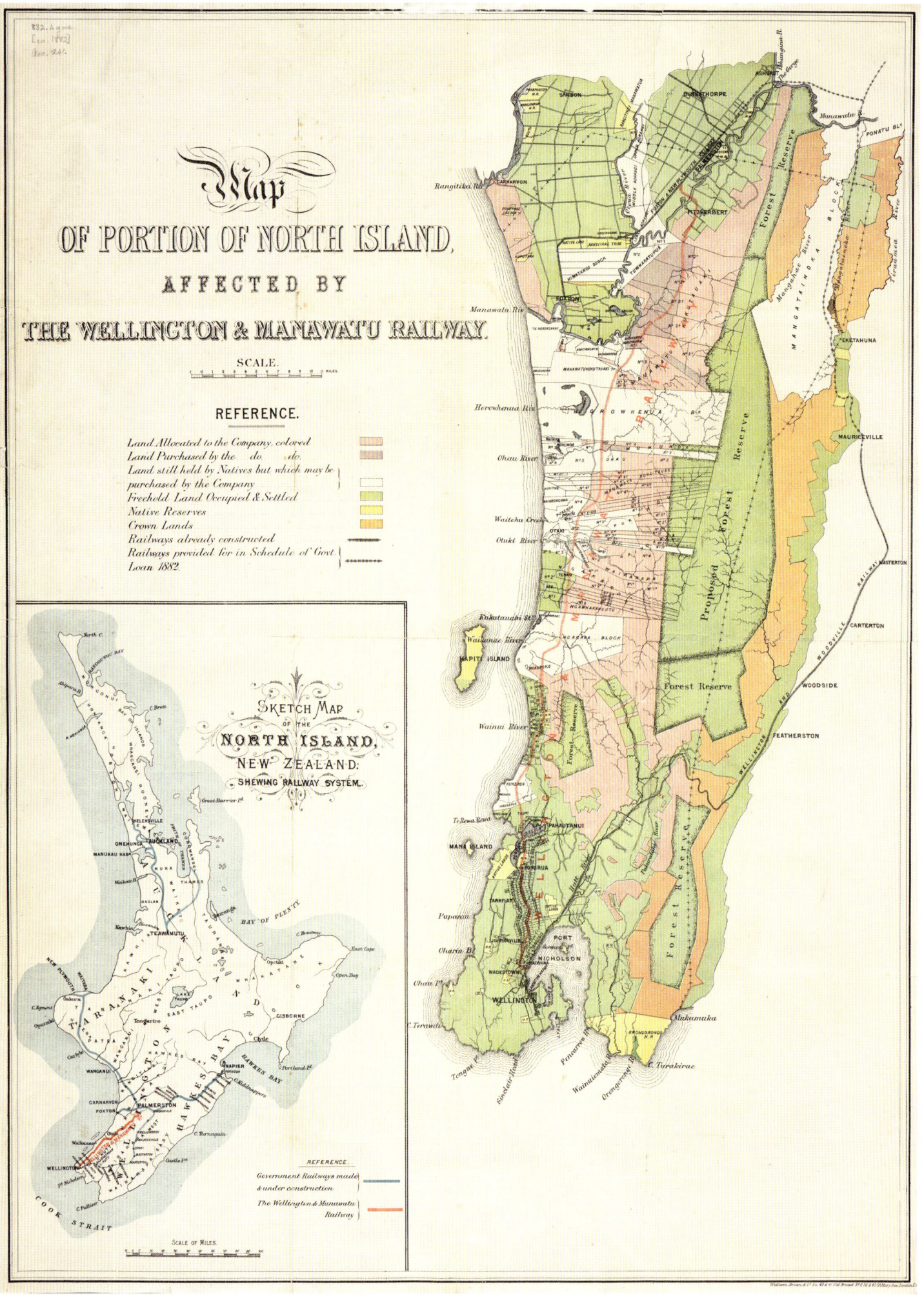

NEW ZEALAND RAILWAYS

Wellington and Manawatu Railway Company Ltd, 'Map of Portion of North Island Affected by the Wellington & Manawatu Railway', 1883.

This 1883 map very much emphasises the role of land use in the development and financing of rail construction and, more particularly, the interplay of company land and lands under other ownership (including Maori and Crown lands, as identified in the coloured legend). The main map indicates the complexities of this situation and its linkage to rail plans. Thus, the key shows the different forms of ownership along the line northeast of Wellington. In turn, the insert map covers the North Island and depicts the railways currently completed or under construction and the Wellington and Manawatu line. The latter was unusual because it was a private company (founded in 1881) that constructed and operated the line until 1908 when the New Zealand Government Railways acquired it. The government had originally intended to build a line for this route – indeed, a short section to Wadestown had been built in 1879, but the September 1879 general election resulted in a new ministry that wanted to cut expenditure and so in 1880 it decided to abandon the line.

In response, a Wellington initiative floated the company and signed a contract with the government in May 1881. This entailed major grants of Crown land (210,500 acres in total) to offset construction costs, while the company agreed to buy what had already been built and to complete the line within five years.

Under the Railways Construction and Land Act of August 1881, joint stock companies could build and run private railways, but these had to be built to the standard New Zealand rail gauge of three feet six inches and connect with the government's railway. The company did so at Palmerston Green, rather than at Foxton as the government had wished, which represented a longer route and therefore a more expensive one.

The landholdings shown on the map helped to provide revenue as well as usage for the line, which made considerable profit and this led to good dividends, the first of which was paid in 1891. The line ran for 83 miles and was difficult to construct, with the terrain north of Wellington initially mountainous. Subsequently, there were rivers to cross, such as the Waikanae, which required a three-span timber truss, as well as peat swamps.

Businessmen linked to the company, notably William Hort Levin and John Plimmer, had towns named after them, which was a frequent occurrence in the late nineteenth century when railways created towns as much as served existing ones.

GEOPOLITICS AND WAR 1885–1918

The unique Western experience of creating a global network of empire and trade, one that Britain came to dominate, was based on a distinctive type of integration between economy, technology and state formation. This experience led to a major and profitable development in the intensity of relations between parts of the world, which was an aspect of modernity. Rail was important to this, as well as in creating efficiencies within core areas of the global system, notably Western Europe (see pages 118–125) and the American Northeast and Midwest (see pages 106–113), and areas that would ascend to prominence. The latter included Japan (see pages 116–117), Argentina and also Western colonial areas that had achieved *de facto* independence, notably Australia and Canada. In effect, as with the American West, their rail systems were also part of the core – in that they shared the commercial and geopolitical interest of the latter. This linkage was also expressed by the use of core finance in expansion, which came in particular from London, Paris and New York. Yet, as shown by the drives for rail-linked modernisation seen in Japan and, less successfully, China and Turkey, it would be a mistake to think simply in terms of a Western capitalist core driving developments elsewhere. Instead, there was considerable autonomy in the process. This reflected a variety of factors, including the degree to which 'the West' was not a united force, but, rather, deeply divided, not only with rival states – especially Germany opposing Britain – providing competition, but also with competing companies. There was also the determination of non-Western governments not to be directed by their need for Western capital and expertise (see pages 114–115). In part, this replicated the earlier success within Europe in avoiding British control.

The competitive framework changed with technology, finances and politics. Steamships were highly significant for not only long-range routes but also shorter-range ones. Moreover, they changed with time, indicating that the challenge of the new did not begin with the internal combustion engine or powered air flight. Steamships became more effective, not least because improvements in engine design and technology increased power and cut coal consumption, and thus the need for coaling. In the 1860s, high-pressure boilers were combined with the compound engine, and in the 1870s the triple-expansion marine engine was introduced; although it was not introduced into a major ship successfully until 1881. The triple-expansion marine engine was to be followed by the water tube boiler, and in the late 1890s steam turbine-powered ships. Railways provided a different aspect of this relationship between parts of the world, one that faced more difficulties than steamships.

Politics were explicitly to the fore, both in international competition and in domestic matters. The latter can be harder to recover historically. The first came to the fore in 1914 with the First World War; and the situation and development of railways in that conflict is covered in the close of the chapter.

The point that has to be taken away is that the history of rail cannot be written without discussion of war. This is true for other conflicts as well as both world wars. To treat peace as the norm is mistaken and reduces the significance of rail in recent history. Moreover, geopolitical consideration of rail in peacetime frequently related to rail's wartime possibilities. Indeed, the Trans-Siberian line (see pages 84–87) demonstrated this in the 1904–05 Russo-Japanese War. At the same time, the discussion and often visual presentation of geopolitics could make such lines appear necessary and inevitable, which downplays the element of choice. Russia would have found it easier to expand its rail system near its European frontiers and this could well have stood it in better stead than the Trans-Siberian Railway. To a degree, it was as if the transcontinental railways in North America drove forward the emulation by Russia. Rail both seemed to fix new acquisitions and interests, and to serve as a base for additional ones.

This factor was underlined in the decades prior to the First World War due to a lack of clarity as to probable opponents. Thus, for Britain, the major supporter of colonial rail schemes, it appeared likely prior to the mid-1900s that Russia and/or France rather than Germany would be the foe. This led not only to sensitivity about their rail plans, but also concern about how best to respond. As Germany emerged as a more likely foe, so anxiety switched to German rail plans, notably in the Ottoman Empire, and the suggestion that these threatened British interests in the Middle East and on the 'Route to India'. As a reminder that rail has to be considered alongside alternatives, this anxiety focused on the Berlin to Baghdad railway plan, and indeed an extension to Basra (see pages 94–97), but the British 'Route' was itself totally maritime, with the Suez Canal replacing the hitherto need to round the Cape of Good Hope.

Yet, however fascinating, thinking in geopolitical terms underrates other core spheres and requirements of rail, most notably local and regional services, and the quest for profit. The latter interacted with engineering issues and operating problems, both of which could be described in terms of gradient.

In the 1890s the emphasis in these equations still very much appeared in terms of rail, not only maintaining existing services but also constructing new ones. This was seen in the most 'mature' system, that with the longest development, Britain (see pages 126–131). Thus, in southwest England, there continued to be expansion – for example, with branch lines to Lynton (1898), Lyme Regis (1903) and Appledore (1908), and indeed the Torrington to Halwill Junction line was not built until 1925. However, that was a work-creation scheme, and 1903 saw a sign of change when the Great Western Railway (GWR) began a bus service from Helston to the Lizard as an alternative to a new railway. This, the first motor bus service in Britain, was matched by 1914 by other services. Change was very much in the air prior to the outbreak of the First World War.

Following pages: A railcut through hard rock during construction of the western portion of the Amur Railroad section of the Trans-Siberian Railway, 1908–13. Ultimately, it was this herculean construction project that effectively united the European and Asian parts of the Russian Empire.

RUSSIA ACROSS ASIA

Stanford's Geographical Establishment, 'Stanford's Map of the Siberian Railway, the Great Land Route to China and Korea', 1904.

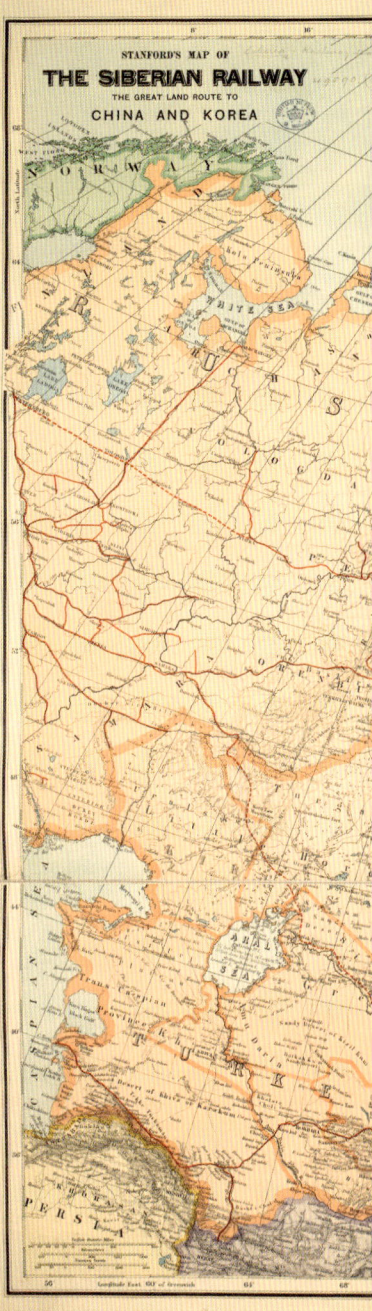

Stanford's, the source of this 1904 map (the eastern section of which is shown here), was the key cartographic publisher of late Victorian Britain. Edward Stanford, the founder of the firm, was to die that November. Born in 1827, he had a youthful interest in maps and in 1852 became a partner in the London map shop of Trelawney Saunders, which he took over in 1853 as Stanford's. Stanford's became the leading publisher and seller of maps in London and that at a time when the British Empire provided a major and changing topic for mapping. Maps from Stanford's were associated with accuracy and clarity.

As with the maps produced in Edinburgh by W. & A.K. Johnston, there was an engagement with the issues of the present day, offering an overlap with newspaper maps on current issues, which frequently included railways. The 1898 Johnston 'Map to Illustrate the Question of the Upper Nile' included the recently built British railway across the bend of the Nile that was designed to support the advance on Khartoum. So also with the 1899 Johnston map on 'The Transvaal Question' in southern Africa.

The Trans-Siberian Railway fired the imagination of commentators in large part due to its scale. This was totally different to the transcontinental railways of North America and was seen in that light. This was particularly so in Britain, where concern about Russian expansionism was longstanding and had become more acute from the 1850s, with war the result in 1854–56 in Crimea. This conflict showed that Russia could be a formidable opponent. There were fresh Anglo-Russia war panics thereafter, notably in 1878 and 1885, with conflict coming close in each case. The situation then appeared to deteriorate further for Britain, with Russia allying with France, which created renewed anxieties for Britain, and notably so at the time of the Fashoda Crisis of 1898, an upheaval over the control of southern Sudan.

It might appear surprising that British commentators should be so anxious about the Trans-Siberian Railway. The pivot of concern in 1854, 1878 and 1898 had been Russian entry into the Mediterranean from the Black Sea, thus threatening what British commentators saw as the 'Route to India' via the eastern Mediterranean, more specifically Egypt. Indeed, this concern helped lead to the establishment of the British positions in Cyprus (1878) and Egypt (1881). The latter was followed by the British commitment to the Suez Canal, which enhanced the significance to Britain of the Mediterranean and represented a defeat for French interests in Egypt.

A second axis of British concern was that of a Russian advance across Central Asia so as to threaten not only British India (which included what is now Pakistan) but also interests in Afghanistan. This concern had nearly led to war in 1885, although, as Robert, 3rd Marquess of Salisbury, then Secretary of State for India, pointed out in Parliament in 1877, in a remark more generally true of rail and geopolitics:

> '...a great deal of misapprehension arises from the popular use of maps on a small scale. As with such maps you are able to put a thumb on India and a finger on Russia, some persons at once think that the political situation is alarming and that India must be looked to ... use a larger map.'

The Trans-Siberian Railway posed different strategic concerns. Most immediately, it appeared to challenge the British interest in China and, with that, to offer the risk that Russia might become the dominant power there, not least because Russian ambitions in Manchuria would press on the capital, Beijing. With the exception of Hong Kong, Britain's interest in China was not territorial but commercial. As a result, British policymakers were concerned about the expansion of foreign, notably Russian, control in China as this was likely to limit British commerce as well as increase Russian influence over the Chinese government. Secondly, there was the fear that Russia would become a stronger maritime presence in the Pacific and thus again challenge Britain in a new area of imperialism

and one, moreover, where Britain had existing interests to protect in the shape of Australasia, British Columbia and a range of Pacific islands.

In 1858–60, Russia had annexed the area between the Amur Valley and the Pacific and in 1860 Vladivostok was established as a naval base, followed by growing Russian interest in Sakhalin and the Kuriles. This expansion was a menace to Japan, which Britain saw as a potential strategic partner.

Thus, this transcontinental railway offered what was clearly seen as a threat. As the map makes clear, the railway entailed a Russian presence in Manchuria, because the route to Vladivostok in Russian territory only – in other words on the northern bank of the Amur – was a far longer one. The construction of this line greatly concerned Japan and encouraged it to go to war with Russia in 1904, a war in which it achieved major success.

INTO AFRICA

Geographical Section of the General Staff, 'Mombasa–Victoria (Uganda) Railway and Busoga Railway', 1916.

The line from Mombasa to Lake Victoria was an exercise in British imperialism that lacked commercial logic but that was designed to make sense of the recently established British colony of Uganda, which had become a protectorate in 1894, but without the maritime access of Britain's other African colonies. Uganda therefore posed some of the issues of access, control and influence seen with the Afrikaner (Boer) republics of Transvaal and the Orange Free State, and also with the establishment, over the following decades, of colonial control in both Sudan and northern Nigeria. In all cases, rail was seen as part of the solution, which only underrated the degree to which rail could also pose its own problems.

Pursuing the interests developed by the Imperial British East Africa Company, founded in 1888, there was a wish to protect both the trade route of the Nile and Protestant natives. The territory initially consisted largely of the kingdom of Buganda, but treaties with Toro in 1900 and Ankole in 1901 extended this control. There was also a concern with thwarting expansion by other powers, a concern that became more significant as the British became anxious about German plans in East Africa as well as French power spreading eastward from West Africa, which was the root cause of the Fashoda Crisis of 1898.

Established to the east of Uganda in 1895, the East Africa Protectorate was the basis for Kenya, with its capital Mombasa until 1905 and then Nairobi. British settler interest developed from 1902, with the land in the Kenyan Highlands regarded as promising and the East Africa Syndicate becoming a key player. The settlers pushed hard for a railway and because settlement was seen as a way to anchor British interests, there was government support for it, which was crucial, as public funds were central. Indeed, looked at in one light, the decision for the railway was taken relatively lightly and without an adequate assessment of costs and benefits. This led, at the time and subsequently, to casting round to find reasons to support investment.

The building of the railway began in 1896 and in 1901 it reached Kisumu on the eastern shore of Lake Victoria, with the line opened for goods and freight in 1903. Port Florence, the name chosen by the British, reflected its chosen role for shipping to Entebbe in Uganda, and the first steamboat belonged to the Uganda Railway company. Built in Paisley, Scotland, in 1890, the ship was moved in kit form to Mombasa, then taken to Kisumu where assembly was completed in 1901. Two other ships were added in 1902 and 1904. The movement of locomotives to the line was also both significant and difficult. The line relied in part on local workers and in part on Indian labour, which was important more generally to British railway activity around the Indian Ocean.

The ambition underlying the scheme can be seen on the map where the proposed line is marked in dotted red lines, a line that was not actually built. On the north shore of Lake Victoria, as shown on the map, there is a line from Jinja north to Namasagali. Port Bell to nearby Kampala is the other line shown, and again it supplemented the steamer links on Lake Victoria. It took 12 hours for steamers to sail from Kisumu to Port Bell, which was named after Henry Hesketh Bell, who in 1905 became Commissioner of the Uganda Protectorate, subsequently becoming governor. A keen believer in the potential of railways, Bell pressed for a line from Jinja to Kakindu and then on to Lake Kioga north of Namasagali and, secondly, a line from Kampala to Lake Albert, which is seen in the northwest corner of the map.

This railway construction should not be seen in isolation. In part, there was a linkage with the expansion of British power southwards from Egypt into Sudan. This expansion had proved abortive in the 1880s, with the humiliating failure to relieve besieged Khartoum in 1885. In contrast, the British succeeded at the Battle of Omdurman in 1898 and subsequently captured Khartoum. In 1899 the British followed up with success in southern Sudan. This created an axis of possible linkage with the developing British presence in Uganda and Kenya, one from Alexandria to Mombasa via Khartoum and Lake Victoria. That was

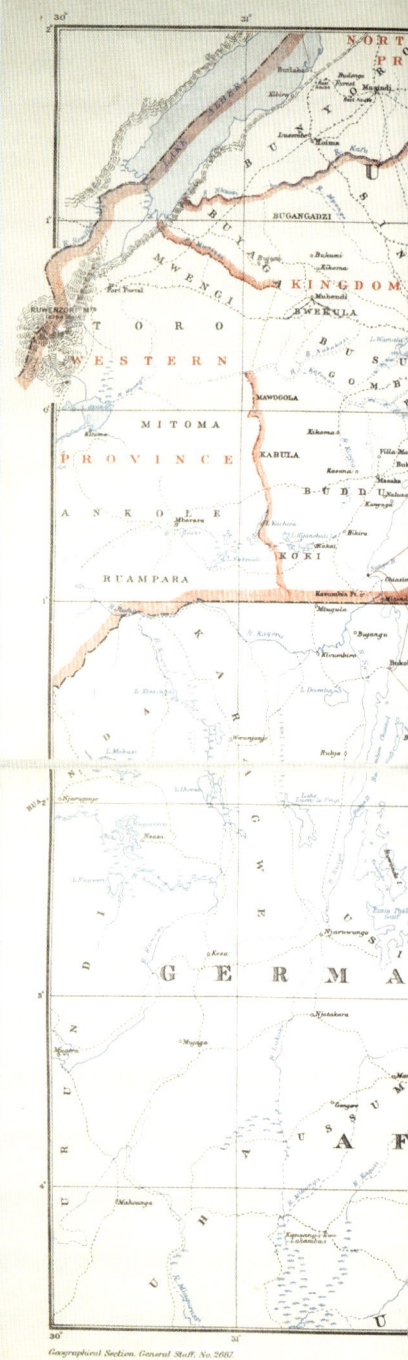

a variant of the Cape [Cape Town] to Cairo railway envisaged by Cecil Rhodes, which he sought to advance in talks with Emperor Wilhelm II in 1899. In practice, more-restricted lines were far more viable, as the *Manchester Guardian*, a critic of Rhodes, argued on 27 March 1899. Kampala was not linked to the coast until 1931 when a line was completed around the north shore of Lake Victoria and the Nile was bridged at Jinja.

RAILS FOR RICHES

Dennis Edwards & Co., 'Railway Map of South Africa with the Routes to the Diamond & Gold Fields', 1895.

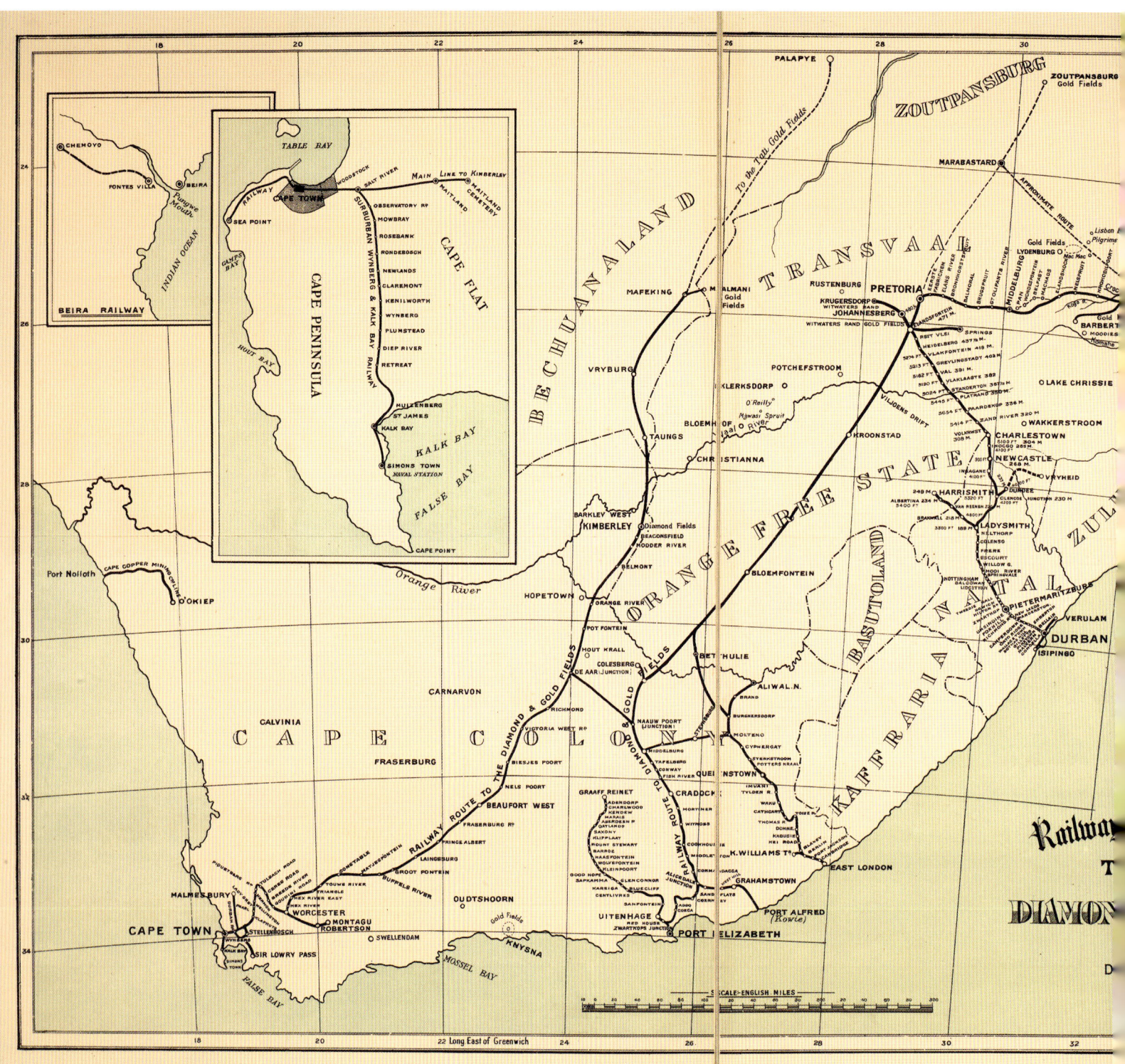

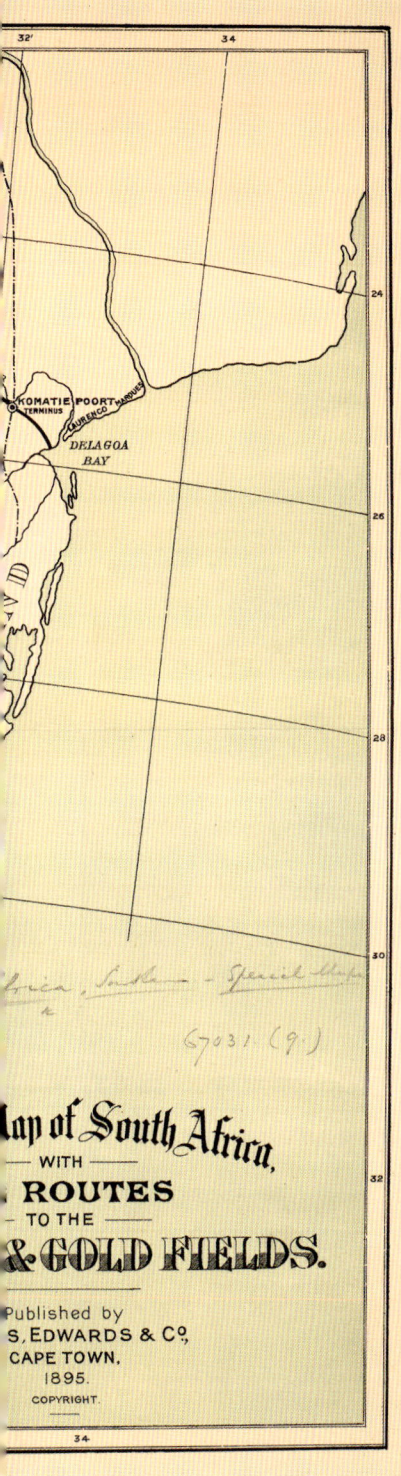

Published in Cape Town by Dennis Edwards and Co., a major publisher whose works included *The Gold Fields of South Africa* (1890), this map shows the significance of rail links in southern Africa. More notably, the map reveals the situation of the Afrikaner (Boer) republics – the Orange Free State and Transvaal – with regard to the British colonies and, more specifically, the leading ports: Cape Town, Port Elizabeth, East London, Durban and the anchorage of Delagoa Bay (now Maputo Bay) in the Portuguese colony of Mozambique, where Beira, which became the outlet for Rhodesia, is also situated. As the map (see inset, top left) shows, Beira is the natural outlet for Transvaal. There had been rival claims to it by Britain, Portugal and Transvaal from the 1820s, but in 1875 the French president, Patrice de MacMahon, as arbitrator, ruled in favour of Portugal. As a result, the railway station on Delagoa Bay in what was to become the city of Lourenço Marques (now Maputo) was to sit alongside Praça MacMahon. In 1874 Transvaal decided to build a line and gained a Portuguese government concession. Delayed by the First Boer War with Britain in 1881, the scheme was relaunched with the Netherlands South African Railway Company founded in 1887 and by 1894 the 350-mile-long line was complete, with a key element being the bridge across the Komati River completed in 1891. As a reminder of the continuing role of politics, the Portuguese seizure of the railway led to a dispute with Britain, which intervened on behalf of the shareholders, and the matter was settled by arbitration.

The British wanted the key outlets to be in their colonies. In 1895 the Cape Colony sought to win the trade by cutting the rates it charged, but Transvaal's government charged it very high usage charges. This was the background to the attempt by Cecil Rhodes to overthrow the government of Transvaal by means of an invasion, the Jameson Raid, combined with a rising by the non-Boer population of Transvaal. The failure of the raid in turn contributed to the further breakdown in Anglo-Boer relations and that led to the outbreak of the Second Boer War in 1899.

South African railway history had a relatively late beginning, with the first lines, both short, from Cape Town in 1859 and Durban in 1860. Begun in 1859 by the Cape Town Railway and Dock Company, by 1862 the Cape Town to Wellington line had only reached the nearby Eerste River and Wellington, 45 miles away, a year later. In 1864 the Wynberg Railway Company linked Cape Town and Wynberg.

The situation changed with the Kimberley diamond rush in 1871 and self-government for Cape Colony in 1872. As in Australia, Canada and New Zealand, self-government was important. John Molteno, the first Prime Minister who served until 1878, was very keen on expansion, which required in particular telegraphs and rail. Cape Government Railways was founded and a distinctive narrow gauge of three feet six inches introduced, which made it easier to cross mountainous terrain. Railways were begun inland from Cape Town, Port Elizabeth and East London, all towards Kimberley. The first crossed the Hex River Mountains and the semi-desert Karoo region, reaching Kimberley by 1885. The second met this Cape Western line at De Aar, and the third reached Queenstown in 1880. This helped spur on Natal Government Railways, established in 1877 with the takeover of the private Natal Railway Company. It reached Charlestown in 1891 and Johannesburg by 1895. The Second Boer War (1899–1902) saw the rail companies of Transvaal and the Orange Free State taken under British Army direction. After the war, the companies were merged into the Central South African Railways. In turn, in 1916, this was merged with Cape Government Railways and Natal Government Railways.

WEST AFRICAN STRAINS

British War Office, 'Map of Freetown–Baiima, Sierra Leone, to Illustrate Messrs. Shelford & Son's Report, June 1904', 1904.

The construction of a railway in Africa posed major problems, as did its mapping. Heliozincographed at the Ordnance Survey, this map is from a 1904 official report on British colonial railways in West Africa and it shows the line built in 1896–1903 from the capital at Freetown to Bo, as well as the southern route of a proposed extension to the Liberian frontier at Kenewa. In the event, that route was followed but not quite that far, because the frontier was not reached. Instead the line only went to Pendembu, while the northern and middle alternative routes were not pursued. The names on the map are not always those used today.

Due to the actions of and pressure from British Abolitionists in the 1780s, freed slaves were settled from 1787 in coastal Sierra Leone. The St George Bay Company established in 1790 was brought under the Sierra Leone Company in 1792, the year in which Freetown was established. In 1808, in turn, Sierra Leone became a Crown Colony.

At this stage, British territorial control was limited, but by the 1880s there was greater interest in expanding a presence in response to rapid and ambitious French expansionism in West Africa. In 1895 a boundary agreement with France was concluded, and there was a need to survey to fix the frontier. The establishment of a Protectorate over native chiefs led to considerable resistance, including opposition to the abolition of king as well as to a house tax to raise colonial revenue. In 1898 this opposition was supplemented as the result of another such tax, which led to serious difficulties that continued in the east near Liberia until 1899.

The Mende, Temne and Loko tribes were prominent in the opposition. Although they are not shown on the map, the map has to be understood in terms of this tribal opposition, because it helps to explain the determination of the British to establish a presence. For example, Pendembu was a town inhabited mainly by Mende. The railway showed power as well as determination: it could be used to transport officials and deploy troops as well as deadly weaponry such as the Maxim (machine) gun. The railway also offered

economic opportunity by providing the means for people and companies to transport products to the port of Freetown.

The railway was unusual in being narrow gauge throughout, and most of the line was built with a 1:50 gradient. This avoided the need for complex engineering and powerful locomotives – as was the case in South Africa with the line from Delagoa Bay to Transvaal. There was also no need to bridge or ferry across major rivers comparable to the River Niger in Nigeria. The 30-pound rail of the Sierra Leone Government Railway limited the axle load to five tons, which underlined its character as a light rail route.

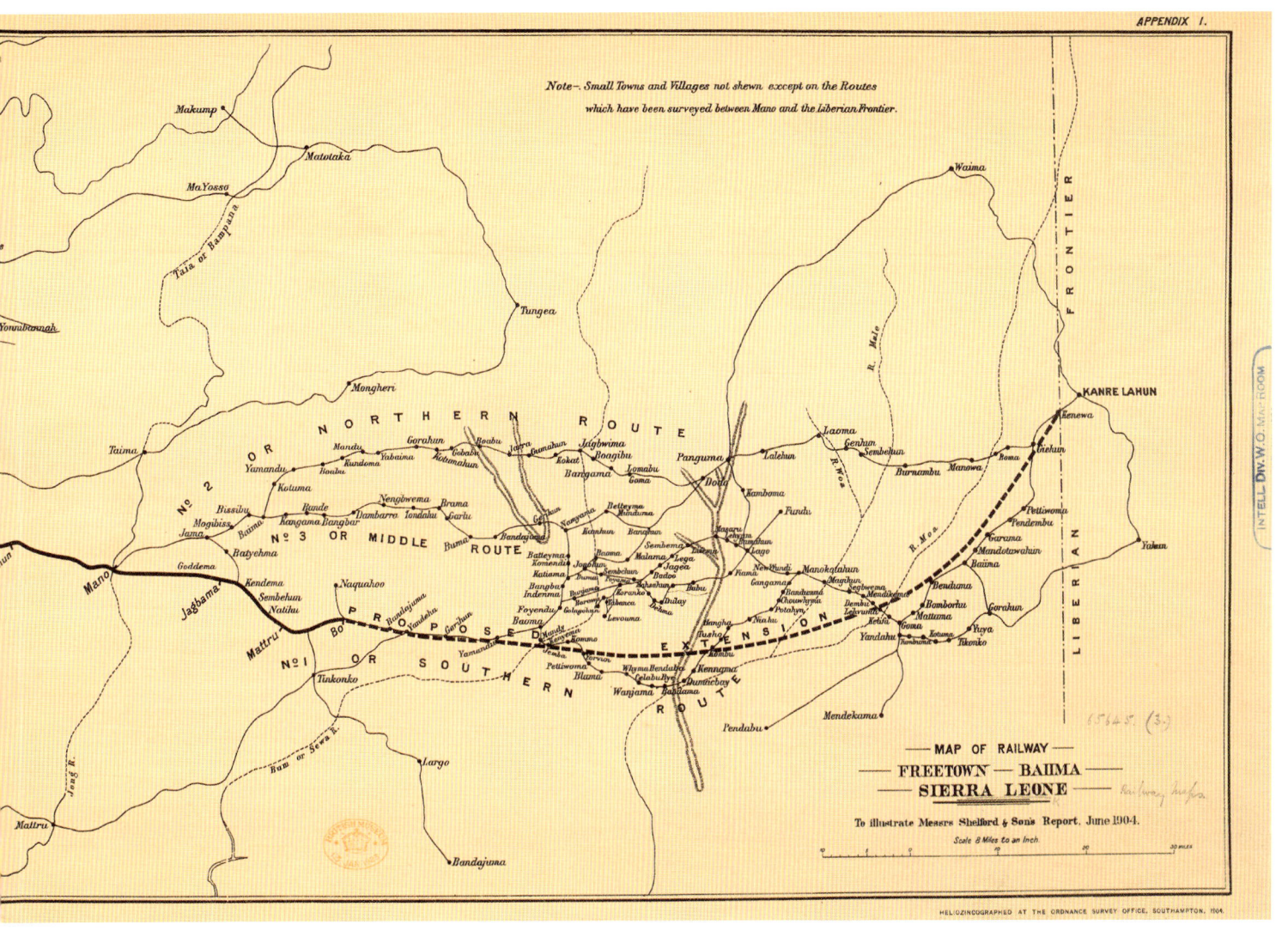

Liberia lay to the east but no line went there to the frontier on the Maro River until 1960. There had been plans in 1871 for the development of a railway in Liberia linked to British loans, but the plans were not pursued; nor were those in the 1920s. In part, this reflected the strength of American commercial interests, notably the Firestone Company, which wanted to focus on Liberia's road links, and on transport to the capital, Monrovia, rather than on links to Sierra Leone.

Nor were there links to the neighbouring French colony of Guinea, although a line was built there in 1904–10, from the port-capital Conakry 345 miles to the city of Kankan in eastern Guinea, which the French had seized in 1891. That line had more logic than the one in Sierra Leone because Kouroussa near Kankan, where the French established a post from 1893, was where the Niger could be reached for trade. On the Niger, the French used steamships to reach their positions downstream, including Timbuktu.

The problems of rail construction and operation in Atlantic Africa were also shown in the Portuguese colony of Angola, where there was the unsuccessful attempt from 1886 to build and operate a line between Luanda and Ambaca.

BERLIN TO BAGHDAD

Justus Perthes, 'The Railway Concessions in Asiatic Turkey in 1914…', 1915, and W. & A.K. Johnston, 'The Short Cut to India', 1909.

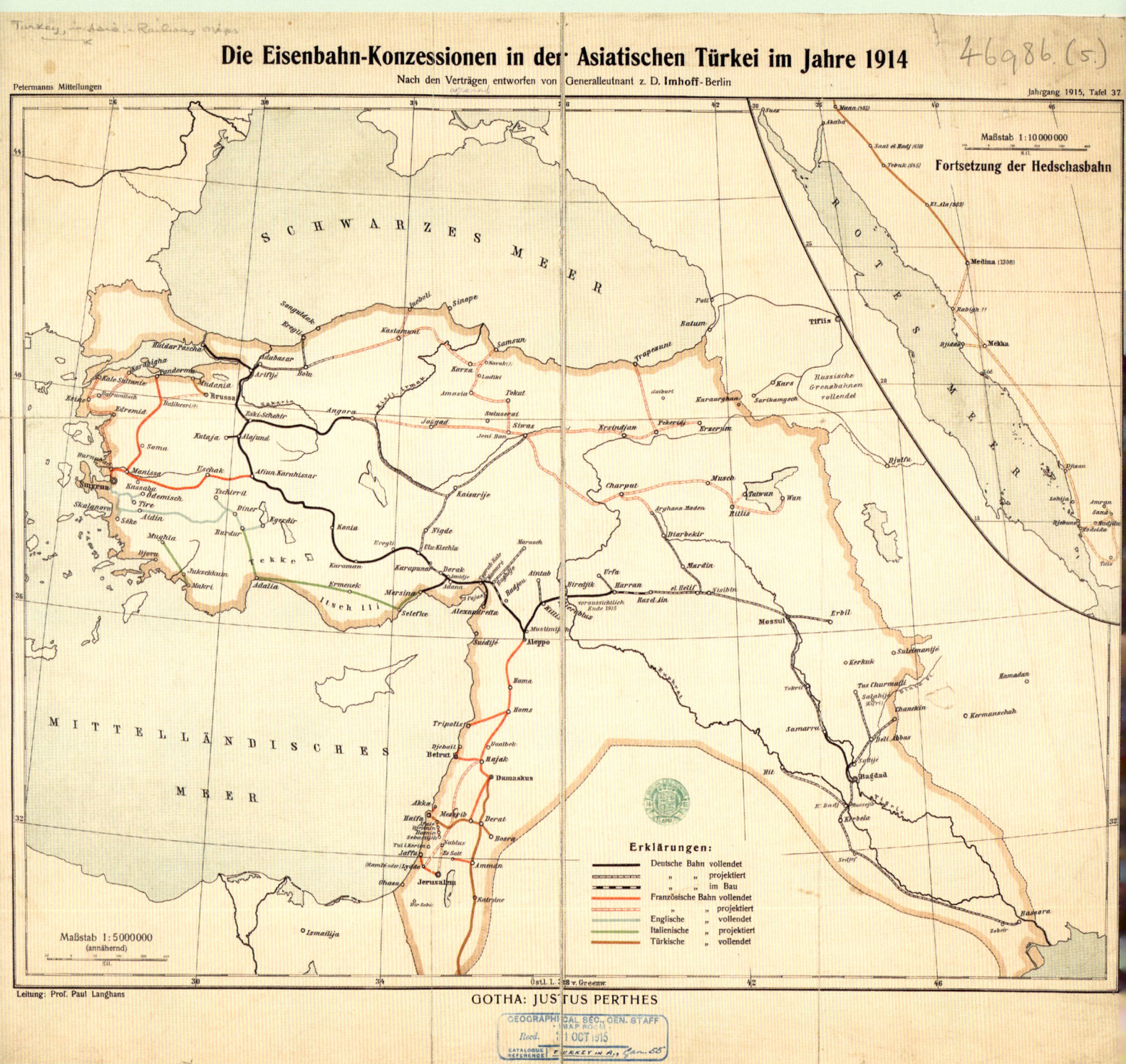

This map (left) of railway concessions in the Asian portion of the Ottoman Empire was produced in 1915 by Justus Perthes of Gotha, the major German cartographic publisher, using information provided by Lieutenant General Karl Imhoff, a German adviser to the Ottoman Army, who became a pasha. This copy was acquired by the British that October by the Geographical Section of the General Staff, indicating its wartime determination to get hold of German maps. The map shows German railways that existed, were projected and under construction; French railways that existed and were projected; British railways that existed; Italian railways that were projected; and Ottoman Turkish railways that existed. In addition, Russian lines near the Ottoman border are shown. The map includes a skilful inset (top right) enabling the display of the Ottoman Hejaz Railway to Medina, and the plan for its southward projection.

The German map does not provide the topography, which is shown in the map (see following pages) produced a few years earlier by the major Edinburgh mapmaking firm of W. & A.K. Johnston to illustrate *The Short Cut to India; the record of a journey along the route of the Baghdad railway* (1909) by David Fraser, a journalist for *The Times*. As well as several Ottoman railways, Fraser's map includes the Russian-built railways in the Caucasus.

The Ottoman Turks turned to foreign capital to develop their railways, and this led to political rivalry. The route reflected strategic considerations: to avoid the possibility of British naval intervention in the Gulf of Iskenderun (also known as the Gulf of Alexandretta) the line was routed inland even though that involved expensive tunnelling. Separately, the Russians did not want a line that threatened their border, which meant opposing the original Ottoman choice of Ankara (Angora in the map) via Kayseri (Kaisarije) and Silvas (Siwas) to Diyarbakir (Diarbekir). Instead, the more expensive route from Konya (Konia) to Adana through the Taurus Mountains was chosen. Any route through mountains was more expensive because of the gradients that had to be addressed, irrespective of the tunnels. In 1906 German interests increased when Deutsche Bank acquired the Mersin–Tarsus–Adana railway. By the end of 1914 significant sections of the line were finished or under construction, but the major tunnels across the Taurus and the Amanus mountains were incomplete.

The line was seen by British commentators as strategic competition, similar to how the British had viewed the Trans-Siberian Railway when Britain and Russia were rivals. As with the latter there were serious engineering difficulties, the Taurus being the equivalent of Lake Baikal. In practice, there were more grave problems confronting Ottoman Turkish rail construction: whereas the Russians in Siberia could draw on the wealth and resource base of their empire, notably tax revenues and an active metallurgical sector, both were absent in the case of the Ottoman Empire and that increased its dependence on foreign capital and expertise.

Furthermore, internal political stability was a major issue. The Young Turks movement that overthrew Sultan Abdul Hamid II in 1908 was divided, leading to a serious power struggle until 1913. Moreover, the Ottoman Turks lost Libya following an Italian invasion in 1911 and were heavily defeated in the First Balkan War of 1912–13. This was not the best basis for bringing to fruition any significant infrastructural project. So also in 1913 with the focus on the Second Balkan War and from 1914 on with the First World War. To foreign commentators, such as the British, it looked like a military timetable drove the geostrategic dimension of rail construction. However, in practice, although a military timetable might seem to make such construction necessary, it also hit at investment in rail. Like the modernisers in Japan and China, the Young Turks were committed to rail but the cost of war meant other needs took priority.

RED SEA RAIL
Justus Perthes, 'The Hejaz Railway', 1906.

Produced by the major German map publisher Justus Perthes, which had an extensive base-map system that could be readily employed, this 1906 map is more accurate than those that stretched the railway into Yemen because the latter was an aspiration rather than reality. The line from Medina, reached in 1908, on to Mecca is marked as under construction but was not built. Nor was the planned extension to Tâif in Yemen. Inset maps set the line in the wider geopolitics of Ottoman railway plans and also offer the branch line from Der'a to Haifa, which provided a valuable port outlet on the Mediterranean. Begun in 1903, the Haifa–Der'a line was inaugurated in 1905. The line was used in part to deliver supplies for the main Hejaz line, which was originally designed to reach Mecca but stopped 250 miles short at Medina, opposition by local tribes having discouraged further expansion. The Damascus to Medina section was 810 miles long and narrow gauge. Postcards (such as this example, right, from 1910, prepared in Cairo by 'Alī Rida Mu'īn) helped to fund the line. This was a novel approach, effectively crowd-funding from among a theoretically large number of devout Muslims whose religious obligation it was to perform the Haj (pilgrimage to Mecca). However, in practice this means did not provide sufficient funds. There were also donations from prominent Muslims, such as the Shah of Iran and the Khedive of Egypt, as well as the sale of honours.

The map is good at presenting the railway and its links with other transport routes in the region. There was competition with the Suez Canal and the Red Sea (Rotes Meer), which, as far as the Ottomans were concerned, was threatened by the presence of British and Italian naval power.

It is very easy to treat the failure to extend the line to Mecca as a defeat but it can be regarded as a sensible response to the local political situation, indeed part of the always changing basis of co-operation that underlay rule. Instead of extending the line, the government turned anew to the Grand Sharif of Mecca to provide a crucial degree of local support. Moreover, there was a shift of interest towards the idea of building shorter lines, including to the ports of Aqaba, Jidda, Yanbu and Hodeida. However, this idea involved more reliance on the vulnerable Red Sea route and for this reason the Minister of War opposed it.

By 1914 the availability of the line meant the number of pilgrims transported had risen tenfold from 30,000 in 1912. During the First World War the line was attacked by the local Arab tribes, encouraged by British agents, notably, but not only, T.E. Lawrence. These attacks in 1916–18 not only inflicted direct damage but also forced the Ottoman Turks to devote considerable attention and resources to protect the line. In June 1916 Jeddah was captured, in July Mecca and in September Ta'if, although Medina resisted attack that October. In January 1917 Wejh was captured and attacks on the railway began, not least the use of British-supplied mines from February. Locomotives, tracks and bridges were all destroyed that year, and in July Aqaba fell. From there, attacks were launched on the railway east of the Jordan Valley.

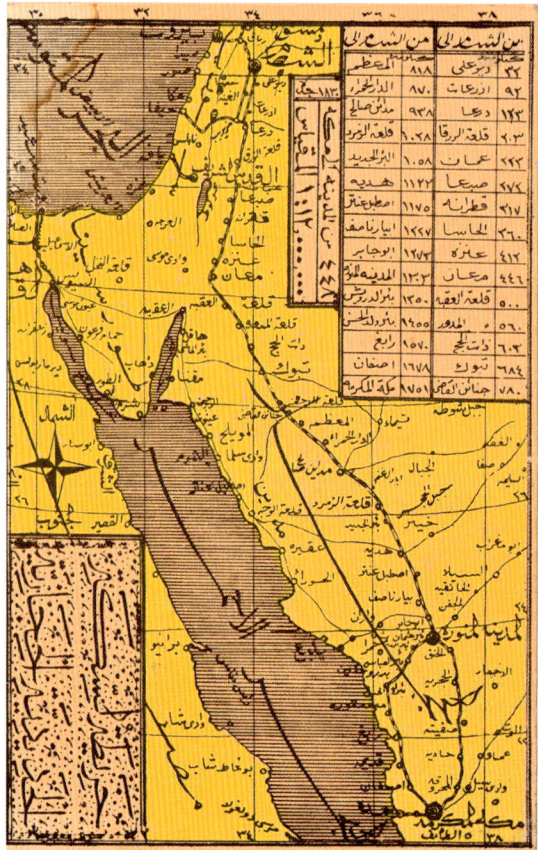

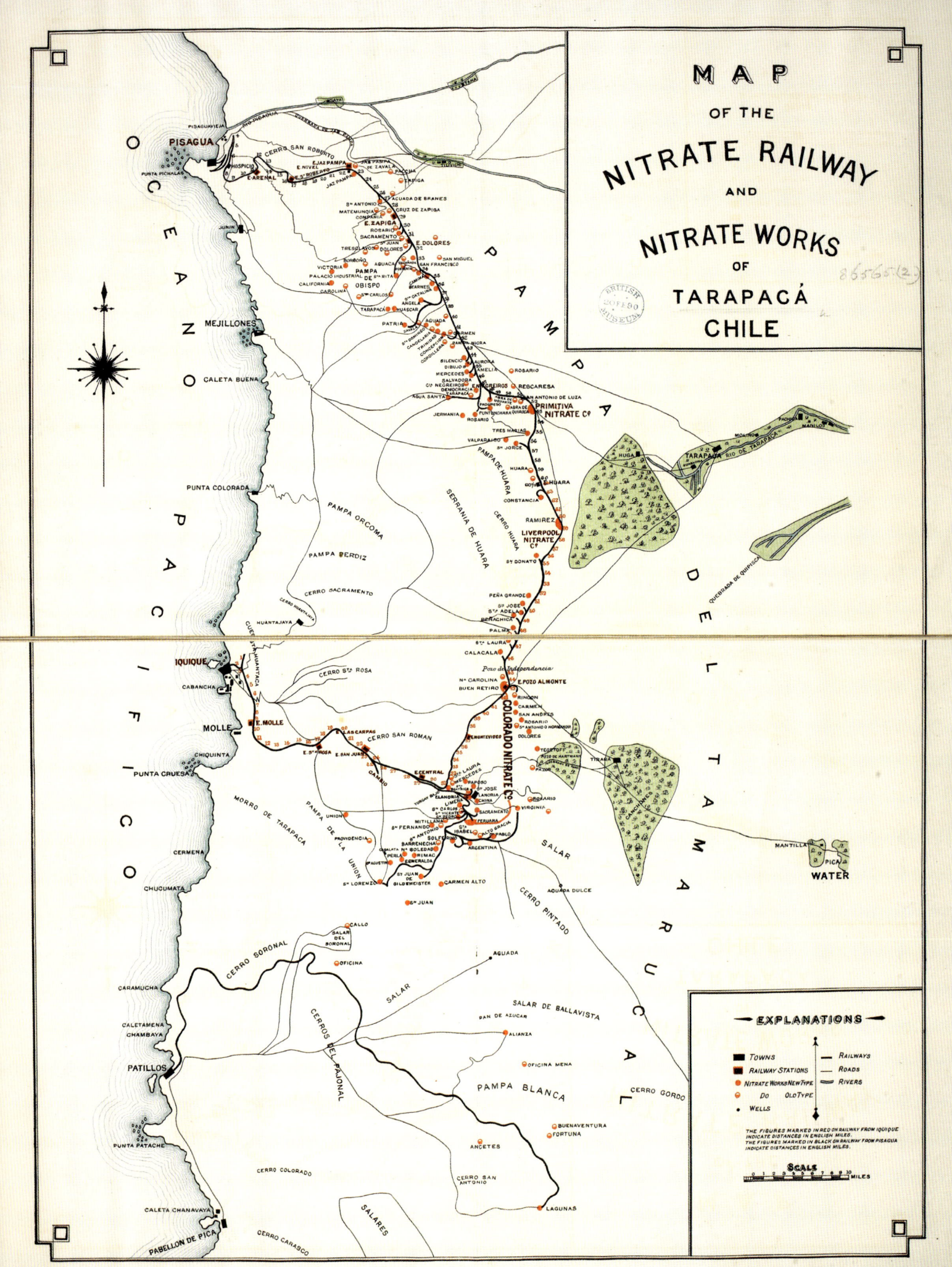

FERTILISERS FOR THE WORLD
W. Morrison & Sons & Mallet, 'Map of the Nitrate Railway and Nitrate Works of Tarapacá, Chile', c.1890.

Published in London, this undated map of the Nitrate Railway (stamped received in the British Museum in 1890) reflects the major expansion of the nitrate industry in Chile in the late nineteenth century as demand for nitrate-based fertiliser and explosive grew rapidly, and within a liberal, globalised economy in which distant, foreign sources could be worked. Chilean nitrates were largely consumed abroad, which ensured a dependence on the steamship, but within Chile itself the movement relied on rail, which provided the bulk capacity that could not be offered by mules.

Chile's victory over Bolivia and Peru in the War of the Pacific (1879–83), a conflict that owed much to rivalry over the nitrate industry, meant that the Tarapacá Department was occupied by Chile in 1879 and ceded by Peru in 1884. The key landing in the area shown in this map was a Chilean one of 9,500 troops at Pisagua on 2 November 1879. The Chileans then marched south towards Iquique, defeating the Bolivians in the Battle of San Francisco on 19 November, after which Iquique fell without resistance. San Francisco is shown on the map 34 miles southeast of Pisagua: the Chileans had advanced along the railway. After the battle, the retreating Peruvian forces fell back to Tarapacá (marked in the green area, right) where, on 27 November, although they defeated the Chileans, their exposed position had no supply lines to Bolivia or Peru, forcing the Peruvians to retreat and cede the province to the Chileans.

Control attracted inward investment – notably from Britain, where between 1882 and 1896 over 30 joint-stock companies were founded in order to benefit, raising over £12 million pounds. This investment was linked to the development of British-owned railway companies, notably the Antofagasta and Bolivia Railway Company. The key decision Chile made was to return into private hands the nitrate industry that had been nationalised by the Peruvians in 1876. The map has marked on it the interests of major British companies, which included the Liverpool Nitrate Company (1883) and the Colorado Nitrate Company (1885), both set up by Cornishman and nitrate engineer Robert Harvey.

The map accompanies the scale by a useful system of recording the distances on the railway from Iquique (figures in red) and Pisagua (figures in black) respectively. This underlined the central significance of the railway. That is also the reason why the map shows wells, because railways needed sources of fresh water. Nevertheless, there were successful rail companies, not least the Anglo-Chilean Nitrate and Railway Company, which operated from 1890 and held a monopoly on nitrate rail transport in the Tarapacá area shown in this map.

The company was founded in 1886 by John Thomas North, 'The Nitrate King,' who had co-established the Liverpool Nitrate Company with a capital of £1 million. To the south of this area, the more far-flung Antofagasta and Bolivia Railway Company was a narrow-gauge railway begun in 1873, initially with mule-power until steam locomotives were introduced in 1876.

SUB-CONTINENTAL NETWORK

J.H. Trott, 'Map of Railway Systems in India, Burma and Ceylon', 1909.

This 1909 map of the rail system in British South Asia was compiled for the Indian Railway Department Board by J.H. Trott, the Station Master in Bilaspur. The best of the Indian railway maps, this work went through several editions, including Calcutta-published ones in 1911 and 1927, the latter by the Calcutta Chromotype Company. The Railway Board published an edition in Simla in 1914 and another in Moradabad in 1917.

The map illustrates each railway in a different colour. Contrasts are readily apparent. Gauge (five feet six inches, three feet three inches, two feet six inches or two feet) is successfully indicated by differences in the width of the lines. Lines under construction are marked, as are through or trans-shipment junctions. Numbers between the stations are the distances in miles. Major roads, mountains and rivers are also marked.

Inset maps show the vicinity of major cities including Agra, Calcutta and Madras. The river services of Bengal connected with railways are another inset, because in the northwest corner of the sub-continent were the British and Russian railways that approached Afghanistan, an area of strategic concern. There were, of course, differences between particular editions. That shown here includes an inset map of the Provinces and Railways, which was somewhat unnecessary, whereas the 1914 and 1915 maps included a map of the proposed railways from Britain to India (through Asia) as well as a map of the world's major rail and steamship routes; the 1911 and 1912 versions offered the former, but not the latter.

Marked on the map, Bilaspur remains a major rail junction to this day. The Bengal Nagpur Railway arrived there in 1888 and the railway station and yard were constructed in 1890.

The contrast is readily apparent between the railway density in the Punjab and also in the Ganges Valley and the less-intense coverage in the Central Provinces and in Rajputana. Economic activity was a clear reason for this contrast.

Alongside this map of the network as a whole, there were other maps of particular networks, as with the 1911 one produced in London by Waterlow and Company as a chromolithography, a colour lithographic technique, showing all the routes of the Great Indian Peninsular Railway (GIPR), a map supported by an illustration of one of its trains leaving Bombay's Victoria Terminus. That map, designed for display in train stations and on billboards, differed from the Trott one because it used two colours – red for GIPR lines and black for other lines. Coalfields were also shown, as were projected lines. An inset showed Bombay lines. GIPR became state-owned in 1900.

Although Indian expertise, resources and labour were each highly significant, Britain played a major role in the development of India's railways, notably with finance, expertise, locomotives and security. The *Railway Gazette* of August 1911 noted:

> 'Between these furthest points the system has shot out many trunks and innumerable branches all over the great plains, connecting remote cities and industrial centres. All the chief cities of India are connected by railways. In fact they would not rank among the principal cities if they did not possess a railway station.'

As a result of the establishment in India of the Railway Board in 1901 and its formalisation in 1905, there was greater coherence than hitherto. This was part of a period of reform in the British government of India that owed much to the Viceroyalty of Lord Curzon from 1899 to 1905. In part, it was a matter of his commitment to the rivalry with Russia, which he saw largely in terms of rail competition. It was a theme in Curzon's book *Russia in Central Asia in 1889 and the Anglo-Russian Question* (1889). He had been Under-Secretary of State for Foreign Affairs (1895–98) and was later Secretary of State (1919–24). Indeed, Curzon bridged the key issues of India and geopolitical concern about Russian railways far more than any other individual.

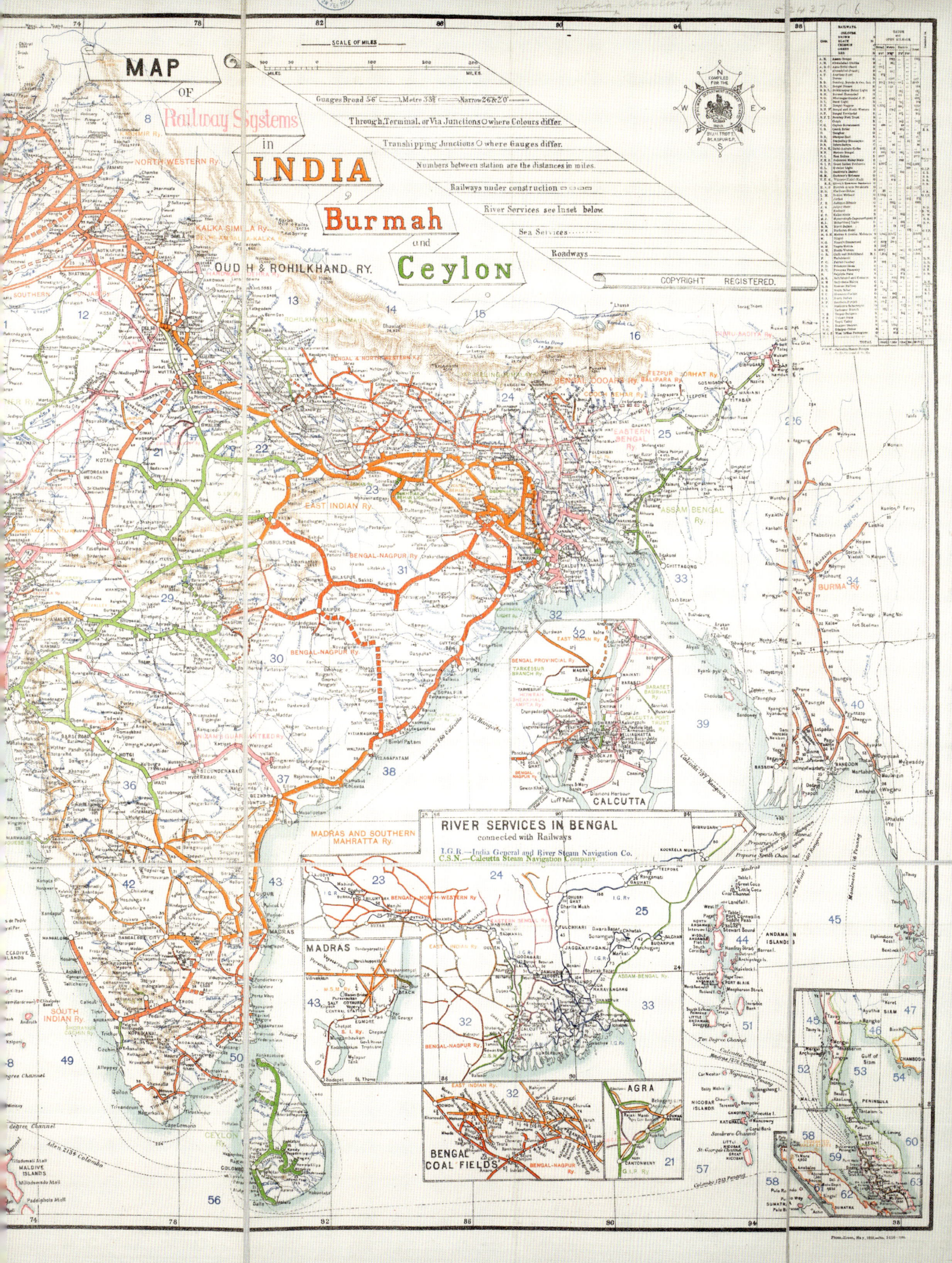

SCENIC WONDERS

New Zealand Department of Lands and Surveys, 'Map of New Zealand Shewing Railways Open for Traffic', 1901.

The South Island (marked Middle Island in the map) saw more mileage in the early stages of rail expansion in New Zealand. This was helped by the fact that the difficult Maori Wars were focused on the North Island and that discouraged activity and investment there, driving up the protection costs of railways. Moreover, the terrain separating the main areas of settlement was more difficult on the North Island, not least because most of that on the South Island was on the flat coastal plain, which, alongside the Canterbury high country, was the prime area for sheep ranches.

By 1877 there were 339 miles of railways in the North Island, but 797 miles in the South, half being the north–south Main South Line from Lyttelton to Bluff. Construction began in the South Island with the Ferrymead line from Christchurch to Ferrymead in 1863, with the line extended to the port of Lyttelton in 1867. What became the Main South Line was begun from Canterbury in 1865 and reached Selwyn in 1867, then stopped. After being resumed it reached Rakia in 1873, and in 1877 was completed from Christchurch to Otago. The line opened to Dunedin in 1878, which was then linked to Invercargill in 1879. In 1878 the first express ran from Christchurch to Dunedin, taking 11 hours. Dunedin Railway Station, New Zealand's busiest with up to 100 trains a day, was completed in 1906. Again ports were crucial.

The Otago Gold Rush, which began from 1861 in Central Otago and was followed by the West Coast Gold Rush in 1864–67, ensured that railway services in the South Island increased, as did steamship ones. There was interdependence as well as competition. The railways were not affected by the storms that hit the steamships, while the gains, liquidity and prospect for more gains arising from the gold rushes encouraged rail construction, which also happened in other parts of the world with gold rushes, notably North America, Australia and South Africa.

British expertise, technology and capital were all important to the development of railways on the South Island. Lines included the Otago Central Railway, which was begun in 1877 but only opened to Cromwell in 1921. Passenger use was not great, but was matched by freight, including the movement of sheep. New Zealand became the prime source of lamb for Britain, with frozen meat first exported there in 1882. The number of sheep in New Zealand increased from about 19 million in 1896 to 24.6 million in 1914.

MAP OF NEW ZEALAND SHEWING RAILWAYS OPEN FOR TRAFFIC MARCH, 1901.

STATISTICS.

Government Railways open for traffic on 31st March, 1901	2,212 miles.
Private companies' lines	99 "
Capital cost of Government Railways open on 31st March, 1901	£17,207,328
Revenue from Government Railways, year ending 31st March, 1901	£1,727,236
Expenditure on " " "	£1,127,848
Passengers carried " "	6,243,593
Number of season tickets issued " "	82,921
Cattle, sheep, and pigs carried " "	2,536,118
Tonnage carried " "	3,339,687
Number of miles travelled by trains " "	4,620,971
Number of locomotives " "	305
Number of passenger-carriages " "	603
Number of wagons and brake-vans " "	10,868
Area of colony, square miles	104,471
Population, estimated at 31st March, 1901	815,349
Chief Cities:—	
Auckland } Population with Suburbs	67,226
Wellington } (Census for 1901)	49,344
Christchurch }	57,041
Dunedin }	52,390

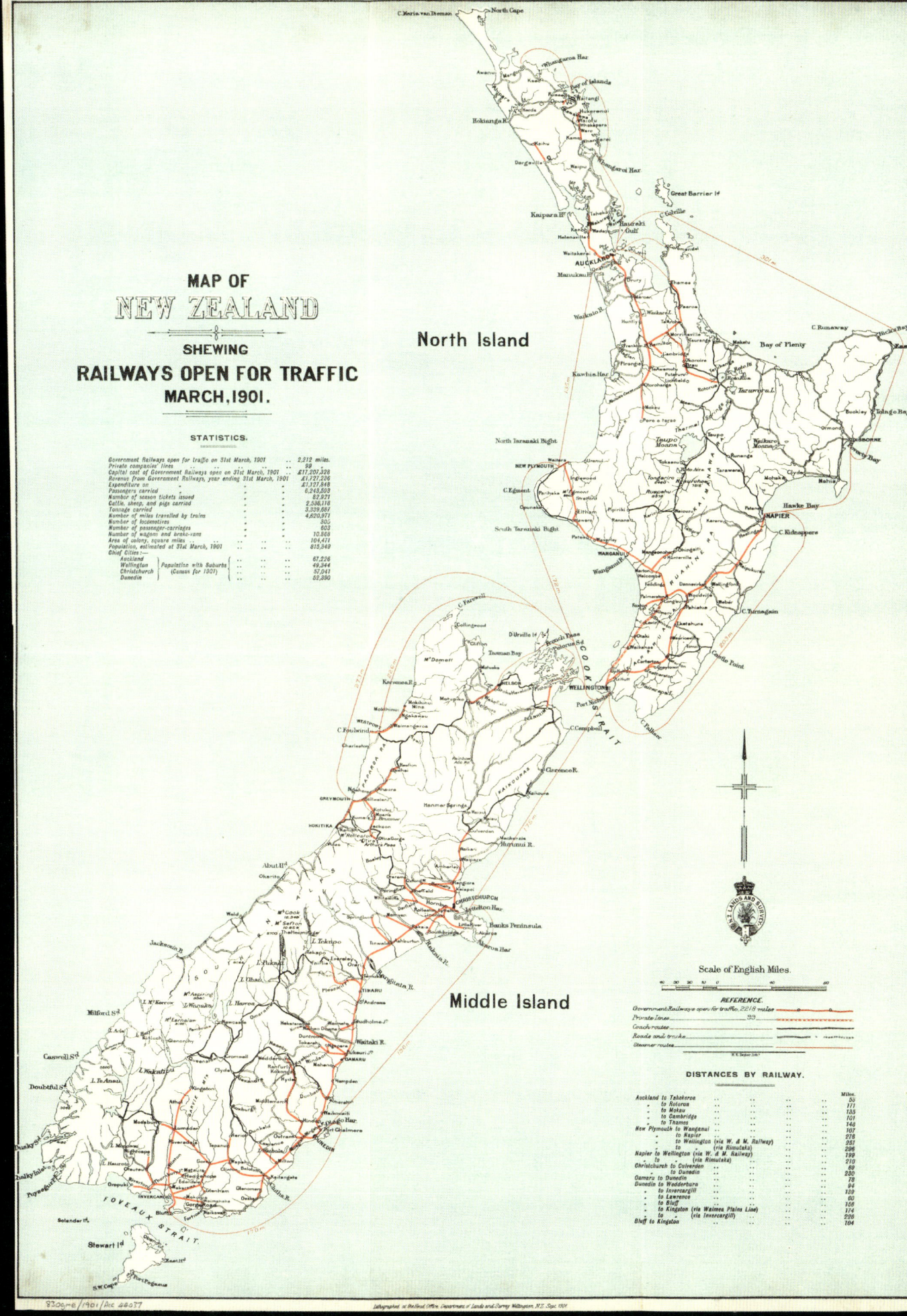

AMERICAN RAILWAYS: THE URBAN IMPACT
D.A. Sanborn, 'Fire Insurance Map of Boston', 1867.

The nature of the railway urban estate appears in this section from *Insurance Map of Boston* Volume 1 (1867) by Daniel Alfred Sanborn (1827–83), a Massachusetts surveyor, who in 1867 established the New York-based D.A. Sanborn National Insurance Map and Publishing Company, which in 1916, after the purchase of competitors, became a monopoly.

Sanborn used a key that combined colours and letters, including blue for stone or concrete buildings, yellow for frame structures, pink for brick and green for 'specially hazardous risks'. This map is the earliest published Sanborn fire insurance map and it shows how the railway estate closely abutted dwellings, with repair shops, coal sheds, warehouses and freight depots all taking up much space. In 1861 a fire near the Carpenter Shop plot on Albany Street, caused by firecrackers on Independence Day, resulted in the loss of many buildings.

There was no central terminus in Boston, but railways came into central locations, with the Boston and Worcester Railroad and the Old Colony Railroad running near Albany Street, where the former became, as a result of mergers, the Boston and Albany Railroad (B&A) in 1867. It is those B&A buildings that are marked on this map. The area on which the rail depots were built owed a lot to the filling in of the bay, including by gravel trains. This reflected the need for much land for rail depots and the problems this posed for urban land use and therefore for the interplay of local politics with financial interests. These are aspects of the hidden history of rail.

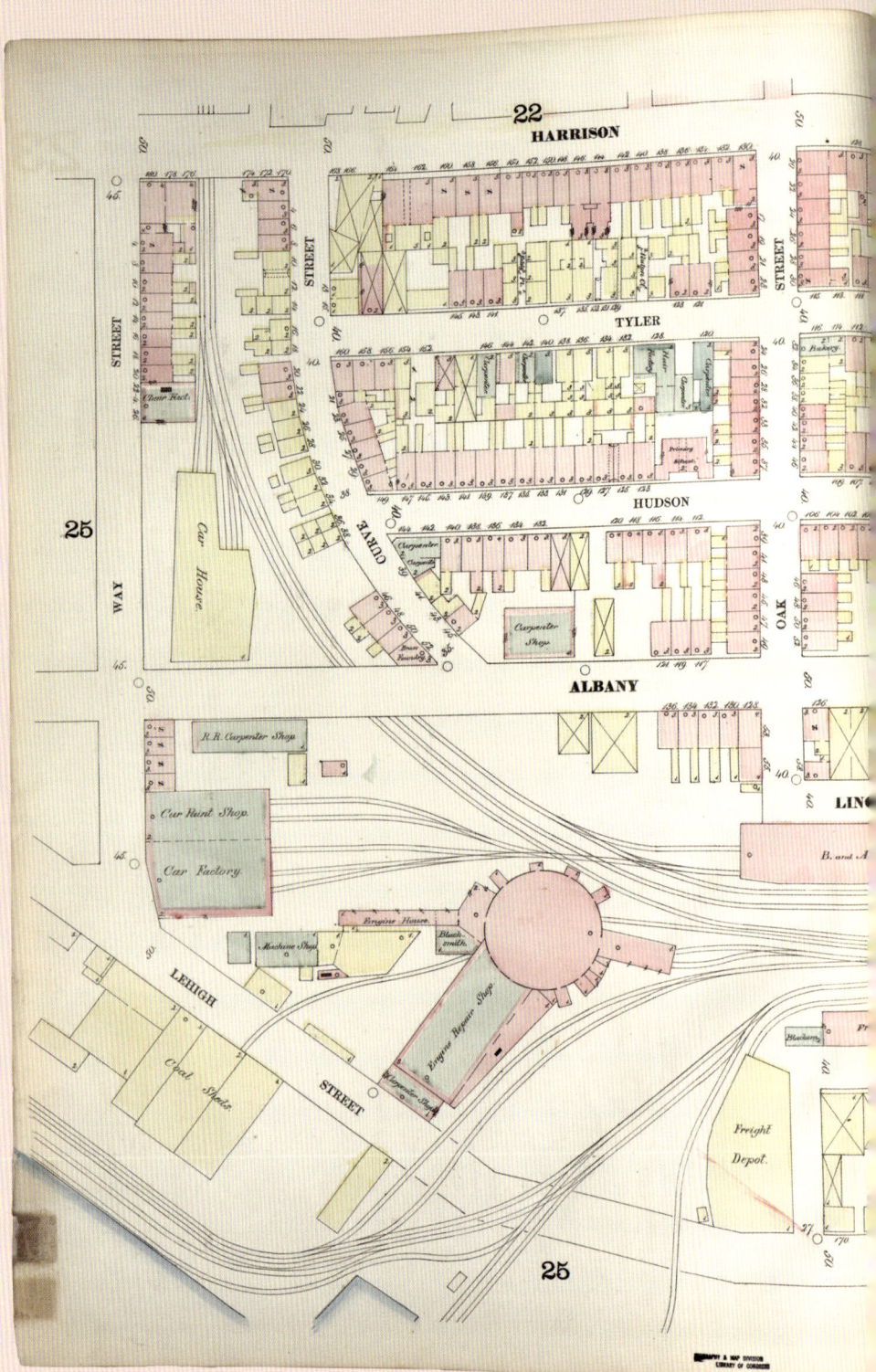

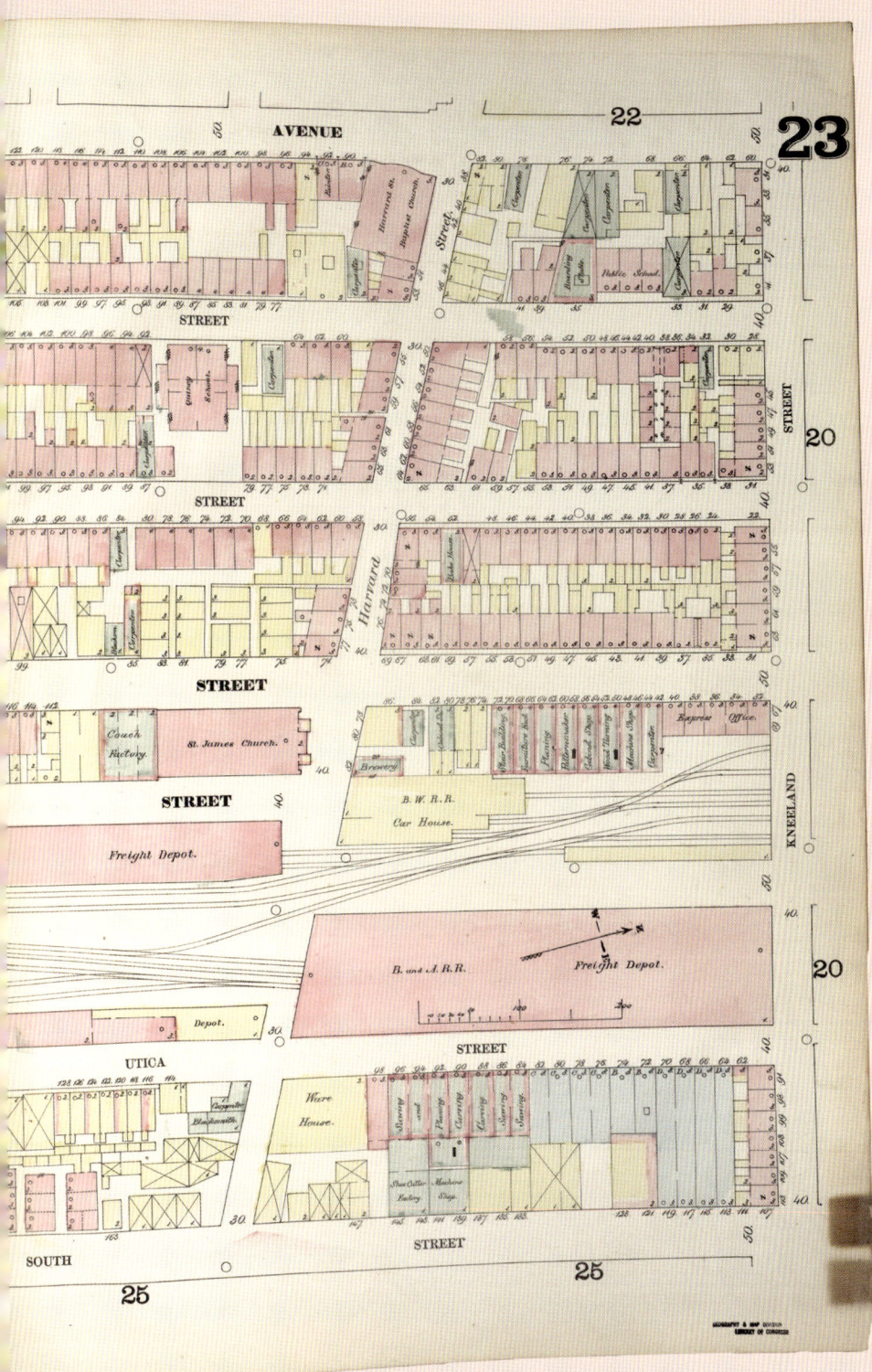

As ever, what is not present in the map is instructive, in this case the segregated housing that was part of both urban life and the railways, because the latter owed much to African Americans. This was true not of skilled jobs such as drivers and machinists, which were unionised, but of those lower-paid roles seen as unskilled and not allowed to unionise, such as station porters, maids and cooks on trains, and manual workers.

Boston's significance as a major centre of rail use was enhanced by the completion of the Hoosac Tunnel route in the 1870s, which connected it to upper New York state. The city was also a major port; aside from the transoceanic services encouraged by a shorter passage than to other American ports, it meant that steamship routes, for example from New York, met rail routes. This process was made very clear in a 1905 map from an Atlantic perspective: *Bird's Eye View of Massachusetts, Rhode Island and Connecticut. A Comprehensive Map of Waterways Traversed by Steamers of the Joy Line between Boston-New York-Providence*.

Sanborn mapmaking was at a peak in the 1920s, but it was hit hard by the Depression of the 1930s and by the consolidation of the insurance industry, which caused profits to fall significantly. Sanborn created its last new map in 1961.

3. GEOPOLITICS AND WAR 1885–1918 107

STATE-LINE STOCKYARDS

Augustus Koch, 'Panoramic View of the West Bottoms, Kansas City,…', 1895.

Probably a recent German immigrant, Koch (1834–1901), a prolific engraver, had served in the Union Army during the American Civil War, including as a map draughtsman. Subsequently, he became a major engraver of views of America, his first being Cedar Falls in 1868 and his last in Montana in 1898. About 110 Koch bird's-eye views survive, including over 20 in Texas. Most were of various cities and towns, presumably because that maximised the number of purchasers. Moreover, there was publication by subscription, to finance the work. These views included Manitowoc (1868), Sacramento (1870), Salt Lake City (1870), Dubuque (1872), Bangor (1875), Virginia City (1875), Nebraska City (1880), New Braunfels (1881), Galveston (1885), Albuquerque (1886), Savannah (1891), Seattle (1891), Ocala (1892), Aspen (1893) and Pensacola (1896).

As this map of Kansas City shows, Koch specialised in very high viewpoints, with the horizon close to the tops of the images and the horizontal dimension not much greater than the vertical, both of which permitted considerable detail. The map of Kansas City's West Bottoms district was that of the economic heart of the city, and the perspective enabled the capturing of considerable detail, which was explained with reference to the letter grid and the descriptions at the bottom, which identify grocers, manufacturers, freight depots and packaging companies. This coloured lithograph of a pen-and-ink map indicated the considerable economic importance to Kansas City of the railway – the city's transport nodality. In the first case, rail was closely linked to the stockyards, located along the Missouri and Kansas rivers and straddling the state line with Missouri, with cattle brought by rail in order to be slaughtered and turned into food in the meatpacking plants shown in the map. This was far more effective than droving. Rail also provided a throughput that ensured economies of scale. In turn, these offered a profitability that encouraged investment, both in meatpacking plants and in the associated rail infrastructure, locally at

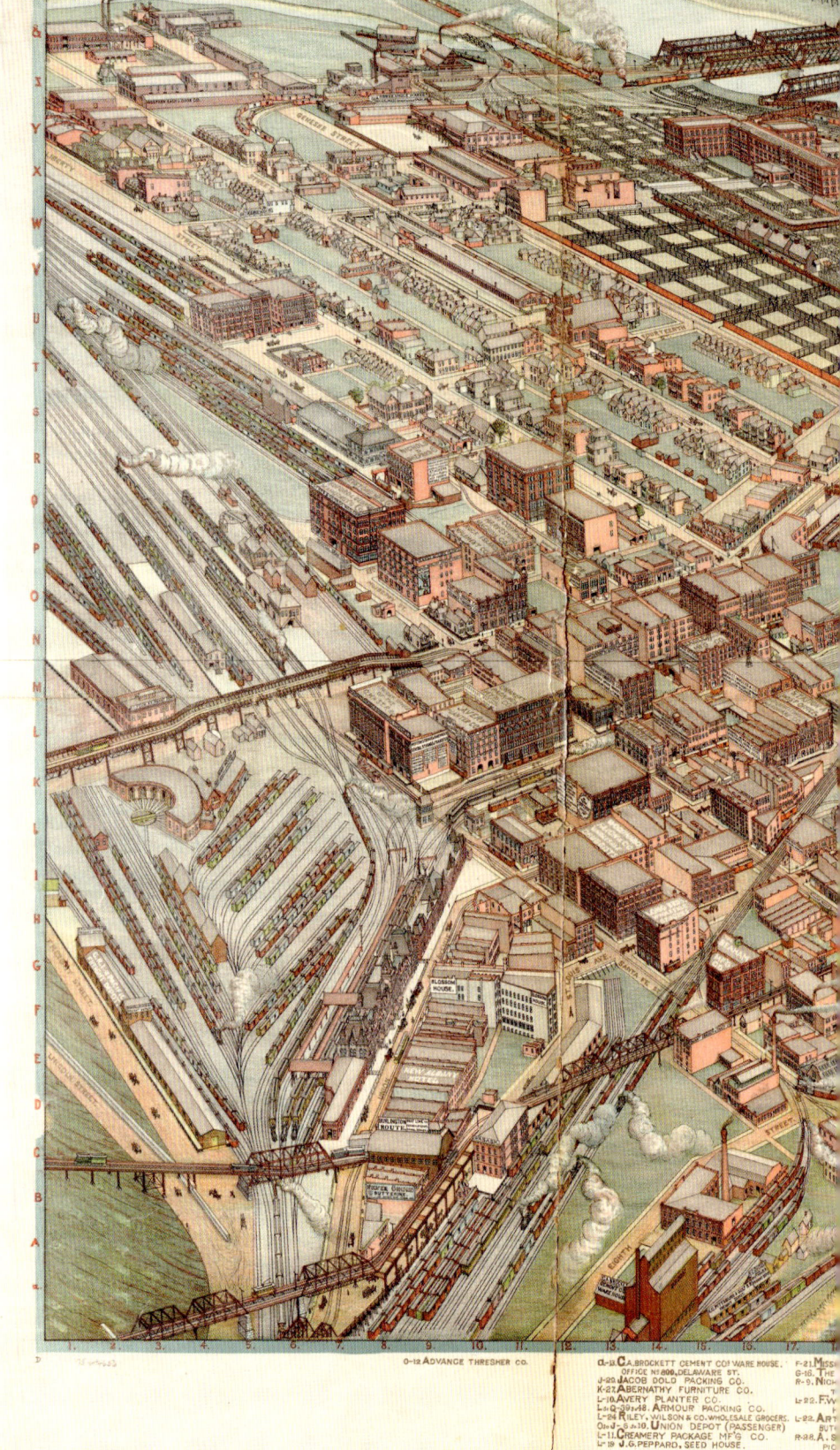

the plants and regionally to move the stock. This infrastructure included considerable investment in rolling stock.

The main competition for Kansas City, which was closer to the cattle, came from

Chicago – the only market in the country that was larger.

There is a separate Koch map of Kansas City as a whole, produced in 1885. Koch spent much of his later life in Kansas City, which accounts also for his maps of nearby places including Kingman and Topeka. This is an example of the personal interests and knowledge of cartographers integrating with commercial opportunities.

ELEVATED TRAINS

Poole Bros., 'Chicago Rock Island and Pacific Railway Station', 1897.

This bird's-eye view of Chicago provides a different form of aerial view, one in which the use of colour drives home the degree of connectivity; the firm red line is the elevated 'Loop', and the Chicago, Rock Island and Pacific Railway's Van Buren Street station is shown in red letters as centrally located and of the railway's three central stations marked it is 'the only one on the Loop'. Other railway stations lack this colour depiction. The elevated railway that looped around Chicago's downtown was designed to connect the competing suburban lines, and it was built and opened between 1895 and 1900. Railways, like skyscrapers, contributed to the importance of the 'Downtown', the central district in which business activity was concentrated. Seen as disruptive, elevated railways were unpopular and uncommon, but underground lines were far more disruptive and very expensive to build. The map also shows the elevated system's interconnections with the 'Cable and Electric Street Car System' (identified by light red lines), as well as the location of several other stations, hotels, important public buildings and theatres.

Moran's 'Map of Chicago and Suburbs of 1892' invited purchasers to look at the suburbs of Eggleston and Auburn, among the advantages of which were accessibility: '...by the Chicago, Rock Island and Pacific, and Chicago and Eastern Illinois railroads, with trains every few minutes. Also the State Street cable line connects with street car lines through the centre of both these residence districts.'

Chicago's first railway, the Galena and Chicago Union Railroad, was chartered in 1836 to build a connection to the lead mines at Galena. Many railways followed, including the Chicago, Rock Island and Pacific Railroad (Railway from 1880), which began operation in 1852 and, despite its title, only reached Joliet; Rock Island (on the Mississippi) followed in 1854. The Chicago, Danville and Vincennes Railroad was chartered in 1865, its main line being completed in 1872.

Chicago was a major national rail hub, the centre of regional rail and a focus for growing commuter activity. As a result, the mapping of its railways was inherently difficult, other than in terms of a mass that looked like a mess. Individual companies handled the visual issue of differentiating between lines but that atomisation did not contribute to an overall impression.

The expanding market of rail information also found Chicago given prominence in general maps – as in Rand McNally's *New Railway Map of the United States and Canada*, which was published to accompany the company's *Railway Guide* (1874). The two were carefully linked with a note on the map: 'To find the population and general description of any railroad or river town on this map, or to find the corporation name of any line of railroad upon which towns are situated, see general index of railroad and river towns in Rand McNally and Co's *Railway Guide*, pages 1 to 23.'

MIDTOWN MANHATTAN

New York Central Lines, 'The Heart of New York: Grand Central Terminal', 1918.

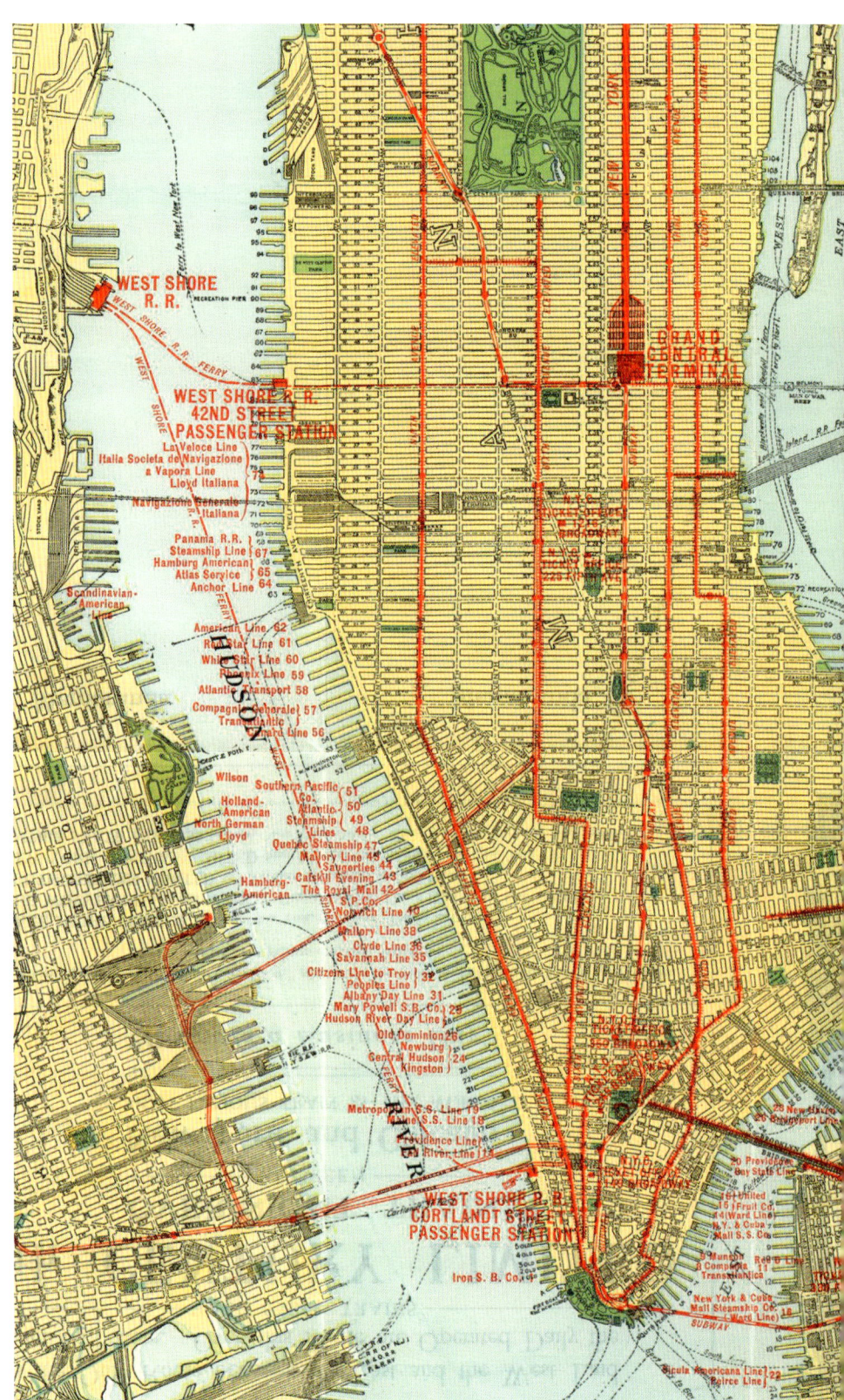

Published by Rand McNally, this map came as part of a guide to New York City issued by the New York Central Railroad. The map, and its inset of the area bounded by First and Ninth avenues from 22nd to 64th street, is designed to show the significance of that company and its railway station to the city. The text announces: 'Only Railway Station on the Subway, Elevated and Surface Lines.' The key very much highlights the station's location, showing hotels, theatres, the New York Central Line, subway lines, elevated lines, surface lines and the Hudson and Manhattan tunnels.

Opened in 1913 on the site of two stations with similar names, Grand Central Station, like its predecessors, was the terminus for the New York Central Railroad, which was established in 1853 as a merger of earlier lines and then greatly expanded. This map appeared against the background of considerable institutional activity, with the operations of 11 subsidiaries merged in 1914 with the New York Central and Hudson River Railroad, re-forming the New York Central Railroad. This encouraged the emphasis on branding seen in this map. The terminal was the basis for the 20th Century Limited, a rapid and luxury service to Chicago that began in 1902, and the Lake Shore Limited service that began in 1897.

The hidden element of this map was the competition offered by the recently completed Pennsylvania Station, which occupied another large site in Midtown Manhattan, between 31st and 33rd streets and Seventh and Eighth avenues. Constructed from 1904, it was finished in 1910 and had 11 platforms and 21 tracks. That station provided direct rail access to New York from the south, as hitherto the Pennsylvania Railroad (PRR) terminated in Jersey City and passengers crossed the Hudson River by ferry, which gave the New York Central Railroad a major advantage. A tunnel had been regarded as not viable due to the pollution arising from steam, but the PRR overcame the problem by adopting a French system of using electric locomotives for the final approach.

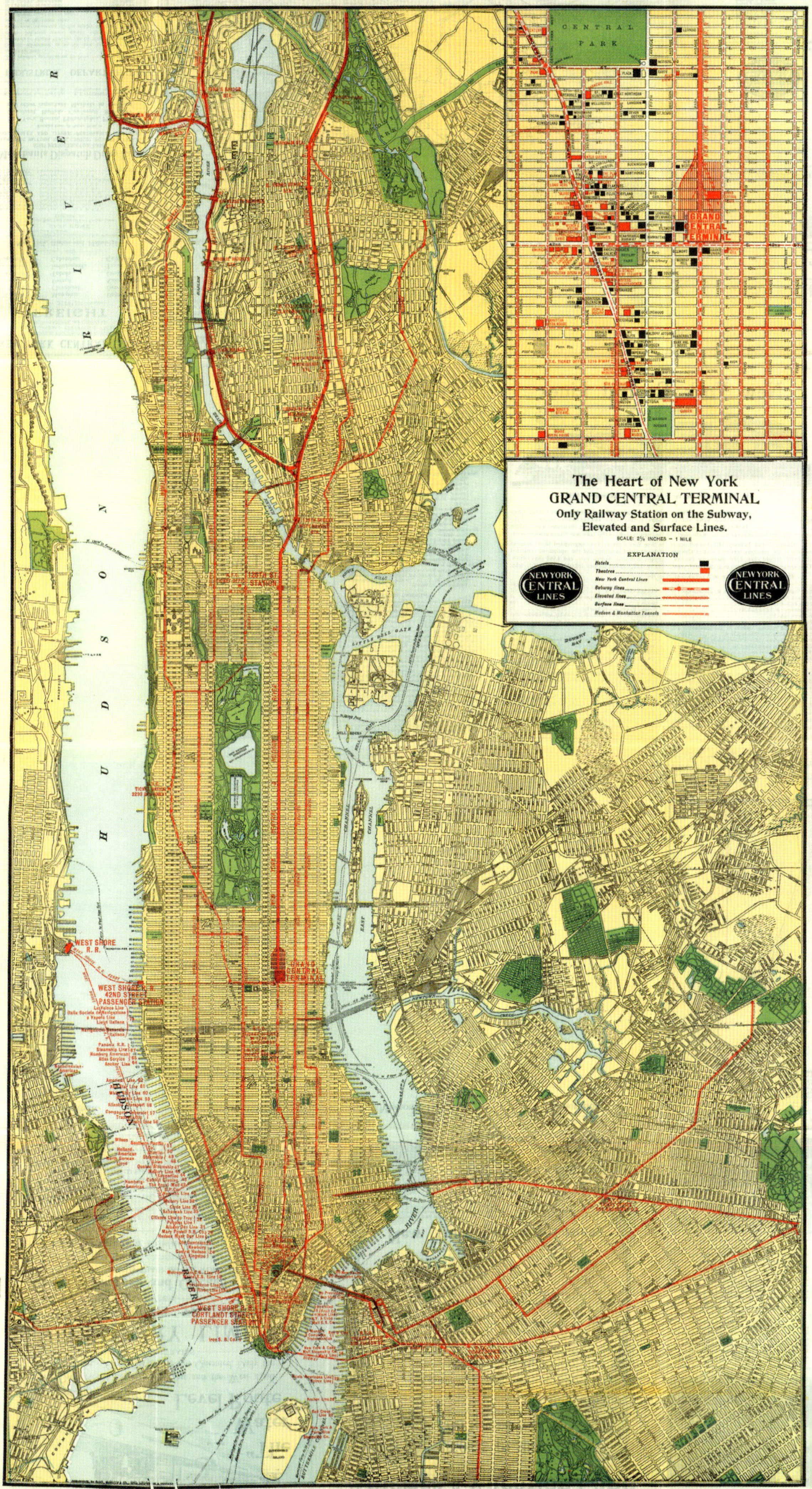

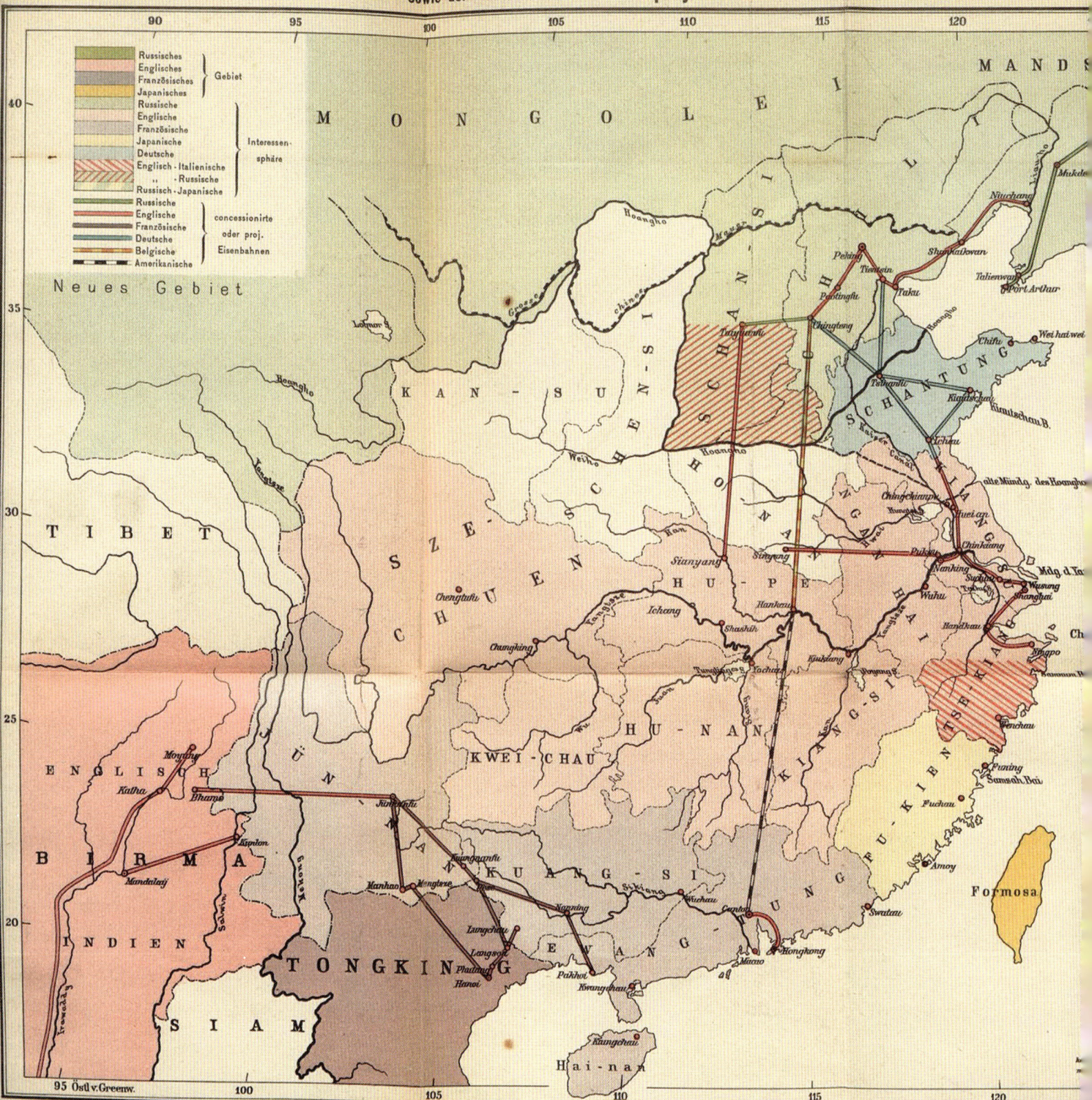

INVESTING IN CHINA

Max von Brandt, 'Industrial and Railway Enterprises in China', 1899.

Published in Berlin, this 1899 map, which records 'the ownership and spheres of interest, and the concessionary and projected railways in East Asia', shows what appears to be the carve-up of China. Max von Brandt (1835–1920) had a longstanding interest in China, having taken part as Prussian attaché in the Eulenburg expedition to Japan and China in 1860–61. In 1861 this led to a treaty between China and Prussia, which was representing the entire German Customs Union. Brandt went on to be the Consul and then Minister to Japan from 1862, moving on in 1875 to become Imperial (German) Envoy in China where he stayed until 1893, becoming the senior foreign diplomat. Brandt published extensively on China, notably from 1895 until 1914.

The map is somewhat crude, not least because it devotes no attention to the terrain, which made the idea of a line from northern Burma, where the British had established control in 1885–86, to Yunnan implausible. There is also the issue of clarity, as the rail lines 'concessioned or projected' ('*concessionirte oder projectirte Esienbahnen*' as labelled in the key) to the British and their French counterparts are not that well differentiated in colour terms, in contrast to the Russian and American lines. Instead, as with the contrast between the German and the Russian lines, it is necessary to read in the information provided by the colour-coding of what the key calls '*interessen sphäre*' (spheres of interest). In some cases the lines show a determination to respond to developing spheres of interest, notably those linked to ports in which there were concessionary positions, but in others there was a 'blue skies' enterprise in seeking to create a new sphere.

Brandt also did not adequately distinguish between realistic prospects, such as the Russian lines in Manchuria ('Mandschurei'), and unrealistic ones, such as an American line from Canton (Guangzhou) to Hankou on the Yangtze. The latter rail line was a means of offering a modern transport route to complement or replace China's ancient, river-based system, but it was also merely a line on the map that bore scant reference to topography or practicality. This 750-mile-long project by the American China Development Company, incorporated in 1895, had too few investors and ended up as a 30-mile line to Sanshui, with the concession cancelled in 1904. In contrast, the Belgian concession for the onward northward section from Hankou to Beijing ('Peking') was bought back in 1909: the line had been built in 1899–1906, but the Chinese government was now opposed to the line being under foreign control. The granting of concessions was intended by the government as a way to ensure foreign investment in economic development, which it was believed would, in turn, produce revenue for the government. This also reflected international competition and Chinese weakness after the Sino-Japanese War of 1894–95. However, rising domestic criticism of foreign control over railways, which reflected how nationalism and localism could be joined, played a role in the unpopularity of the government that culminated in the 1911 Revolution. The Chinese had already been far from passive respondents to foreign pressure, thwarting the Portuguese attempt to build a line from their colony at Macao to Canton, for example.

MODERNISING JAPAN

'The Railway Map of Japan', 1900.

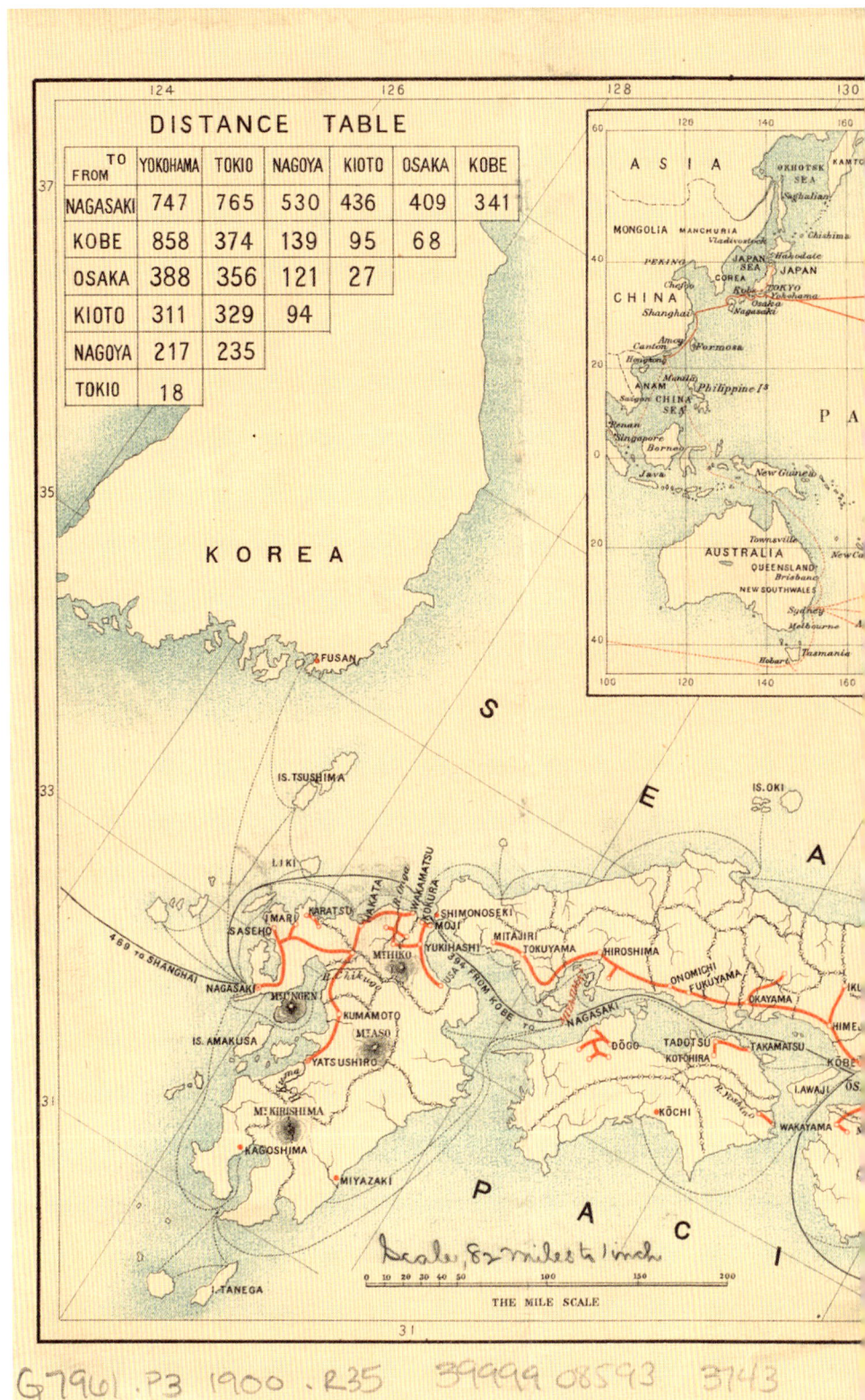

Weakened by the very limited attention devoted to topography, this 1900 map with several insets captures the rapid expansion of the Japanese railway system. Railways were explicitly seen as a key means of modernisation, as well as helping provide for troop movements, and they also reflected the significance of Westernisation at this stage of Japanese development. More particularly, Britain offered an influential model and was a source of knowledge-transfer expertise. In 1872, when the first line (from Tokyo to Yokohama) had been completed, construction had been overseen by Edmund Morel, a British railway engineer with experience in Australasia and British North Borneo. The train, with Emperor Meiji among the passengers, was pulled by a British locomotive and British loans had played a major financing role.

In many senses, there was a natural bifocality: the concentration of population on the main island, Honshu, and notably so between Tokyo and Osaka, favoured rail. However, the degree to which Japan was a matter of islands, and with populated areas on individual islands, which were in practice coastal enclaves that were separated by mountains, encouraged the use of steamships. The map shows many of the main steamship routes with a distance table in miles but the difference between the two methods of transport was not really brought out by including both in a map. Ships are affected more by weather, impacting their speed and reliability.

Maps of Japanese railways of the period also directed attention to the imperial position, with Japan, as a result of victory in the Sino-Japanese War of 1894–95, the ruling power in Taiwan (Formosa Islands, inset) and Korea. The main inset map ('The Toyo Kisen Kaisha Routes') advertises a major new Pacific steamship company founded in 1897 at a time of greatly rising competition in trade between North America and East Asia. Korea was to be an important staging point in the deployment of Japanese power into China.

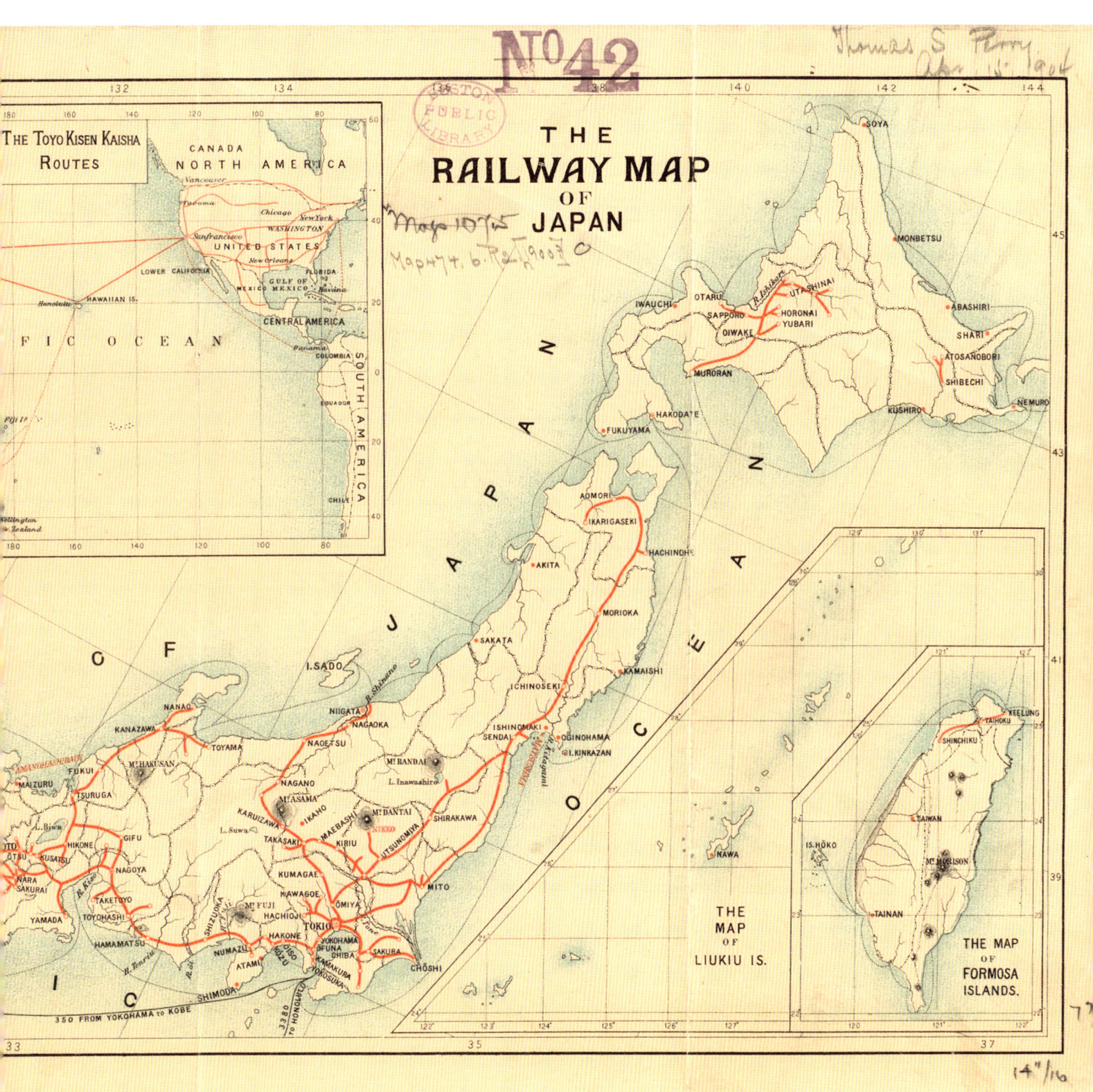

3. GEOPOLITICS AND WAR 1885–1918

NORTH EASTERN RAILWAY

MAPPED IN TILES
Craven Dunnill, 'North Eastern Railway', *c.*1900.

Manufactured by Craven Dunnill and Co. Ltd. in Jackfield, Shropshire, about 1900, this map in 64 ceramic tiles depicts northern England, from the Dee estuary in the bottom left to Berwick-upon-Tweed in the top right, with inset maps of eastern Scotland and of the docks at Hull, Hartlepool, Middlesbrough and on the Tyne. The map includes the line from Beverley to North Frodingham, the 9.5-mile North Holderness Light Railway, which was approved by Parliament in 1897 under the Light Railways Act, designed to facilitate such railways by cutting down their costs. However, the line was never built and was instead replaced in 1903 by a bus service.

Tile maps were produced across the network and although this example is on view at a railway museum (in Shildon, County Durham) they can be seen still at a number of working stations including Beverley, Hartlepool, Middlesbrough, Morpeth, Saltburn, Scarborough, Tynemouth, Whitby and York.

The tile map also records other lines, although they are not distinctively marked. Across the Pennines, the original cross-country Carlisle–Newcastle link of 1838 was not supplemented until the line from Durham to Barrow-in-Furness in 1861 and the Carlisle–Settle line of 1876. The first line took Furness iron east and Durham coke west, helping the iron and steel industry on both sides of the Pennines.

The map shows that the impact of the railway was local as well as national, and far from simply replacing the waterway network railway reached places and areas that did not have access to waterways or only had limited and sometimes only seasonal access. The Durham and Sunderland Railway reached Sunderland in 1836, the Newcastle and Darlington Junction Railway in 1852 and the Londonderry, Seaham and Sunderland Railway in 1854. On the northern bank of the River Wear, branches of the Brandling Junction Railway reached Wearmouth and North Dock in 1839. The significance of port links remained strong. The same railway reached Gateshead in 1839, a year after the Newcastle and Carlisle Railway.

This railway tile map reflects, but also underplays, past and current competition. Thus, the Clarence Railway had offered a shorter route for West Durham coal to the coast than the Stockton and Darlington Railway, and in turn was challenged by the Great North of England, Clarence and Hartlepool Junction Railway. The West Durham Railway was laid from Crook to Byers Green to compete with the Clarence line. The large number of companies led not only to competition but also to different services supplementing each other, as at Carlisle where seven railway companies operated.

Maps provided no guidance to usage: annual passenger numbers on the Tyneside loop rose from six million in 1903 to over nine million a decade and a half later.

At the national level, the mainline system was not completed until the early 1870s, with no link between London and Aberystwyth until 1864. However, as shown on this map, many local and branch lines came later. Thus, in Berkshire branches were built to Faringdon (1864), Abingdon (1873), Wantage (1875), Wallingford (1886) and Lambourn (1898), while in Essex light railways to Tollesbury and Thaxted were opened in 1907 and 1913.

A PROSPERING PORT

Percy Home, 'The Humber, the Company's Docks at Hull & The Railway Connections', 1900–15.

This North Eastern Railway (NER) poster map is by the eclectic artist Percy Home, who also produced numerous sketches of Gilbert and Sullivan productions. Home's aerial perspective has provided an opportunity to highlight particular links. In this case he has enhanced the process by the use of the colour red for the railways. Hull competed for trade as the railway altered, indeed transformed, the economics of hinterlands. In particular, in 1882 the Great Eastern Railway (GER) had introduced a through train from the north of England to the newly opened Parkeston Quay at Harwich. This was an example of the way in which the train created new routes.

This poster map is also a new iteration of the standard aspect of Hull's rail history, namely the link between the port and industrial zones, rather than the rural and small town hinterland to the north and northwest. For long, the focus had been on links with the industrial zones of Yorkshire. Thus, under legislation in 1836, Hull had gained a rail link in 1840. This, the Hull and Selby Railway, was a line to Selby, where it linked with the Leeds and Selby Railway that had opened in 1834. The Hull terminus was adjacent to the Humber Dock. In turn, the Victoria Dock Branch Line opened in 1853. Separately, the York and North Midland Railway (Y&NMR) had expanded to reach Hull by leasing the Hull and Selby Railway from 1845. In 1852 the NER acquired the lease and in 1872 took over the Hull and Selby Railway.

As a major game changer, the very large Alexandra Dock in Hull, which opened in 1885 and was backed by Hull Corporation, served the new Hull, Barnsley and West Riding Junction Railway and Dock Company (later Hull and Barnsley Railway), which was formed in defiance of the NER's attempt to control the local rail system. Steam-powered excavating machines were used. The line focused on coal exports. The Hull Dock Company amalgamated with the NER in 1893, and in 1899 the NER and the Hull and Barnsley Railway agreed to co-operate on building a new dock, the King George V Dock, which was contracted in 1906 and completed in 1914. In part the new dock competed with the Great Central Railway's Immingham Dock on the other side of the Humber, on which construction began in 1906 and was finished in 1912, a dock that involved the construction of three new short connecting lines. Coal exports, as well as iron ore deposits in Lincolnshire to the south of the Humber, were important to the development of rail capacity there, but it could not rival the trade of Yorkshire, notably coal exports, as represented in the use of Hull's docks.

120

THE COAL CENTRE

J.P. & W.R. Emslie, 'Official Railway Map of South Wales', 1910, and 'Cardiff, Cogan, Penarth & Taffs Well', 1898–1905.

The maps of the Lower Taff Valley and of the Cardiff and Penarth docks (following pages) capture the role of export by sea in the development of the South Wales coal industry, and also the use of colour-coding to distinguish different lines (green for Taff Vale Railway, yellow for Great Western Railway, orange for Barry, blue for Brecon and Merthyr, and red for both Port Talbot and LNWR) and to explain the resulting junctions. The handbook of *Official Railway Junction Diagrams*, published by the Railway Clearing House in 1905, based in Euston Square, London, contained maps of the junctions in England, Wales and Scotland, and was used to calculate the apportioning of fares between the various railway companies. The diagrams are dated between 1898 and 1905, and most have added annotations in pencil and ink.

Although the Glamorganshire Canal opened in 1794 made it possible to move coal cheaply, rail proved more economic. The development of the port of Cardiff by the Marquess of Bute and the spread of the railway, especially the Taff Vale line between Cardiff and Merthyr Tydfil (1841), permitted the movement of large quantities of coal. In the decade after 1856, exports from South Wales led to the opening of three new dock developments, all marked on the map (following pages), the Bute East, Roath Basin and Penarth docks, all linked to the Taff Vale Railway.

Production of coal rose from 1.2 million tons in 1801 to 10.2 by 1860, 13.6 by 1870 and 57 by 1913, much of it exported. The peak year for the number of collieries in South Wales was 1910, when there were 688. The growth in production reflected greater demand, but also engineering improvements in drainage and ventilation that made possible the working of deeper seams. The increase in coal freight contributed greatly to the growth in the volume of freight carried in Britain by rail: from about 38 million tons in 1850 to 513 million in 1912.

Other products and industries were also greatly affected. Use of the railway from the 1840s enabled the brewers of Burton-on-Trent to develop a major beer empire, and also helped speed slates from North Wales towards urban markets. The Ffestiniog Railway of 1836 linked the slate mines with Porthmadog harbour, and the rail network in the area improved from 1867. In the 1870s the railway companies opened up urban markets for liquid cheese, encouraging dairy farmers nationwide to produce 'railway milk' rather than farmhouse cheese.

In 1901 a labour dispute between the Taff Vale Railway Company and the Amalgamated Society of Railway Servants led the House of Lords to decide that unions could be liable for damages resulting from actions. The Conservative government, in power until 1905, rejected trade union pressure to change the law but the legal position was reversed by the Liberal government that came to power and passed the Trade Disputes Act of 1906. This was an instance of the Liberals looking to the Labour Party for support. Alongside the miners, the railway workers were to be the most militant of the trade unions, and the two were often united in their activism.

CARDIFF, COGAN, PENARTH & TAFFS WELL

12

THE BUSIEST RAIL CITY

W.J. Adams & Sons, 'The "District Railway" Map of London', 1892, and
J.P. & W.R. Emslie, 'Official Railway Map of London and its Environs', 1899.

Robert Lowe spoke at the lunch served on Farringdon Station on 9 January 1863 to mark the opening of the 'Tube' (London's underground railway) services:

> 'The traffic of London has long been a reproach of the most civilised nation of the world, and the opprobrium of the age. Dr Johnson used to say that if you wanted to see the full tide of human life, you must go to Charing Cross, but Dr Johnson would have to raise his estimation of the full tide, or rather of the close jam of the full tide of human life, many hundred per cent before he could arrive at the state which the traffic of London has now reached.... Through gas-pipes and water-pipes and sewers, ... and ... the Fleet Ditch.... The line has had to worm its way through a complicated and intricate labyrinth under difficulties almost insuperable.'

Opened in 1862 the Metropolitan Line was joined by the Metropolitan District Railway, parts of both systems incorporated into the Inner Circle (now the Circle Line), completed in 1884 and electrified in 1905. The Lots Road power station in Chelsea was built to provide the electricity for the 'Tube' lines. The early lines were built by 'cut and cover' near the surface, rather than the use of a tunnelling shield, the basis of the deep-bore 'tube' tunnels. No other British city had an underground railway until Glasgow opened one in 1896.

Although London was the key centre of England's new rail network, it suffered, like Paris, from the lack of through-routes or a central station. This lack can be seen as a failure of planning akin to that after the Great Fire of 1666, but massive disruption would have been caused by such construction, and schemes such as that by Charles Pearson, Solicitor to the City of London Corporation, in 1851 for a 100-foot-wide tunnel with eight sets of tracks were not viable. This unrealistic plan formed a counterpart to the bold lines on the map provided by so many projectors.

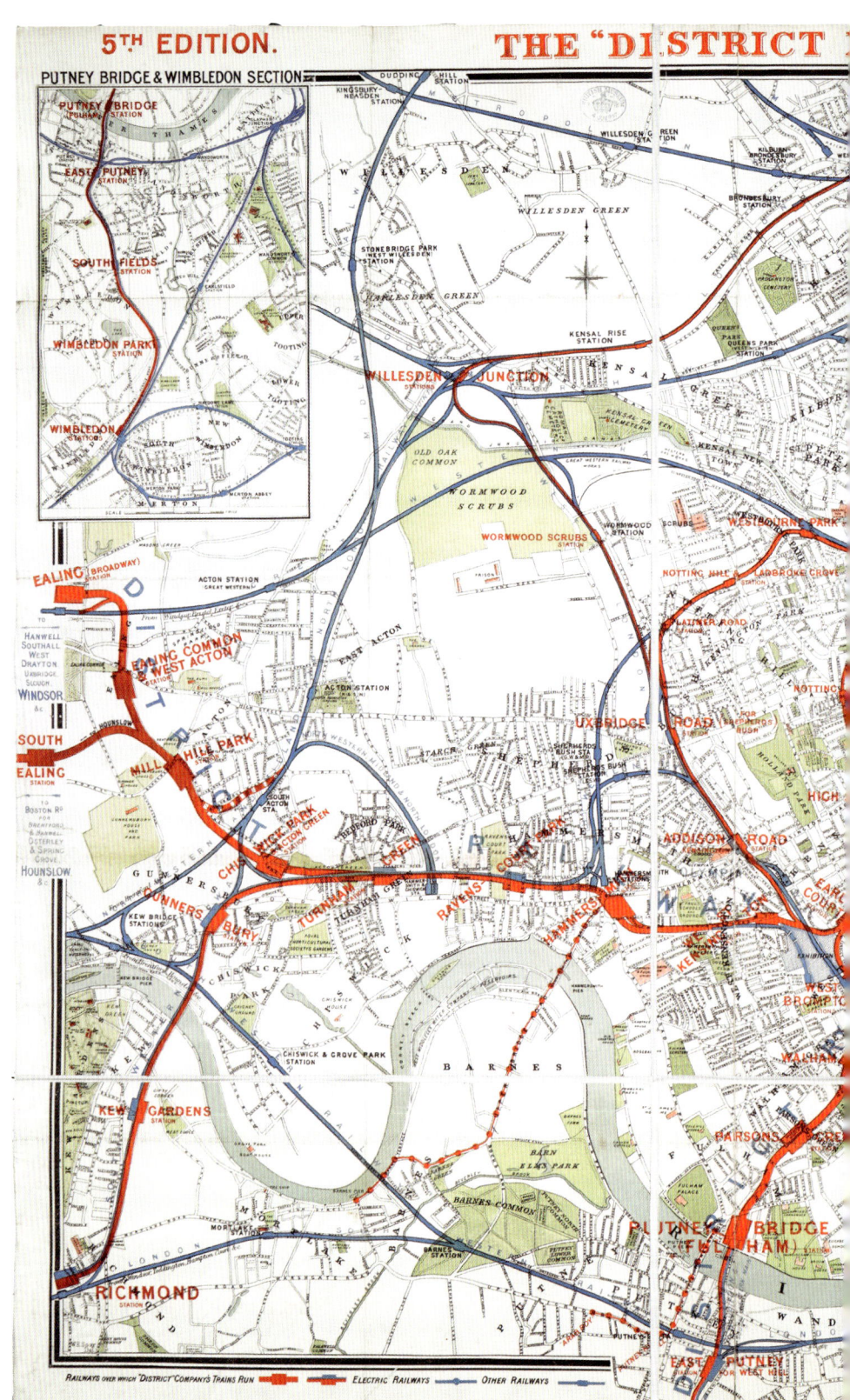

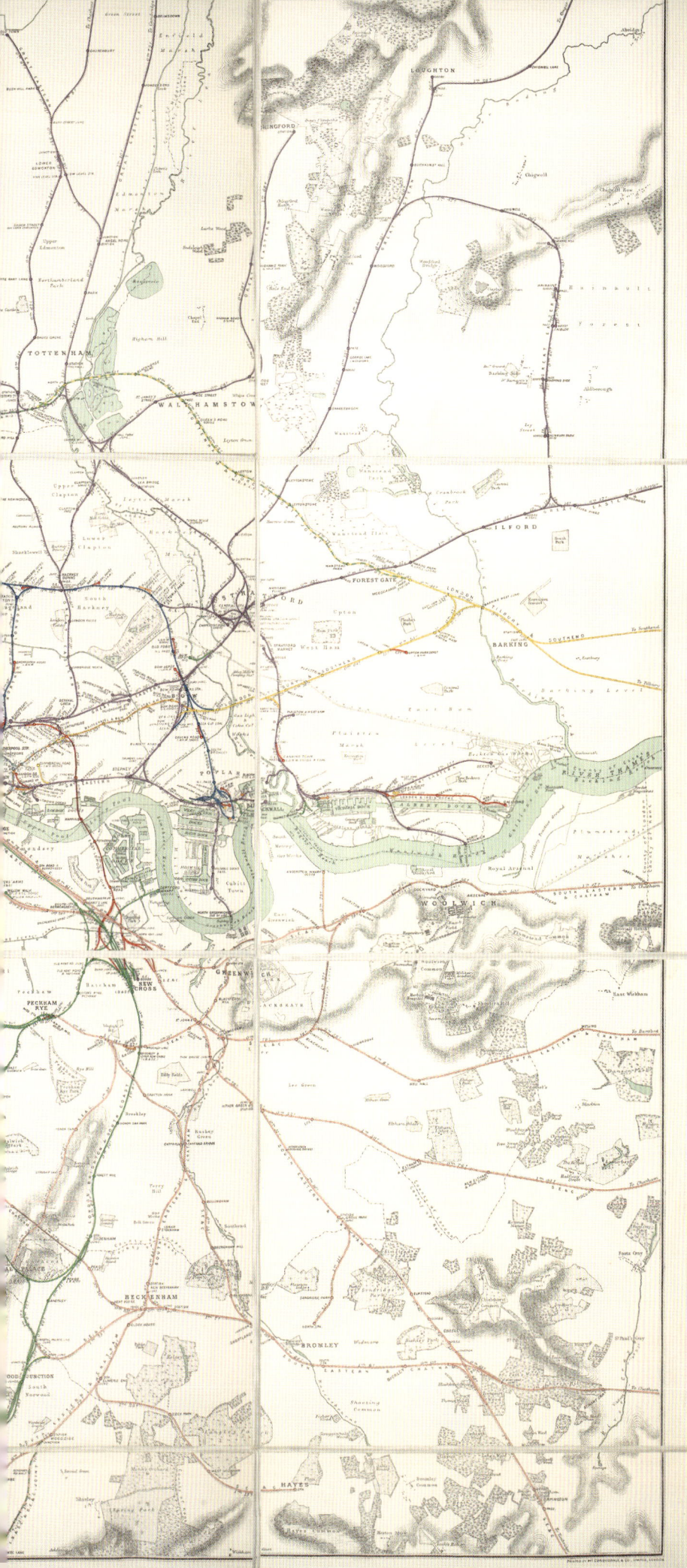

As it was, the construction of London's stations involved massive demolition. The building of Marylebone, the last of the termini, led to the destruction of Blandford Square. As in New York, for example with Pennsylvania Station, the areas destroyed for the London termini, such as Agar Town for St Pancras, were usually inhabited by low-income groups.

As part of the process by which Victorian London continued to develop, some termini did not last. Pimlico, opened in 1858, was a temporary wooden structure that was made redundant when the Victoria Railway Bridge, the first dedicated railway bridge across the Thames, instead made possible the more centrally located Victoria. The 14 termini that remain today were opened between 1836 and 1899. They are a legacy of Victoriana, like so much else in 'modern' Britain. Some were truly impressive: at St Pancras, M.H. Barlow for the Midland Railway used what was, at nearly 250 feet in length, then the largest single-span structure in the world to cover the station's platforms, with a basement underneath with 800 cast-iron columns supporting both the station floor and a store for beer transported to London from Burton-on-Trent. Very differently, railway viaducts ensured that Londoners became used to operating 'under the arches', for example in Borough market.

The railway transformed the region, both within the established built-up area and on its expanding edges. Tottenham, a village on the old Roman Ermine Street, was largely farmland and marshland until the railway came in 1840, with stations established at Tottenham Hale and Marsh Lane. In 1872 a different line with a station at Bruce Grove brought cheap workmen's fares, and Tottenham then grew rapidly, helped by the Cheap Trains Act of 1883. Due to these lines, commuters could go to the dockland areas or central London. Recognition came with rail links: the parish of Tottenham was granted urban district status in 1894.

3. GEOPOLITICS AND WAR 1885–1918 129

RAIL-BORNE SUBURBIA
Knapp, Drewett & Sons Ltd, 'London's Most Healthy Residential Area Served by the Great Central Rly. Co.', 1910.

The Great Central Railway (GCR) was, in 1899, the last major line running into London to be built. This map is a promotional account for the line as a commuter and residential service, and as such it preceded 'Metro-land', a term devised in 1915, with a similar approach taken for the Metropolitan Railway, which in the end took over some of the services of the Great Central. The map uses photographs to suggest the appeal of the area, and ignores the terrain.

Suburbia had spread greatly in the late nineteenth century, with the railways absorbing previously separate villages, but development then had generally not moved far from the stations. In contrast, car transport permitted less-intensive development, although in practice this often meant more extensive estates that were otherwise as densely packed as the basic housing model permitted. In advertisements, cars were pictured against backdrops of mock-Tudor suburban houses, detached and semi-detached (the 'semi'). The Metropolitan Railway County Estates Limited developed railway-owned land into estates, notably at Wembley Park, Harrow Garden Village and in Pinner.

As with the car, the 'semi' expressed freedom: to escape the constraints of living in close proximity to others and, instead, to enjoy space, privacy and respectability. Semis were not the suburban villas of the wealthier members of the middle class, which continued to exist in what were now enclaves, but they captured the aspirations of millions, and offered them a decent living environment, including a garden.

To support the spreading city, the underground railway system became a widespread over-ground to the west, north and east of London; and, to a limited extent, also to the south of the Thames. The 'sub-surface' lines (as opposed to the 'Tube' – the bored tunnels) were already edging well into the country by 1914. The first 'Tube' extension into greenfield areas was the Northern Line extension to Edgware in 1923–24, a project supported by Treasury money to help reduce unemployment. This line was followed by the Metropolitan to Watford, Amersham and Uxbridge in 1925, the Metropolitan (later Bakerloo and then Jubilee) to Stanmore in 1932, the Piccadilly to Cockfosters in 1933, and then another Northern Line branch to Barnet in 1940, where the station was built on the site of Barnet Fair.

Each of these lines organised an important section of London's expansion, while also providing a local pattern of housing and shopping, as well as highlighting areas that were to a degree bypassed by not having 'Tube' lines and/or stations. In this, the pattern of train services was repeated. South of the Thames, the Northern Line reached Morden in 1926. The 17-mile-long tunnel between East Finchley (on the High Barnet branch) and Morden was, at the time of its construction, the longest continuous tunnel in the world. The lines also supplemented conventional rail services. Thus, Ealing, on the Great Western Railway line to Paddington, acquired a District Line service in 1880, the Central in 1920 and then the Piccadilly. The London Underground stations at Ealing Broadway merged with the Great Western Railway station in the 1960s.

The 'Tube' was also responsible for the transformation of London's image, in the form of the map of the system designed in 1931 by Henry (Harry) Beck, a talented electrical draftsman working for London Underground, that depicted the various 'Tube' lines diagrammatically and as straight. Prior to Beck's innovative and arresting work, maps presented by London Underground were designed to be accurate in terms of distance and direction. By contrast, using a topological structure and abandoning scale, Beck, as if imagining that he was using a convex mirror, expanded the central section for the sake of clarity. He also shrank the apparent distance between the suburbs and the inner city, implying that peripheral destinations, such as Morden and Edgware, were within easy travelling distance of central London. Thus, to move there did not appear to be a case of leaving London. Instead, Beck's treatment emphasised the ease of travel into the centre, a visual effect that was encouraged by the use of straight

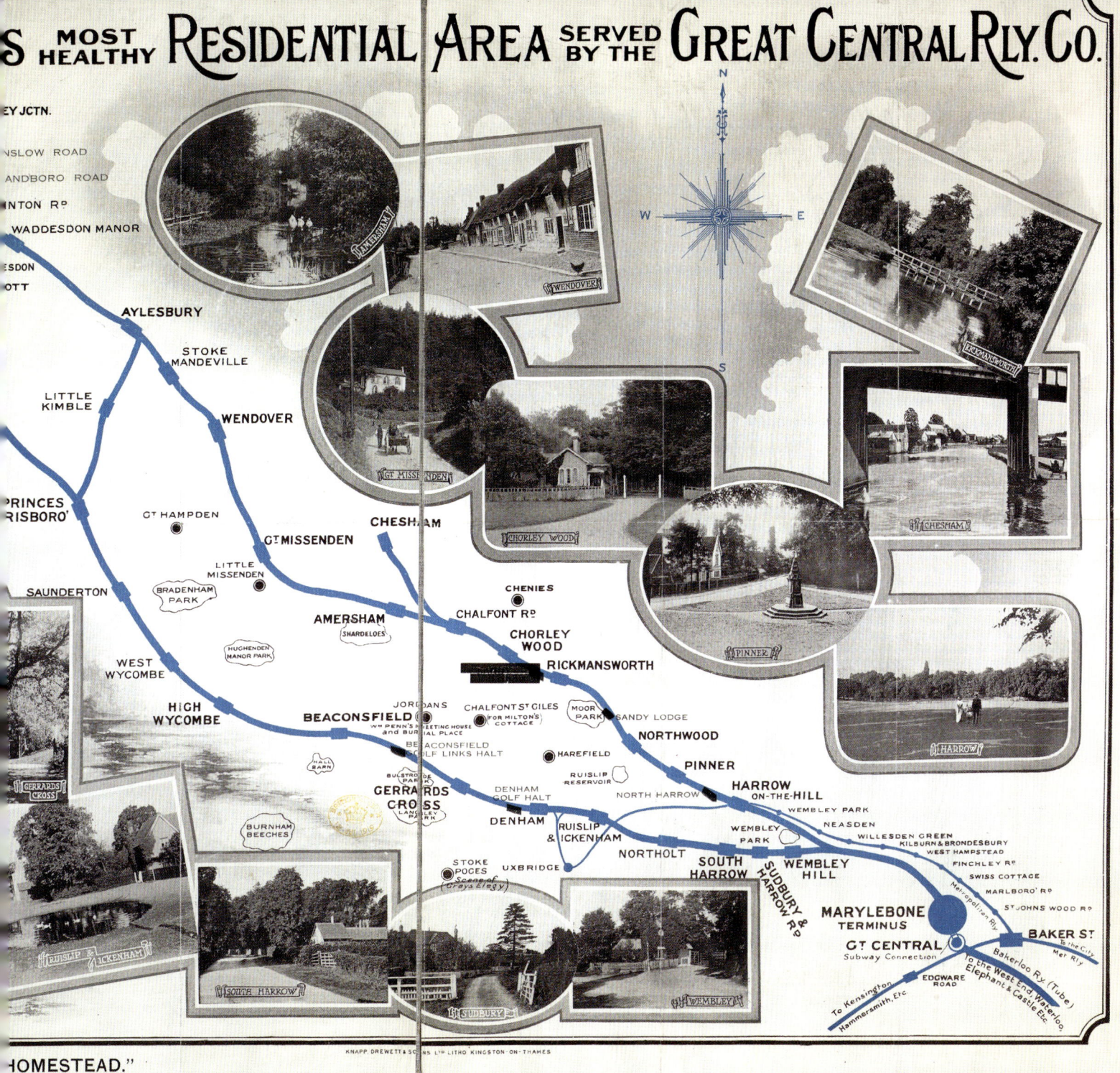

lines for the individual 'Tube' services.

Describing his task as turning 'vermicelli into a diagram', Beck continued to produce updated versions until 1959, establishing the iconic view of the city. However, the London Passenger Transport Board held the copyright and he was embittered by the limited returns he enjoyed.

Different fare tariffs on the Underground played a role in the development of the linked housing system, for, with effective advertising, the 'Tube' was presented as integrating housing, transport and work. For example, fare tariffs helped to ensure that Edgware grew more rapidly than Stanmore, from where the fare to the city centre was higher in the 1930s. As a result, Stanmore was more socially exclusive, or could be seen in that light.

3. GEOPOLITICS AND WAR 1885–1918 131

IRON RAILS AND THE STATE

Ferdinand Walseck, 'Latest railway map of Germany and the neighbouring countries with quick-to-find stations', 1891.

The scale of this large fold-out map captures the German achievement in railways, and the extent of territory that had to be covered. That was in part a matter of the expansion of Germany to 1871 when the Second Reich (Empire) was declared in the aftermath of the annexation of Alsace and part of Lorraine from heavily defeated France. Indeed, the boundaries of Germany and therefore the coverage of railways shown in this map covered both these and the annexation of Schleswig-Holstein after the victory over Denmark in 1864. Thus, Strasbourg and Colmar, both in Alsace, appear as German rail hubs, while Flensburg was no longer in Denmark.

The map is significant in several respects. It did not include many physical features. Rivers, such as the Rhine and Isar, are shown, but not with any focus on them; and there is no attention to terrain.

More specifically, as a classic instance of what the map omits (often a potential interest that can be underrated today), there is an absence of the governmental divisions within Germany. These remained significant after the creation of the Second Reich, with Bavaria, Saxony and Württemberg all kingdoms, and a number of other independent princely territories, including Baden and Hesse. None is shown on the map. Instead, the sole boundary is that of the German Empire. So also with the failure to mark that between Austria and Hungary within the Austro-Hungarian Empire.

Ferdinand Walseck's location in Cologne, a major rail centre (as with Trott's map in Bilaspur, pages 102–103), had no consequences for the selection of material for the map. It was produced in Berlin using lithography, a production technique that contributed to the increase in the quantity and range of maps produced, not least the development of thematic mapping. In lithographic transfers, the design for an engraved plate was transferred to a litho stone on which alterations could be made without affecting the original plate. A much finer line and neater lettering could easily be achieved, and lithography was less expensive than copperplate-engraved maps. It permitted the density of printing seen in this map, not least with the station names, which are closely packed but nevertheless distinctive and legible, and therefore useful for map readers.

Germany from the 1830s initially had railways within the individual states, including Bavaria, Brunswick, Prussia and Saxony. Northern Germany saw more development, in part due to the presence of coal, but also because of a greater population, more industrial activity and close links with Britain, from where rail technology was obtained, notably engines. As in Britain, coal-based energy prices in Germany were low by international fuel price standards, and the coal was easier to use than wood, and this encouraged industrialisation.

The 263-kilometre-long line from Cologne to Minden, opened in stages in 1845–47, was the longest of Germany's early railways and a key to the coal-based industrialisation of the Ruhr. It was also an aspect of the affirmation of Prussian power in the Rhenish territories acquired under the Congress of Vienna in 1814–15, including Cologne. In part, this was a question of Protestant control over Catholic territories. Encouraging an alignment from Cologne to Minden, a major fortified position and garrison town, was part of the Prussian power system. Thus, the rail link was in part a means to deploy the military. The Minden railway station opened in 1847 had the style of a fortress. In 1847 a line was opened from Minden to Hanover, with a further line in that year from Wunstorf on that route to the port of Bremen, which provided an all-German route for Rhenish commerce to reach the ocean.

The network shown in this map bore signs of the plan for a German railway system advanced by Friedrich List. In 1841, he wrote: 'The iron rails become a nerve system, which, on the one hand, strengthens public opinion, and, on the other hand, strengthens the power of the state for police and governmental purposes.'

Germany sat on top of some of the biggest coal deposits in Continental Europe, while the introduction of railways brought particular benefit to parts of

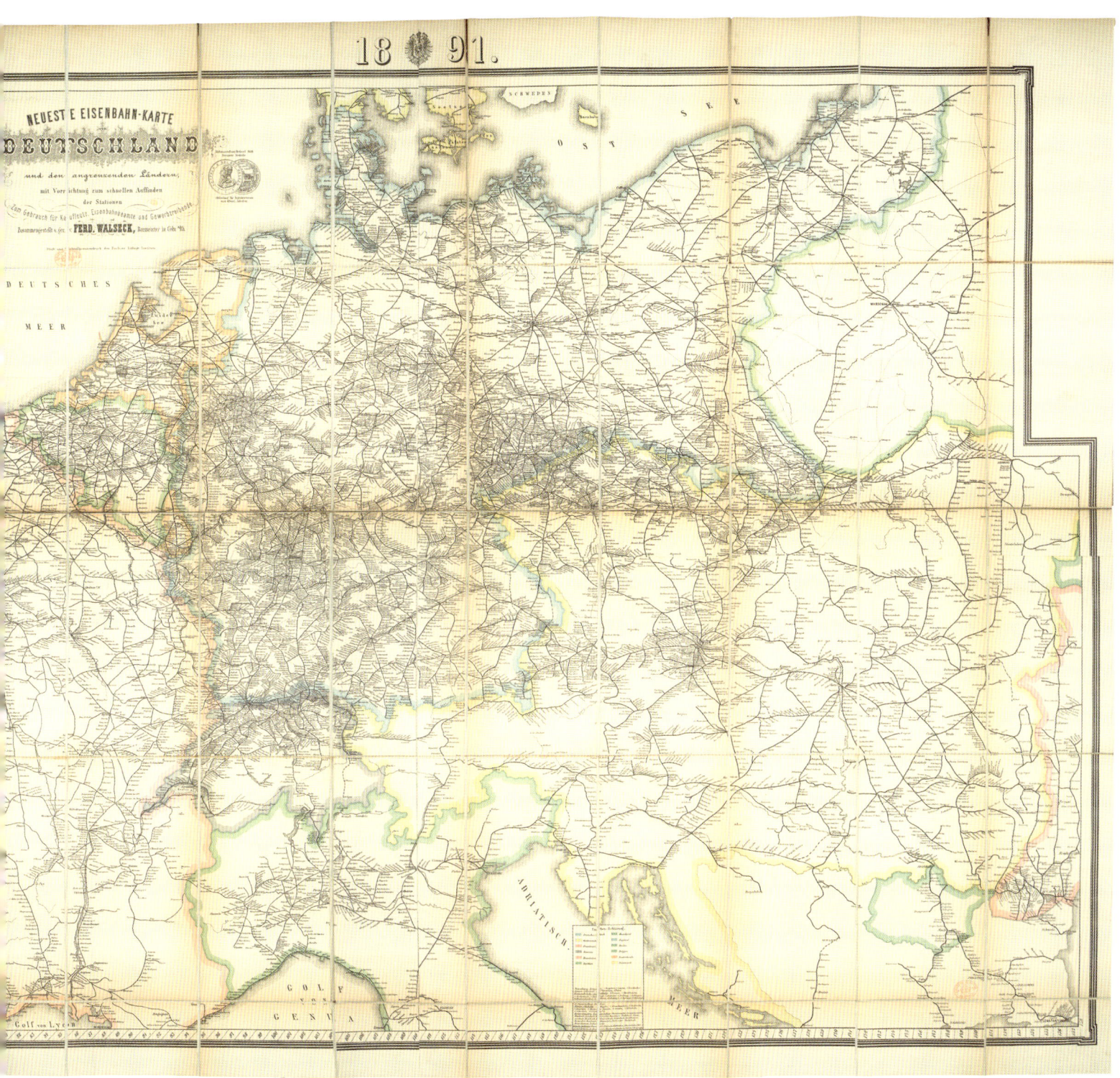

Europe furthest from the coast and from navigable waterways, which advantaged Germany in particular. By 1913 Germany was producing possibly 15 per cent of the world's manufactured goods. The number of kilometres of German railways rose from 11,089 in 1860 to 63,378 in 1913, the annual average output of coal in million metric tons from 41 in 1870–74 to 247 in 1910–14, and for pig iron from 2.7 in 1880 to 14.8 in 1910. The total length of navigable waterways doubled between 1875 and 1914, with steam and electric towing of barges important. The expansion of rail was therefore part of a more general growth in the economy and in infrastructure. The passage of rail over the Kiel Canal between the Baltic Sea and the North Sea, built in 1887–95 and widened in 1907–14, was particularly impressive. Erected in 1911–13, the Rendsburg High Bridge had long elevated embankments and access bridges in order to enable the railway to make the gradient.

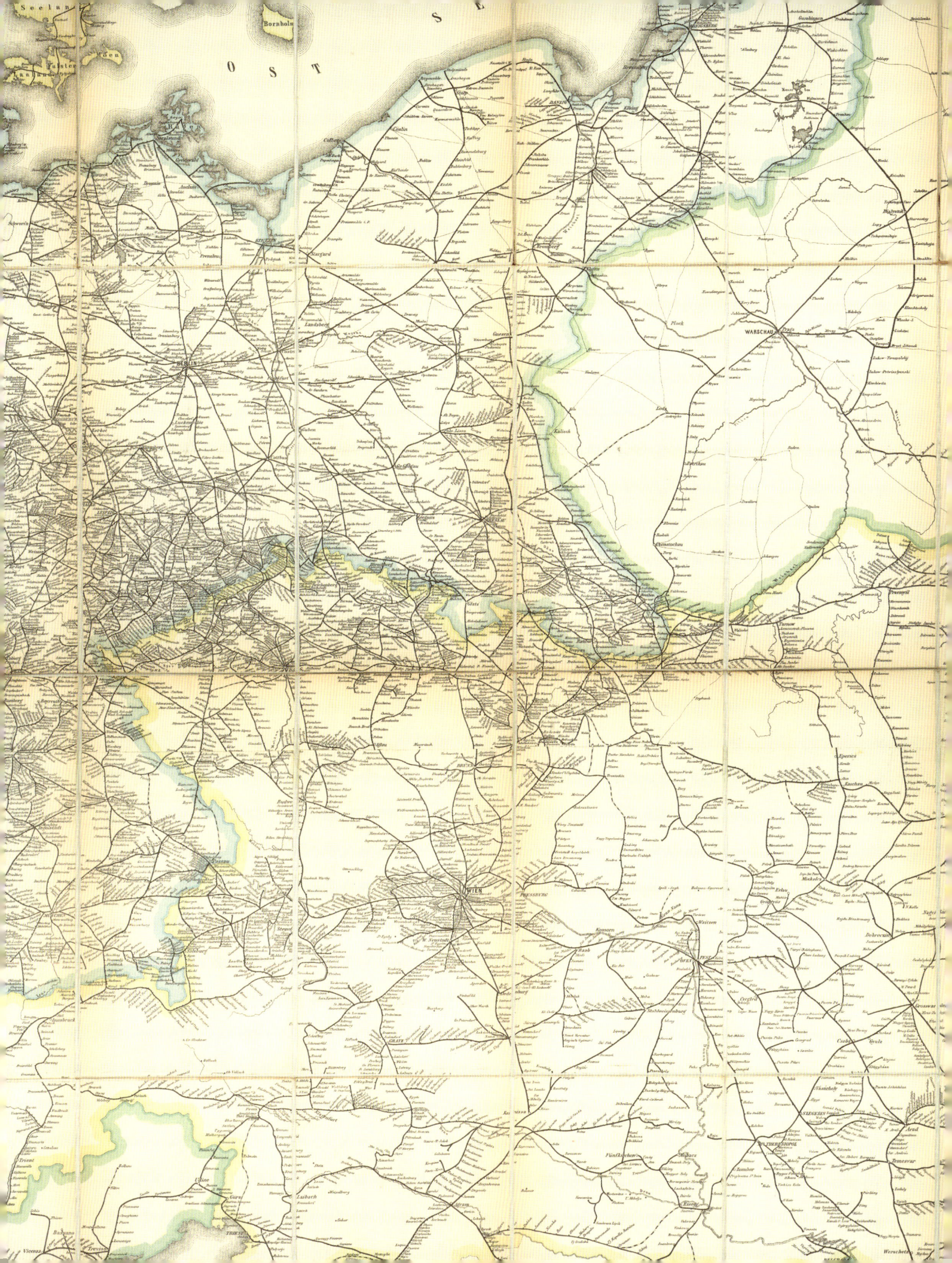

THE AUSTRO-HUNGARIAN EMPIRE AND THE BALKANS

Artaria & Co., 'Artaria's Railway Map of Austria-Hungary', 1913.

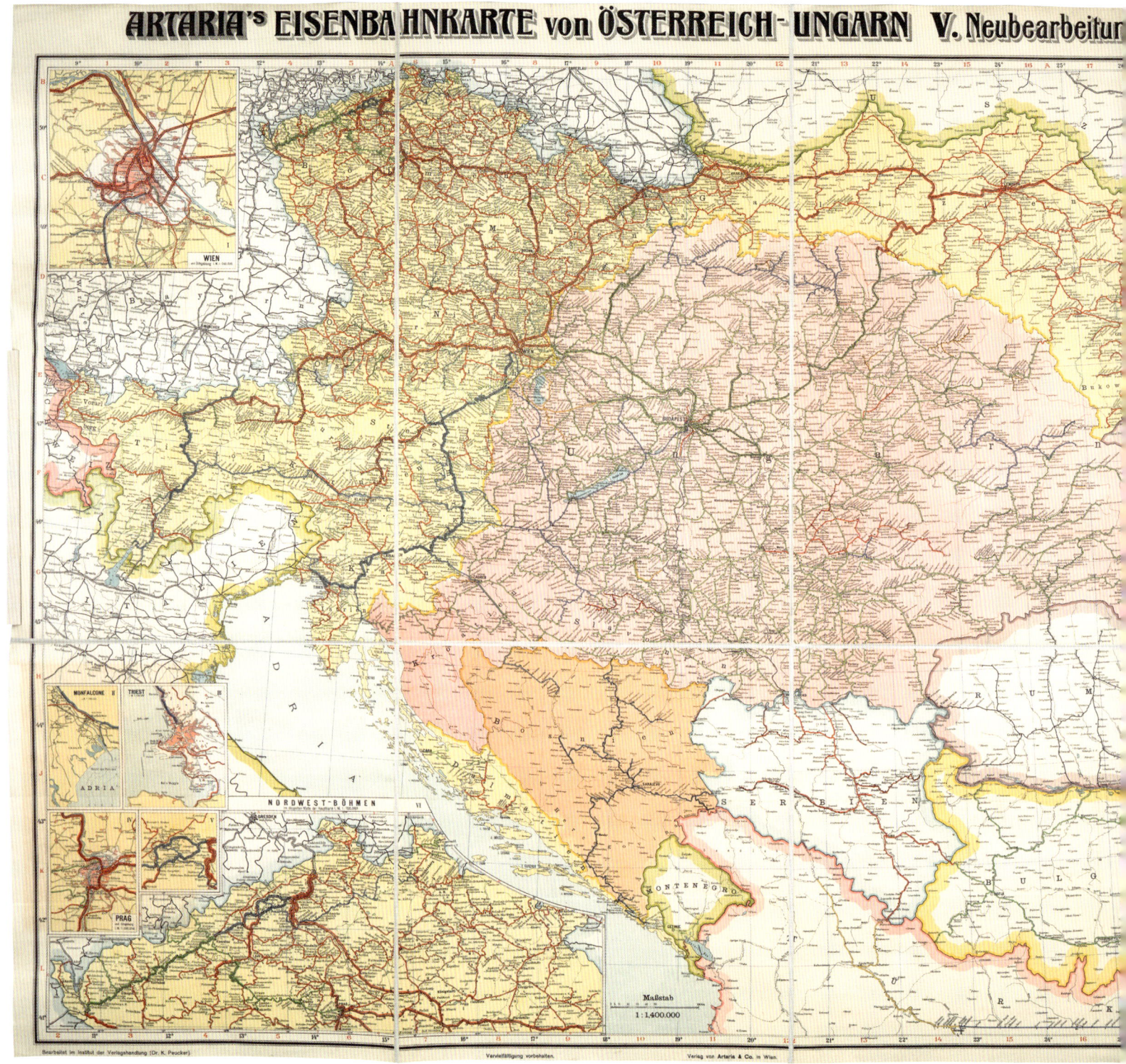

This 1913 map was produced at a time when the Balkans were being remoulded due to a war that had begun in 1912, revealing the difference between the relatively sparse networks in those countries and the far denser situation in the Austro-Hungarian Empire. Even so, the latter, as the map indicates, had major variations; for example, there are few lines in Bosnia, which had been occupied in 1878 and annexed in 1908, while one of the inserts shows that northwest Bohemia has numerous lines, most owned by the private Austrian Northwestern Railway, which had been nationalised in 1908, and others in the major cities, notably Budapest, Vienna, Prague and Trieste. The linkages between the Austrian, Hungarian, Bohemian and Galician (southern Polish) parts of the empire emerge clearly. The use of colour reveals the relationship between different lines. As a result of takeovers, notably in 1906–09, 82 per cent of the Austro-Hungarian Empire's railways were owned by the state (the Imperial-Royal State Railways having been established in 1884).

Initially, links with the water system, notably the River Danube, were important to the development of Balkan railways. In 1860 an English consortium linked Cernavoda on the Danube with the Ottoman (now Romanian) Black Sea port of Constanța, thus avoiding the Danube delta.

The first railway in Romania was linked to the export trade, namely a concession granted in 1865 to an English company for a 70-kilometre-long line from Bucharest to the Danube port of Giurgiu, completed in 1869. Austrian, German and Russian consortia were also significant in Romania, ensuring links to the Austrian and Russian systems, so that by 1880 there were over 800 miles of railway line in the network. In turn, the state took over the railways, creating in 1880 the General Directorate of Romanian Railways, which became the largest industrial enterprise in the country, and in 1889 a monopoly.

Bram Stoker's novel *Dracula* (1897) offered a fictional account of a Balkan journey, one in which the train was supplemented by subsequent horse-drawn carriage transport. Travelling east by train from Budapest into Transylvania (then part of Austria-Hungary, now Romania) via Klausenburgh (Cluj) to Bistritz (Bistrita), the protagonist Jonathan Harker had to rush breakfast at Cluj only:

'...to sit in the carriage for more than an hour before we began to move. It seems to me that the further East you go the more unpunctual are the trains.... All day long we seemed to dawdle through a country which was full of beauty of every kind. Sometimes we saw little towns or castles on the top of steep hills such as we see in old missals, sometimes we ran by rivers and streams which seemed from the wide stony margin on each side of them to be subject to great floods.'

Romania was better placed than Bulgaria and Serbia to attract international investment, not least because of its resources, especially grain, which was much in demand to feed the rapidly expanding population, and due to its proximity to the Austro-Hungarian and Russian networks.

The Bulgarian system was less extensive. The first major line, begun in 1864, was commissioned by the Ottoman rulers. Opened in 1866, the 223-kilometre-long line went from Ruse on the Danube, opposite the Romanian city of Giurgiu, to Varna on the Black Sea. This was followed by the work of the German entrepreneur Moritz von Hirsch (1831–96) who, having made a fortune from banking, purchased in 1869 the concession for railway building in the European part of the Ottoman Empire and sought, with his company the Chemins de fer Orientaux, to build a line from Vienna to Constantinople. Completed in 1888, the line provided, from 1889, the route for the Orient Express.

The route was planned to avoid Serbia, which was seen as hostile by both Austria and the Ottoman Empire. Work was delayed by Ottoman financial problems and by the conflict in the Balkans between 1875 and 1878. Instead, in 1883 it was agreed to build via Serbia and thus connect with the new Serbian National Railways. This was an aspect of the extent to which there was a degree of integration of national systems in Europe that was not matched in the larger and more environmentally challenging continents of Asia, Africa and Latin America.

Lines in the Balkans were to play a role in the strategic planning and propaganda of the First World War. Thus, in the United States, the Committee on Public Information, emphasised the German geopolitical threat in its 1917 map-poster 'Why Germany Wants Peace Now', which showed occupied Serbia and Romania as adding substance to German geopolitical plans on the Berlin–Baghdad rail access, specifically railways from Nis via Salonika to Athens, and from Berlin via Lemberg (Lviv) and Constanța to Varna, with the extension from the latter to Constantinople not as yet built.

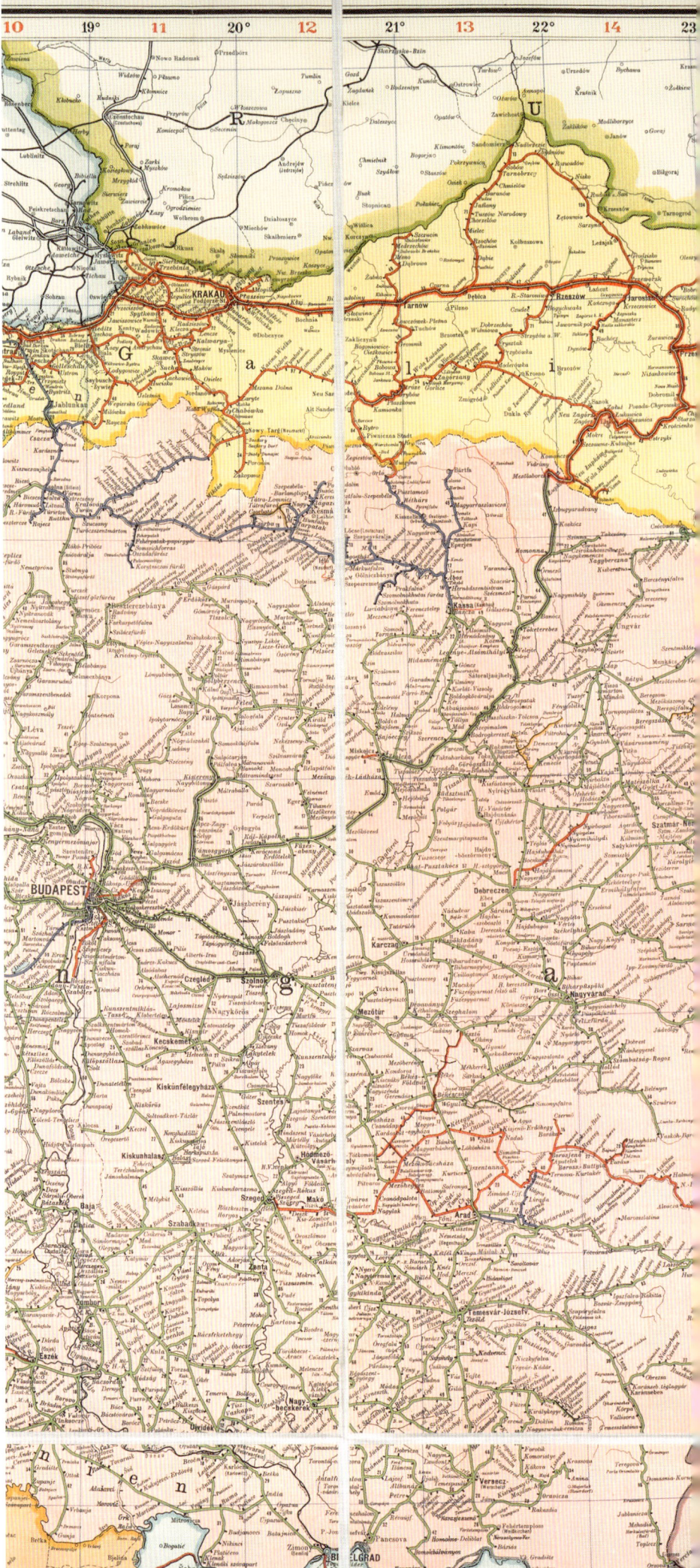

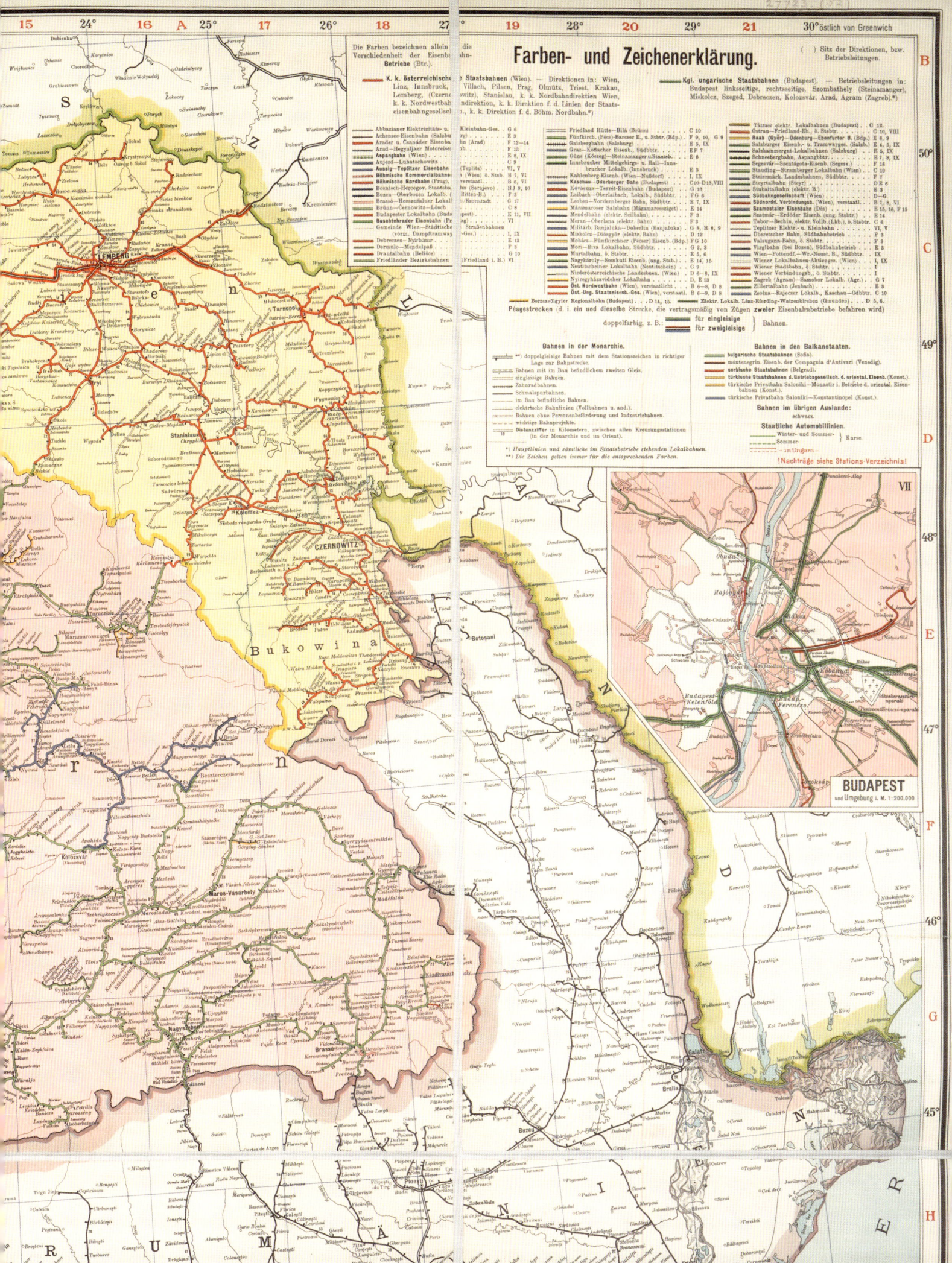

SWISS RAILWAYS
Jules Décor, 'Official Kilometre Map of the Swiss Railway', 1888.

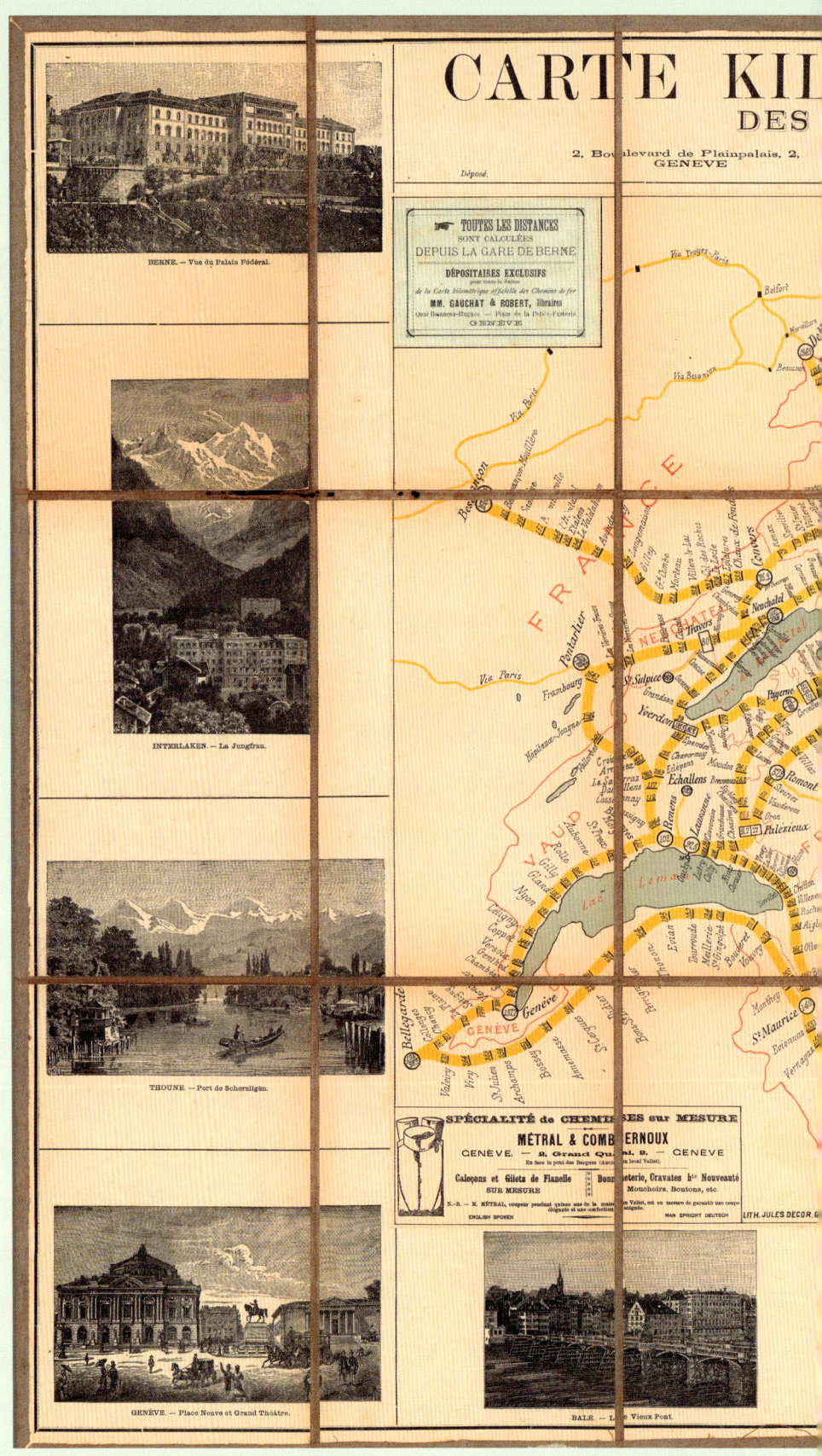

This map depicts the Gotthard Pass route, which opened in 1882, but not the line under the Simplon Pass, completed in 1906; that the Simplon Pass line is not even projected suggests that the map is pre-1900. The map was printed as a chromolithograph and published by Jules Décor in Geneva, where it was segmented and laid on linen, allowing the traveller to fold it up and take it on the road. This was very different to the standard form of map made for display in stations and on billboards.

Swiss railways were constructed and operated by private companies until 1901 when the system was nationalised under the aegis of the Chemins de fer fédéraux suisses (CFF)/Schweizerische Bundesbahnen (SBB)/Ferrovie federali svizzere (FFS) or (Swiss Federal Railways). The map uses a colour scheme of yellow and red with stylish fonts. The numbers given are distances in kilometres from Berne.

Two advertisements are found below the map, both located in Geneva. Framing the map section is a series of engravings of major buildings and picturesque locations throughout Switzerland, including the Gotthard Pass.

IBERIAN RAIL

D.J.M. Serra, 'Plan of Barcelona and its Surroundings', 1891.

This plan of Barcelona, approved by the city hall, makes plain the impact of rail on the urban cityscape. Railway construction developed rapidly in Spain after the Railway Act of 1855. Railways were a symbol and reality of national integration and convergence. Across the country, from Madrid to Barcelona, major stations were dramatic buildings.

Serra responded to city plans proposing the redevelopment of Barcelona along modern lines, notably the Cerdà Plan of 1860. Relief, shown by hachures, indicated the Collserola mountain range that influenced rail links, which to a considerable extent stayed in coastal areas. The city walls had constrained expansion until their demolition in 1854; thereafter the city began to incorporate adjoining municipalities. The rail links were designed to cross the grid development that was emphasised in the major plans of the period. In 1888 the Universal Exposition provided a second impulse for expansion. While roads were central to the grids, Ildefons Cerdà incorporated railway lines, although he thought they should go underground in the centre. He had offered, in 1851, a vision of the railway as modernisation, and that image affected plans such as this one by Serra: '... when railways have become generalised, all European nations will be one city, and all families only one, and their forms of government will be the same.'

The first railway in Spain was built from Barcelona to the fishing town of Mataró, 19 miles away. Hence Barcelona had a major railway station from 1848, the Sants station, and it was progressively expanded in the late nineteenth century as the city became the base of the service along the Catalan coast, one that was at once to be local, regional and national. As with much construction, high explosives played a role, in this case dynamite being used to help drive through the Montgrat Tunnel. Five million pesetas was raised and the railway began service in 1848. Rapidly becoming prosperous, indeed it paid a dividend of 23 per cent in its first year, the line helped in Catalan industrialisation.

The neoclassical Estació del Nord followed in 1862 in order to serve the Zaragoza line. The role of Barcelona as the link point between much of Spain and France underlined its significance for transit. The line with Madrid was opened in 1859.

Most of the expansion in Spanish railways was from 1855, notably in 1855–64, 1878–84 and 1893–99. This episodic character was typical of rail construction. It reflected, in particular, the availability and price of credit. In the case of Spain, credit from neighbouring France was decisive, so that by 1914 French companies, notably the Norte Company (Madrid–Irun on the French frontier) and the M.Z.A. (Madrid–Zaragoza–Alicante), controlled 85 per cent of the Spanish system, having bought the Catalan and Basque-financed lines to add to their own. French expertise, capital and iron were all important, but competition, heavy debts and the purchase of small lines with their own debts helped limit profitability, which ensured that the renewal of equipment and track was inadequate. The M.Z.A. never provided a dividend, while Andalusian lines lacked profitability, and the Norte paid nothing from 1865 to 1873 and from 1891 to 1906. The economy did not provide the necessary volume nor the resulting profits that could be used for investment. Moreover, Spain lacked sufficient coal, as well as cheap iron. In particular, it had fewer resources than Britain, France, Germany, Belgium, America and Russia.

In Spain, the first special rail tariff for sheep was in 1899, and migrant workers and wine were also moved by train. The presence or absence of a station became crucial to the fate of towns. However, much of Spain was poorly connected, let alone served, with minimal links to Galicia. As a result, from 1904 narrow-gauge lines were built with state encouragement, in order to fill the gaps, notably in northern Spain; only to find there were not sufficient goods to transport. The railways were also affected by political instability and trade union activism. In 1912 there was a major rail strike, followed by a strike with revolutionary intentions in 1917 that divided rail workers.

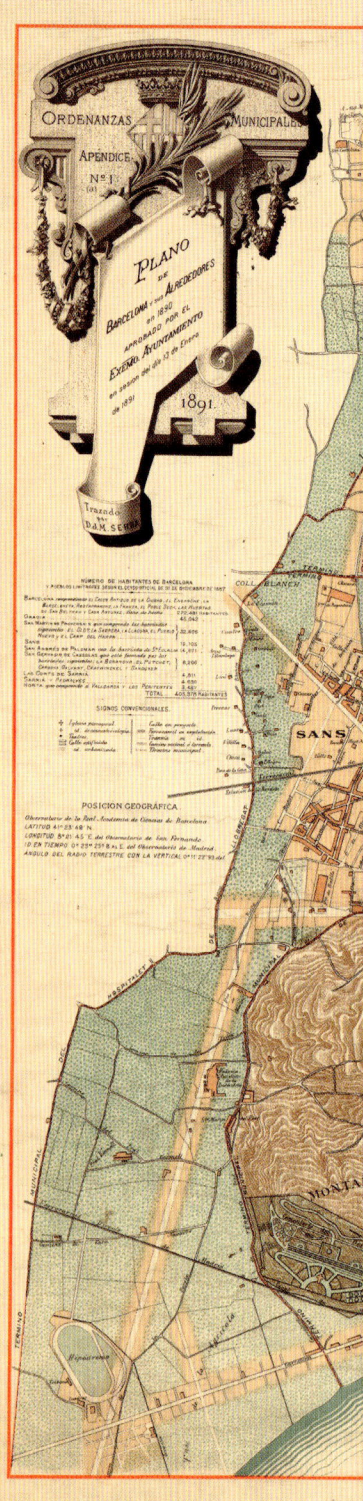

There was growth and modernisation in the 1920s. The railway network was then expensively renewed, with freight trade greatly rising partly as a result. The contemporary expansion of the road network showed that the two were not inherently incompatible.

However, in the 1930s there was recession, insolvency and a lack of new construction and locomotives. The competition of lorries was therefore far more effective, such that, by 1933, Madrid received 64 per cent of its fish by lorry, compared to 4.5 per cent for Paris. Today the fish arrives by air.

PORTUGAL

Gazeta dos Caminhos de Ferro de Portugal, 'Map of the Portuguese Railways on 1 January 1895 in the Mainland and Overseas', 1895.

The first railway line in Portugal, that from Lisbon to Carregado, was only 36 kilometres long. It was commissioned in 1853 and opened in 1856 in a lavish ceremony presided over by Queen Mary II, and the network then spread, albeit far more slowly than in wealthier France or Britain. In 1864 the line reached Gaia, on the south bank of the River Douro, linking Lisbon with Oporto, the second-largest city, with the ferry replaced in 1877 upon the completion of the D. Maria Pia bridge, designed by Gustave Eiffel. In contrast, there was no bridge across the far broader Tagus at Lisbon until 1966; instead, ferries had to be used.

Rail within Portugal was expanded – in accordance not only with technical considerations (for example, a resort to narrow-gauge railways) but also economic, financial, military and political considerations – with profit and national improvement all playing a role, while Portugal's adoption of the Gold Standard in 1854 eased foreign investment. There was, as in Britain, no general railway plan. In 1887 a line along the Douro that had begun in 1875 was completed. In 1888 the Lisbon-based railway newspaper *Gazeta dos Caminhos de Ferro de Portugal* was launched, which in 1895 produced this map of Portuguese railways. The map includes inserts of the railways in Mozambique (the bottom two maps on the left) and Portuguese Guinea (now Guinea-Bissau) and Angola. The emphasis on rail is increased by using red for the lines, while also including lines under construction. The emphasis on northern and central Portugal is clear, as is the poorly served nature of the northeast and the south, with only one line to the Algarve. Towns such as Braganza and Silvas were not on the railway.

There was also a degree of integration into the broader international system, first through lines and then by means of through services, for there was often no automatic or even rapid progress from one to the other. In 1866 Lisbon was linked into the system via the Portuguese rail junction of Entroncamento (south of Thomar) and Badajoz in Spain. In 1887 the Sud Express first ran from Paris to Lisbon via Madrid, and in 1888, taking advantage of the ferry links, the service was run twice weekly from London, becoming a daily from 1907. This long-distance train provided an attractive alternative to the stormy sea passage across the Bay of Biscay.

In Portugal, as elsewhere, townscapes were changed as stations were built and lines driven through. The stations were major works. Finished in 1887, Rossio Station in Lisbon served destinations including the royal summer palace at Cintra. The station was built in a Neo-Manueline style with Moorish-style horseshoe arches. The original Oporto railway station was followed in 1903 by the more central São Bento train station, finished in 1916 on the site of a monastery. It was enhanced in 1930 by about 20,000 tiles that show the history of transport as well as historic battle scenes, including Henry the Navigator's conquest of Ceuta in 1415.

The development of a train system increased the need for coal, but Portugal produced neither coal nor iron, which increased its dependence on Britain. The total of kilometres built increased steadily as the decades passed: to 1,177 in 1880, 2,083 in 1890 and 2,898 in 1910. The Caminhos de Ferro de Portugal (State Railways) was created in 1892, but most railway lines remained in private hands. In 1975 the Companhia dos Caminhos de Ferro Portugueses (CP), which had run most of the railways since 1927, and all the network bar the Cascais line from 1951 onward, was nationalised, and in 1976 the Cascais line was included. Today, in a pattern that is true of many European states, the overall map of the Portuguese railway network closely follows that drawn between 1853 and 1910.

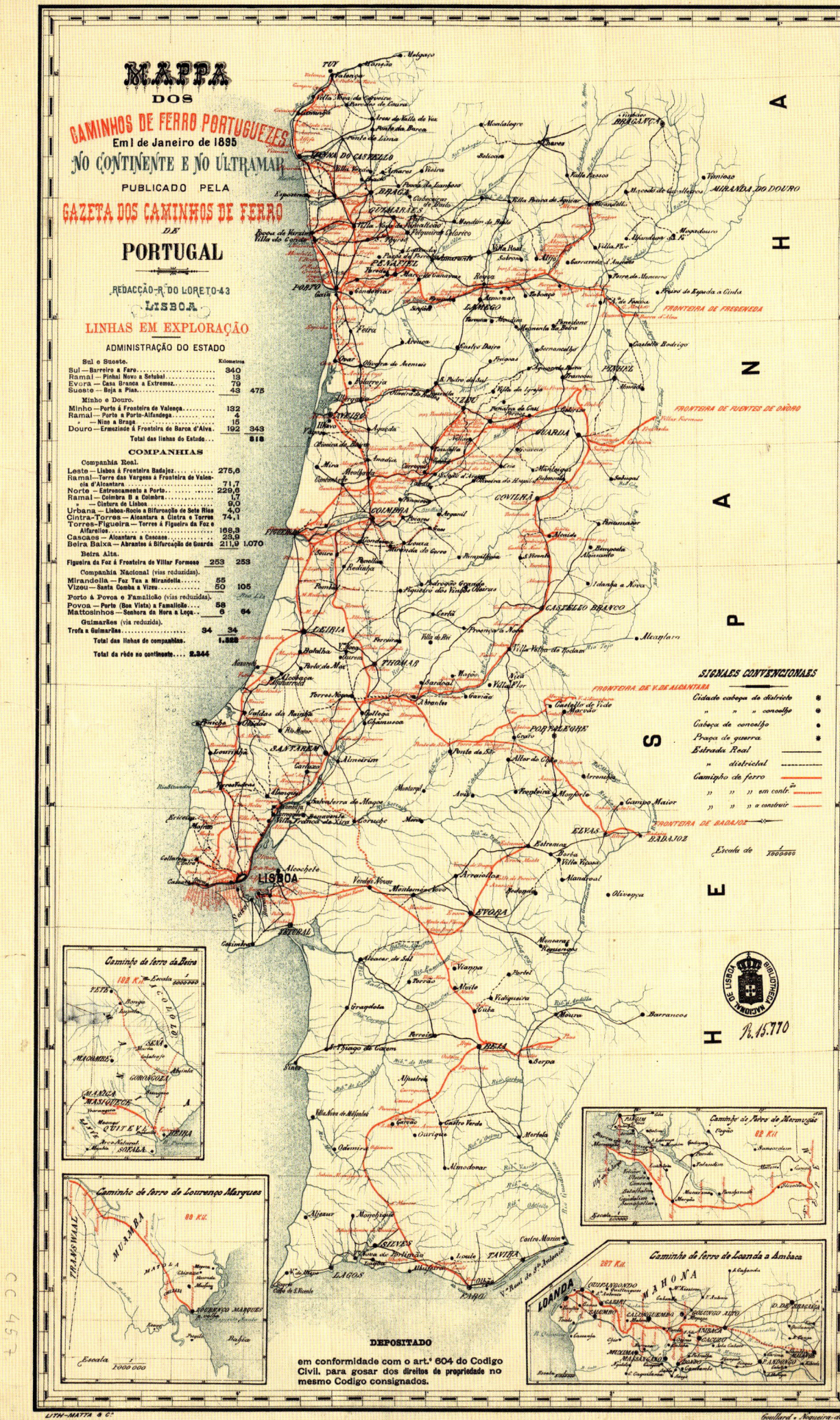

FRANCE AND BELGIUM

Paul Vidal de la Blache, 'France: Railways', 1894, and L. Mols, 'New Map of Belgium', 1851.

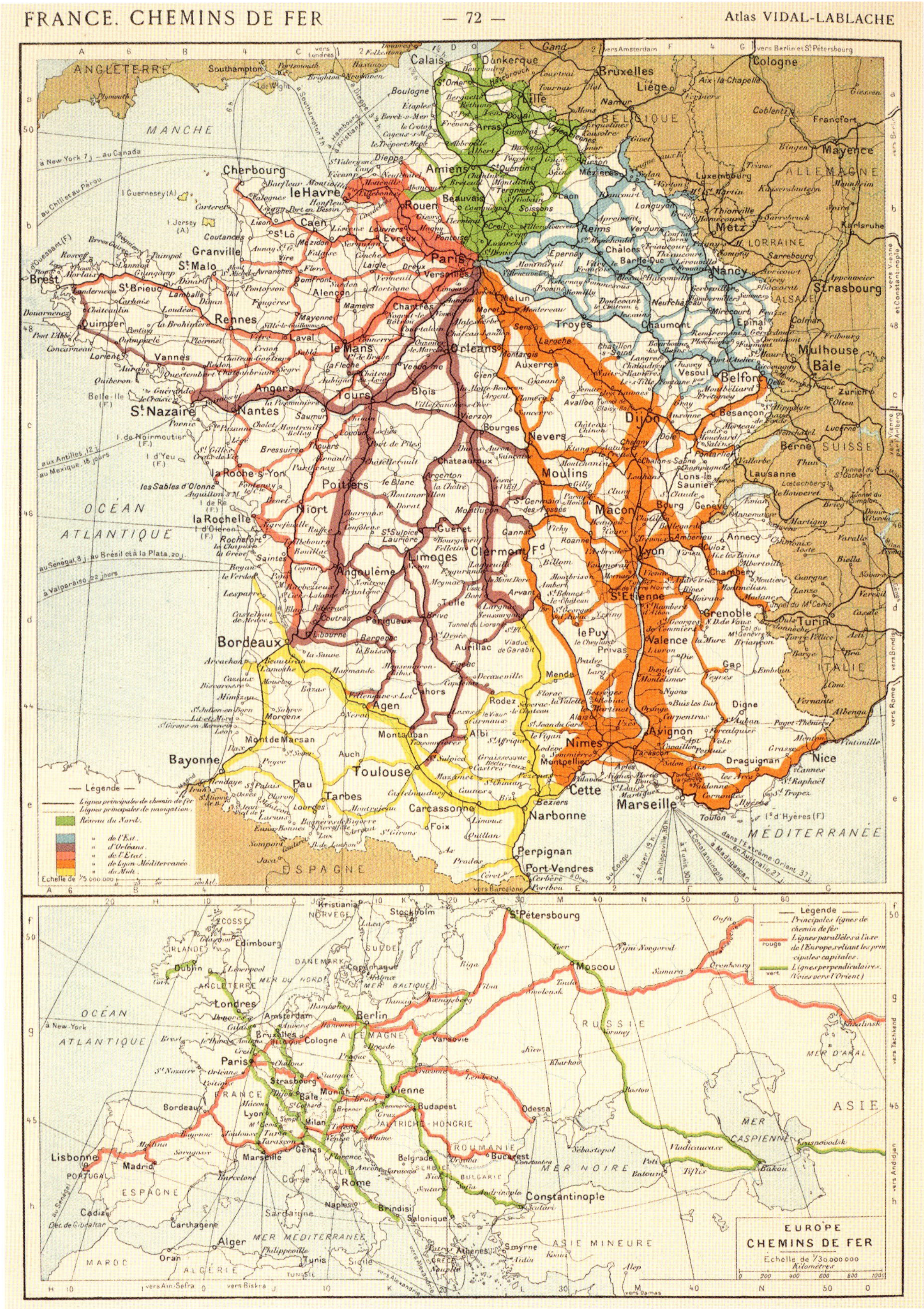

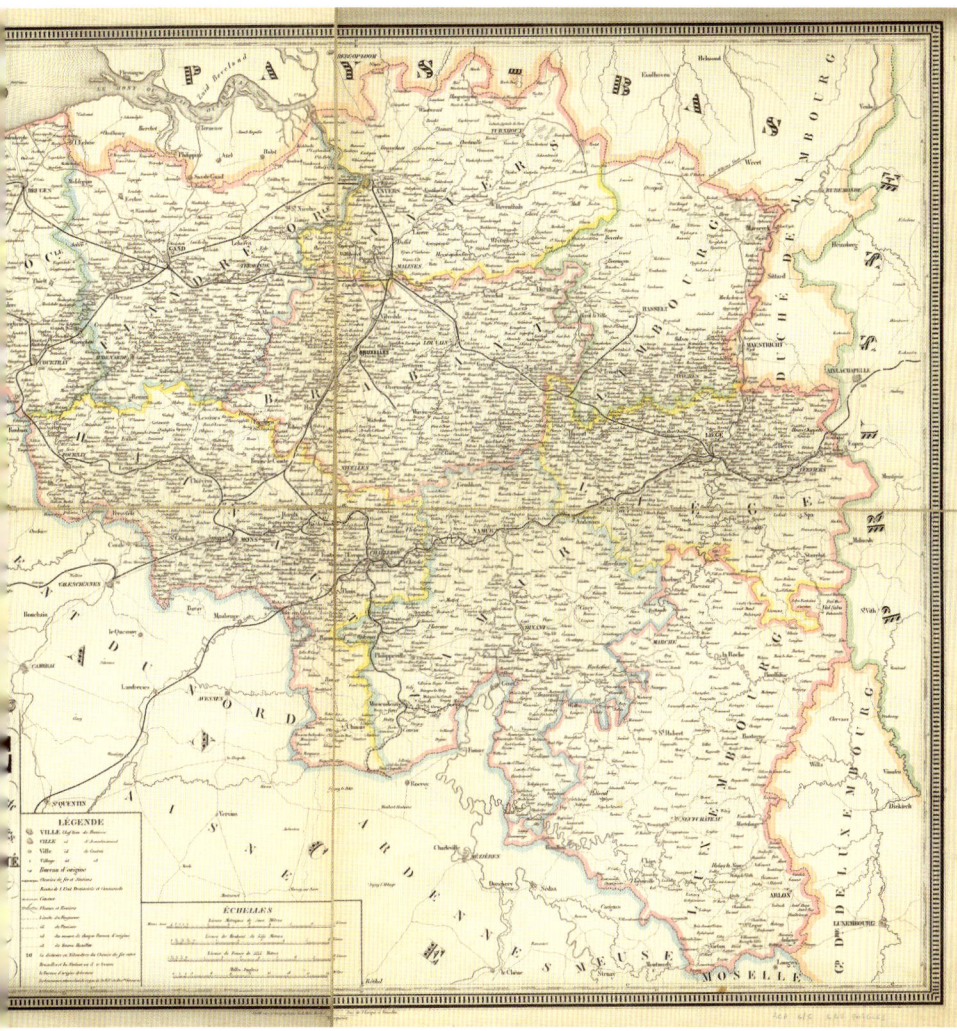

There is a more complex situation to organise than in England and Wales, as France is four times the size. De la Blache presents a system of major provincial hubs, including Amiens, Orleans, Le Mans and Limoges, and notes that the railways then carry nearly two-thirds of national freight, and that only in northern and eastern France is there a major use of canals. Coastal trade is therefore less significant than in Britain.

In 1914 France had about 37,000 miles of rail, about one-third of it narrow gauge. Developed largely from the 1880s, the latter was classified by law as local interest rather than general interest. It was in part a variant of the more general branch network. About 12,500 miles of 'lines of local interest' were built by 1930, in part a matter of pork barrel politics but also in pursuit of the Third Republic's wish to bring equality through modernisation, and thus to lessen the appeal of reactionary politics. Taking forward earlier themes such as Horace Vernet's 1838–47 mural in the Palais Bourbon, the poet Pierre Lachambeaudie (1807–72) expressed this in his 'La Vapeur':

> 'Thus, in hamlets deprived of
> The abundant, civilising waters running all around them,
> Does poverty, ignorance, and envy reign.
> The iron arteries of the beneficent network
> Over every piece of soil affected and crossed,
> Carry life through the social body.
> Every city is becoming a vast reservoir.
> From which comes great flows of wealth and knowledge.'

In a work first published in 1894, Paul Vidal de la Blache, the leading French geographer, mapped the country's rail system with the lines proportional to the freight carried. This infographic approach was typical of the innovative cartography that railways attracted. The colour-coding reflects the merger in 1857 into six major companies of the previous larger number of concerns, a situation that remained in force when the map was being drawn. De la Blache points out that the system of radiating lines is affected by the high ground of the Massif Central in south-central France, which led to detours between Toulouse and both Paris and Lyon. The limited economic benefits of the Massif Central acted as a deterrent to rail construction there.

In contrast, the earlier map of Belgium, which had chosen public ownership in 1834, shows that by 1844 it had the core of its railway system, with one track running west–east, from Ostend to the Prussian frontier, and the other north–south, from Antwerp to the French border, the two crossing at Malines/Mechelen. The role of ports in this system was more generally significant to the development of early railways. Like Britain, Belgium had plentiful coal and this made the financial choice for rail rather than canal easier.

3. GEOPOLITICS AND WAR 1885–1918 147

ITALIAN STATE RAILWAYS

Federico Sauer, 'Italian Railways in Operation, Under Construction and in Planning', 1908.

Federico Sauer's map is a detailed but bold and clear map of the Italian system, providing coverage of all the lines and stations, with the distances between stations recorded in kilometres. Steamship services are also included, which was a norm for rail maps of the period. In the case of Italy, it also reflects the need for a national transport service to provide routes to the major islands of Sardinia and Sicily. Sauer colours the various provinces but no guidance to the terrain is offered.

The inserts provide rich supporting information. They include a map showing the principal rail communications between Italy and the rest of Europe – a map that emphasises the significance of northern Italy. Another map gives the regional organisation of the Italian railways. There are also maps of areas with dense networks, especially near Naples, north from Milan and the Italian lakes. The title emphasises that all the stations are included, which creates a formidable task of fitting in a mass of information.

Italian railways had a grim year in 1908 as three trains crashed at Acquabella bridge with nine fatalities: a Milan to Rome service carrying a number of politicians hit a stationary train and then a train from Bergamo crashed into the wreckage. In 1905–06 the state bought the private railways, adding over 8,000 miles of lines, of which about 15 per cent was double tracks offering more capacity. At that point, the railways were in difficulties because of a marked lack of investment, with 738 of the 2,664 steam locomotives more than 30 years old. The situation for rolling stock was also very bad.

Riccardo Bianchi, an Italian engineer and the first Director General of Ferrovie dello Stato (Italian State Railways), a post he held from 1905 to 1915, had to confront marked variations in practice and regulations. He pushed hard to invest in new locomotives and rolling stock, as well as create an expanded network. The map in part reflects the energy he was conferring on the system.

ADELAIDE TO BRISBANE

A. Combes, 'Railway Map of Part of Australia Showing Through Connections from South Australia to Queensland and Proposed Strategic Railways', 1916.

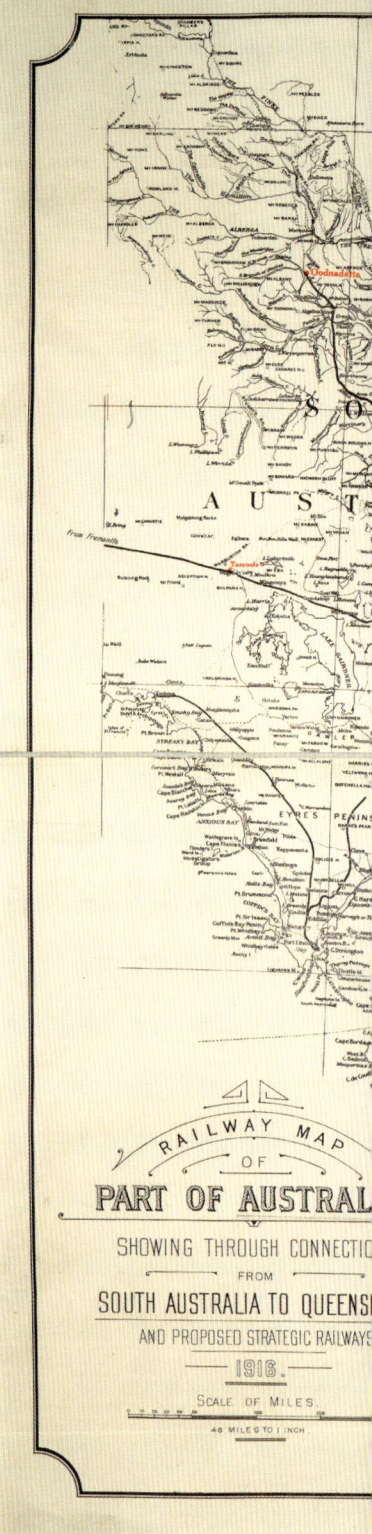

Compiled by the Lands & Survey Branch of the Department of Home Affairs in August 1916 to accompany a 1915 report, this map covers the most densely populated part of Australia and illustrates the proposal for new 'strategic railways'. Using a scale of one inch to 48 miles and making an effective use of colour, the map shows existing railways and those under construction. Overcoming distance was an aspect of 'strategic' because cost was found in terms of time, the use of coal and wages. Moreover, avoiding the coasts made the railways less vulnerable to amphibious raiding – for example, near Newcastle, New South Wales. Such raiding might appear implausible, but there were invasion panics in Australia as well as concern about the bombardment of coastal lines. In practice, the avoidance of the coast was difficult because the key population centres were all on the coast, as were many industrial sites, such as Newcastle with its steel works. The relationship between interior sources of raw materials and urban ports was a key element in the development and operation of many railways worldwide.

An endorsement accompanying the map, signed by the Minister for Home Affairs, explains that it replaces 'an inaccurate and misleading map' inadvertently prepared by the Commonwealth Railways Department that had been presented earlier in the year to the House of Representatives.

The map provides information of proposed routes across southeast Australia, one the existing railway marked (with a solid route) on the line through Serviceton (between Adelaide and Melbourne), Albury (between Melbourne and Sydney) and Wallangarra, and second the proposed route (marked with a dashed line) by way of Terowie, Hay and Waranary. Terowie was the trans-shipment point at the railway break-of-gauge between the broad-gauge line to the south and the narrow-gauge line on to Peterborough. Hay was a major transport hub, both locally and regionally, with a bridge over the Murrumbidgee River, and a link by river steamers to Echuca, which from 1864 had a rail link to Melbourne. Hay was linked to Sydney by rail in 1882, which hit river traffic.

The first Australian railway had opened in 1831, a short gravitational railway designed to service a coal mine in Newcastle, New South Wales. The early systems were state based, with their circumstances essentially dictated by finance. Thus, whereas in the state of South Australia, railways were publicly owned from the start, this happened in New South Wales and Victoria only because the companies proved financially weak. As a result, the Sydney Railway Company became the New South Wales Government Railways. The first Australian steam-powered railway, from Melbourne to Port Melbourne, opened in 1854.

Despite British advice for a standard gauge, New South Wales and Victoria built to different gauges. This reflected the absence of a central government able to provide control and also the degree to which individual states had a very strong sense of autonomy. However, as between independent countries, for example Spain and France, this made trans-shipment at break-of-gauge points a major issue. This occurred at Albury on the Main Southern Railway between Sydney and the state border with Victoria, after which the North East Railway continued to Melbourne. The Main North Line was the original main line between Sydney and Brisbane, Queensland. The Broken Hill Line was opened to Broken Hill, New South Wales, in 1919, with a missing link finally finished in 1927. This ensured a service through from Sydney to Adelaide, South Australia, involving two breaks of gauge. In contrast, in 1932, the completion of a double-deck rail/road bridge at Grafton over the Clarence River ensured that the standard-gauge line from Sydney to Brisbane was opened, replacing the *Induna* rail ferry operational since 1925.

The gauge issue – linked to particular states rather than to specific functions – remained one that caused problems for far-flung services. A key development was the appointment in 1922 by South Australian Railways of the American William Webb

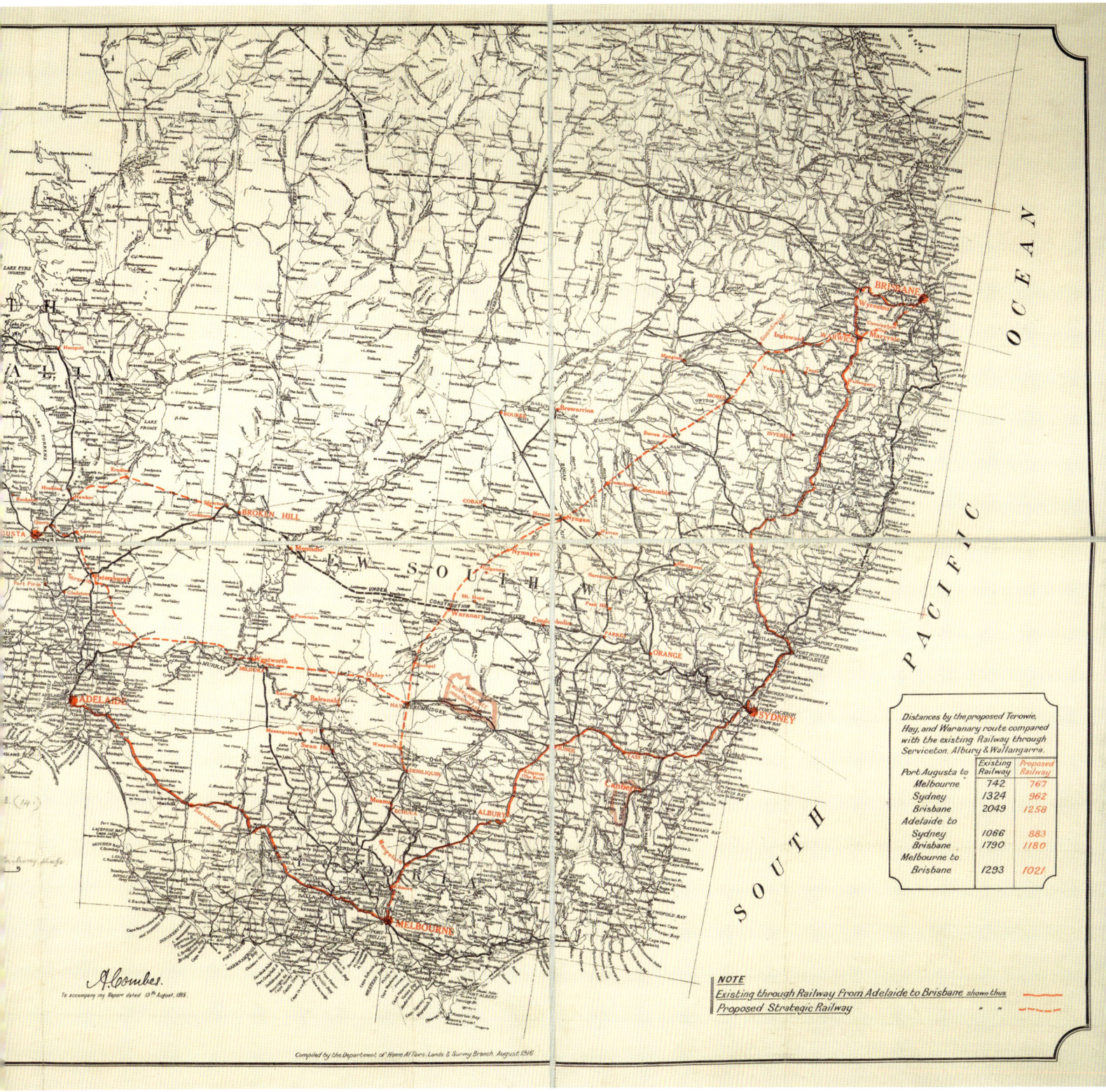

as Chief Commissioner. This led to large, standardised locomotives, which arrived from 1926, as well as steel-bodied freight wagons. Stronger tracks and bridges carried the heavier trains. There was also the replacement of the old railway workshop.

As with other maps of the period, the coverage here is of 'government railways' and not of private lines, which in Australia included not only mining and timber lines but also those linked to the Queensland sugar industry.

THE FIRST WORLD WAR
Geographical Section of the General Staff, 'Third Army Railways', 1918.

This superbly detailed map captures the range of rail support, notably by gauge, but also with reference to whether the track is double or single, under construction or proposed, as well as the surname of the Traffic Officer in control of the station. The thickness of the double-tracked broad-gauge lines helps direct the viewer's attention. The Third Army, Britain's largest army, and then headquartered in Albert, had played a major role in the fighting on the Western Front in late 1917 and early 1918, including the resistance to the German attack on the Somme in early 1918. Made by the Geographical Section of the General Staff working with the Ordnance Survey, the map covers the area from Arras to Peronne.

International competition, a factor in the initial spread of railways, became even more important in the run-up to the outbreak of the First World War in 1914 when attitudes towards railways and their capabilities, in both the infrastructure of empire and as a vital tool of modern warfare, developed a politico-strategic and technical logic which, to a degree, imposed itself on international politics. A 'railway mentality' fostered an offensive approach to war by making mobilisation and logistics seem more plannable and therefore biddable. War by timetable appeared possible. National rail systems had actually been developed earlier in part for the mobilisation of troops. Thus, France financed the building of railways in the Russian-ruled part of Poland in order to help the mobilisation of its ally against Germany.

Once war began, the impact of war on railways, and vice versa, was seen in a massive demand for rail transport for the movement of forces and supplies, which included construction material, notably wood and sandbags. At the tactical level, trench railways, narrow-gauge light ones, were important to the support of front-line units. Experienced engineers and workers were deployed from home countries. The weight of heavy guns and their shells were significant in the rail provision needed for supporting units.

Rail density varied considerably between the fronts, being affected by prior provision, wartime construction and damage. The last was a particular problem for the Hejaz Railway in Arabia. The situation was best on the Western Front, and was most extensively mapped. In contrast, the situation on the Eastern Front was one of inadequate railways (especially in Russia), which affected the density of forces that could be readily maintained and the ability to sustain offensives – as in 1915 when the Germans suffered from that situation in Russian Poland. This was exacerbated by the Russian destruction of bridges over the River Vistula. The dependence of rail systems on bridges is particularly to the fore in wartime.

Alongside destruction, there is also the building of railways both close to the (changing) front lines, and also for strategic reasons linked to the movement of supplies, notably (on the Eastern Front) from the port of Murmansk to Petrozavodsk, which enabled Russia to avoid some of the problems created by its no longer being able to gain access by sea via the Baltic and Black seas, due to conflict with Germany and Ottoman Turkey respectively. Labour for this project was provided by prisoners of war, including those involved in the Kazakh rising of 1916 against conscription.

THE OPERATIONAL LEVEL
Ordnance Survey, 'Railway Map of the Western Theatre of War Showing Broad-Gauge Lines Only', 1916.

This map shows most of Belgium and part of northern France, in essence the areas facing British forces as opposed to those facing French ones further south. The rail lines are linked to facilities including railway ammunition parks, especially at Mons and Namur, and gas factories (because poison gas was moved by rail). Such maps were based on intelligence from aerial reconnaissance, which included the use of photography, as well as espionage: there was a particularly important cell in Luxembourg reporting on German rail movements through that major rail centre between Germany and Belgium.

The British used the Ordnance Survey (OS) not only to map the railways in their own section but also more generally, thereby providing a range of information from the tactical to the strategic. Thus, in 1916 the OS produced for the Geographical Section of the General Staff this map with double and single track distinguished by the thickness of lines, and lines under construction also shown. The map was black-and-white only because that was all that was required for clarity, and reader interest was assured.

Maps had to be carefully interpreted, not least on how effectively railways could be used. This was amply demonstrated during the German invasion of Belgium and France in 1914. The French could readily use their rail system to move troops from eastern France to support Paris. In contrast, the Germans faced the problems of taking over the rail systems of conquered areas, including different gauges of track and the lack of a lateral system to match that of the French. The extent of the German advance towards Paris provided the French with a relative advantage stemming from the focus of their system on the capital.

Control over the railways was introduced rapidly after the outbreak of war. Thus, in Britain the government took over control of the railways in 1914. Railways were regarded as significant and vulnerable. 'They could do a tremendous lot of damage in an emergency by ... destroying railways and telegraphs' observed Maurice Hankey, Secretary to the

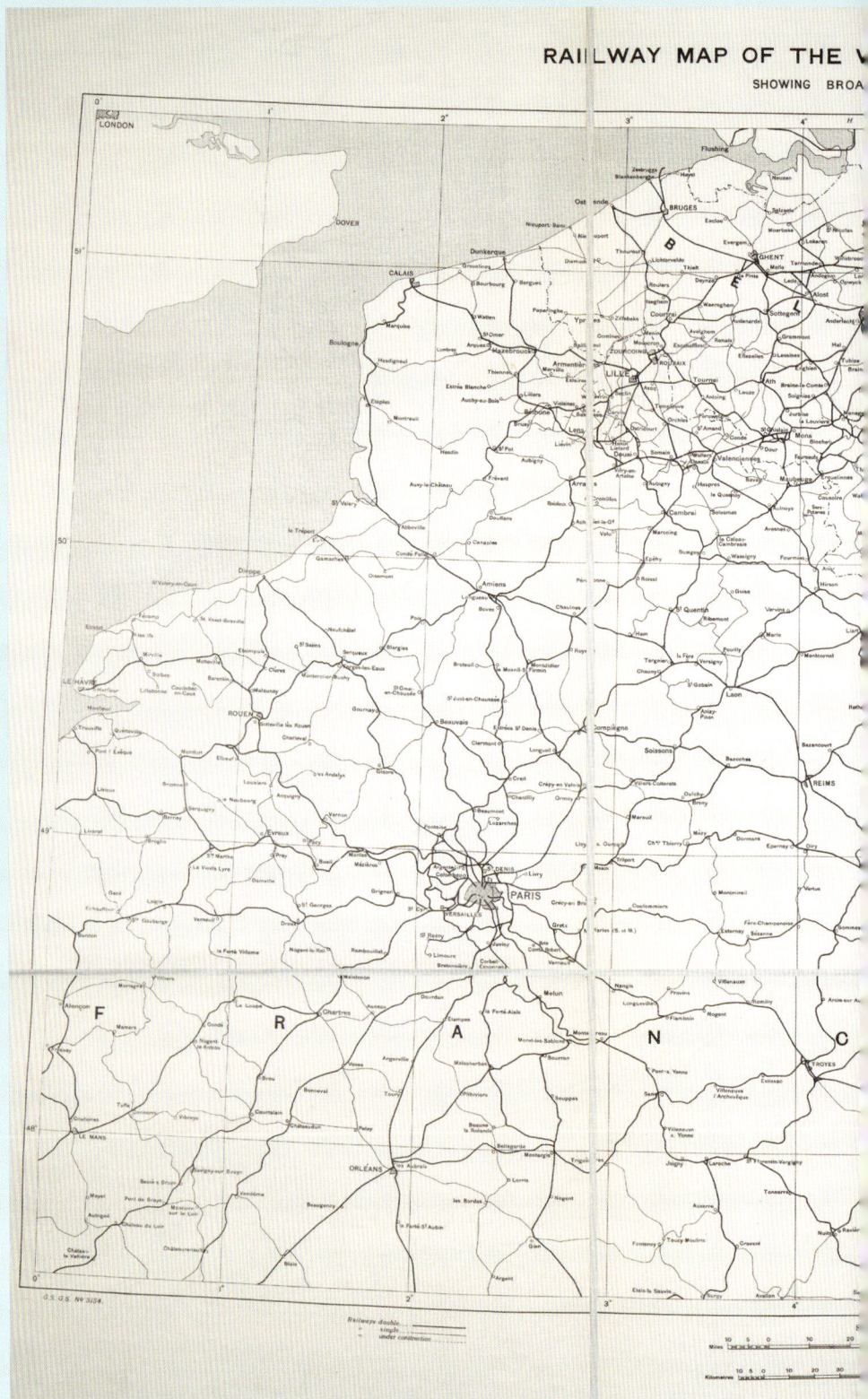

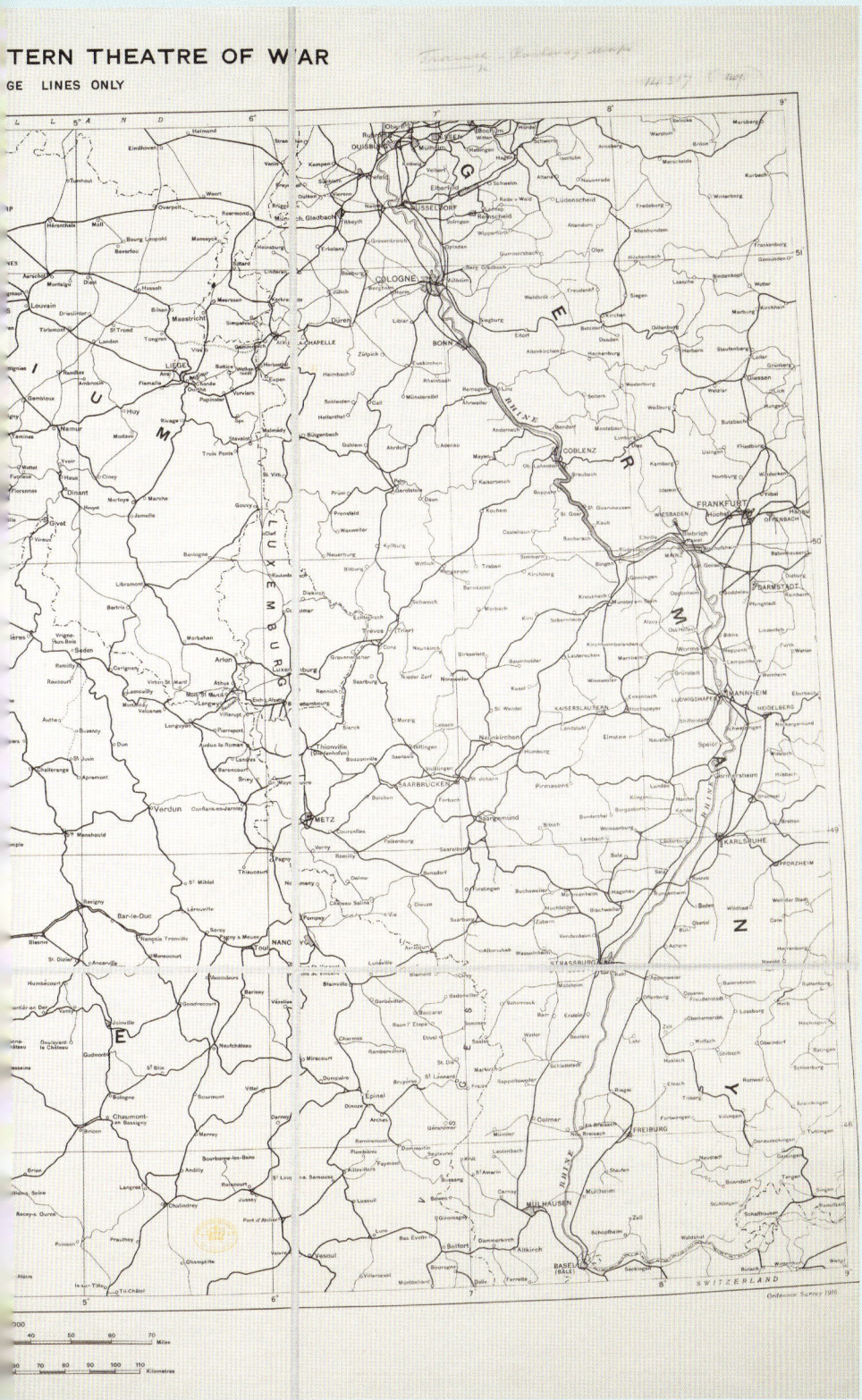

British Committee of Imperial Defence, in 1914 when writing about 'aliens'. Aside from concern about sabotage and espionage, there were repeated issues over the availability of sufficient coal to run the railways, as well as coal of sufficient quality. France had lost key coal-producing areas to the invaders, which made the provision of British coal, always necessary, particularly important. Convoys were first introduced in February 1917 to protect, from German submarines, ships carrying coal to France, in response to calls from the French government.

Germany did not have a comparable reserve, and in 1917 German coal production fell due to a lack of miners. Coal shortages and the lack of workers affected the rail system from 1916, which was also under great pressure due to the focus of steel production on armaments. The cumulative impact of such shortages was a running down of the German economy and its growing atomisation, which hindered attempts to better coordinate and direct production.

ALONGSIDE ROAD 1919–39

There has never been a 'normal' in rail history, for its present has constantly changed, and with that hopes and fears of the future – and accounts of the past. Indeed, the idea of a successful and/or stable past is in part a response largely to the uncertainty of the present. The two decades that followed the First World War exemplified these points. The most recent years of world war were so tumultuous that it was easy to hope to go back to a pre-war 'normality'. But there had been no such calm, and partly as a consequence the years from 1919 saw a reworking, albeit in new circumstances, of the earlier difficulties, not to say disorder. In the world of railways the decade prior to the First World War had witnessed serious labour disputes, as in Britain; serious pressures on company finances, as in the United States; and political change threatening existing ownership and related investment, as in China. Each of these were widespread issues – indeed problems – anew from 1919 onward, for management and owners. In addition, there were more general difficulties in the world economy and for the global liberal economic order that rail exemplified, and on which it greatly relied, affecting not least investment, expertise and equipment.

At the same time, the rail network expanded in many countries, notably in Africa (see pages 160–163) where imperial rule provided a context for the building of lines to the benefit of the imperial powers. The most significant investor in rail was Britain, but both within the empire (see India on pages 170–171) and outside (see Argentina on pages 162–163) there was opposition to British control over rail.

The shock of the new can be discussed in terms of competition from air and road, but it was more clearly political, with changes resulting from the formation of states, such as Poland and Czechoslovakia, and in countries that underwent significant domestic transformations, for example the Soviet Union, China, Mexico and Spain. Far from this crisis for rail diminishing as a new order was grounded, it was to accelerate with the Great Depression that

'Toward Los Angeles, Calif.', by Dorothea Lange, 1937, shows two Depression-era travellers hiking along the highway and about to pass by a billboard advertising the comforts of Southern Pacific Railroad.

began in 1929 (see pages 166 and 180). This economic shock dramatically cut revenues as activity declined, reducing the amount of freight that was carried and the rates that could be charged. As a result, the money available for investment reduced, and this was linked to a major fall both in maintenance and in the replacement of locomotives and rolling stock. In addition, economic contraction meant a reduction in government revenues. This increased the pressure on railways as the alternative to nationalisation was of scant value if there was no money for investment.

During this period there were changes in the production methods for maps, but these did not transform the finished product. From the late 1920s, cartographers reverted to engraving; however, instead of metal plates, they used glass ones before moving, in the 1940s, to transparent plastic ones (see also page 162).

ACROSS AFRICA

A.J. Clevely, 'Africa: Story of the Cape to Cairo Railway & River Route', 1922.

This map of post-war African railways in the 1923 book by Leo Weinthal, *The Story of the Cape to Cairo Railway & River Route, from 1887 to 1922*, reflects in particular the degree of progress achieved in accomplishing the plan for a railway from Cape Town to Cairo in order to link British colonies, and thus strengthen governmental, military and economic links. In opposition, the French had been interested instead in a west–east route, as, further south, had been the Portuguese.

A Cape to Cairo railway was pushed by Cecil Rhodes, the ambitious Prime Minister of the Cape Colony from 1890 to 1896. Construction began at Cape Town, going northward to Kimberley, Gaborone and Bulawayo, which was reached in 1897, with a bridge opened in 1905 over the Zambezi near Victoria Falls providing a linkage to the line already laid between Livingstone and Kalomo in what is now Zambia.

The ambitious project could not be brought to fruition prior to the First World War due to Germany establishing a colony in what became Tanganyika, where the Germans had completed a line between Tanga and Moshi in 1911 and then the Dar es Salaam line in 1914. The colony was gained by Britain as a result of the war, but the task of extending the network was too great, due to the forest terrain, the economic need for eastward rail links to Indian Ocean ports, the rival possibilities provided by air and road services, and financial considerations, which had been highlighted in 1907 by Percy Girouard, in a report on the transport policy for Nigeria. When discussing cotton exports from Nigeria's northern region, he observed that 'the line or lines should be built and worked under local control, economy of construction being secured by a low standard in structures, building, etc. without sacrificing hauling capacity'. He added:

'...the probable staple exports of Northern Nigeria will not bear heavy rates.... I have no doubt of the feasibility of constructing a cheap pioneer line on an alignment which can be improved as traffic justifies it. It will be a line of easy gradients through what I must call a very rich country, and, though too much must not be expected in the commencement owing to the underdeveloped nature of the inhabitants, its promise for the future would appear distinctly bright. The line can undoubtedly be built for, say, £1,200,000, or £3,000 per mile, and at such cost rates of carriage should be possible which will permit of the export of the staples of the country with profits remunerative to both producer and exporter.'

Bridges were frequently as major an issue as the line. Thus, just as with the Zambezi at Lake Victoria, the Niger at Jebba had to be bridged, and until the latter was finished in 1916 trains were ferried across the river en route from southwestern Nigeria to the north. So also in the southeast with the line from Port Harcourt over the River Benue at Makurdi, where the railway bridge was completed in 1932. Other bridges linked Lagos to Ebute Metta on the mainland.

In colonies there was a desire to benefit the economy by developing exports. Road travel offered scant competition for rail. In Nigeria construction began from Lagos inland from 1896, reaching Ibadan in 1900 and Jebba in 1907, with a branch line from Zaria to Jos, the latter among tin fields. The railway from Baro on the Niger to Kano, 366 miles away, was constructed in 1908–11. In southeast Nigeria, the railway from Port Harcourt reached the colliery at Udi in 1916, thus providing a local supply of coal for the trains.

As an important aspect of any Cairo to Cape route, the railway system in Sudan was initially built in the 1890s to help British conquest, and in the early twentieth century it was greatly extended, notably in 1906 to reach the new harbour facilities constructed on the Red Sea at Port Sudan (as indicated in Clevely's 1922 map 'Egypt and the Anglo-Egyptian Sudan showing the Valley of the Nile'), with a new line built in the mid-1920s from central Sudan to the port (under construction at the time of the map). This expansion was linked to the attempt, beginning in 1900, to use the Gezira plain for the cultivation of cotton, which led to the construction of an expensive irrigation scheme officially opened in 1926. This produced a cash crop designed to further the British imperial economy and was therefore an example of the process by which distant regions were more extensively integrated into the global economy. Prior to irrigation, the Gezira had instead been used to grow grain for nearby Khartoum. The purpose of Britain's small and restricted Colonial Development Fund (CDF) of 1929 was to facilitate such improvements.

In 1935 the landlocked British colony of Nyasaland, now Malawi, saw the opening of the extension of the line from Blantyre to Salima, as well as the completion of a bridge over the Zambezi in the neighbouring Portuguese colony of Mozambique, which ensured the opening of an all-rail route to the port of Beira. This provided a parallel to the line from Transvaal to Lourenço Marques.

The previous year, in the French colony of Côte d'Ivoire (Ivory Coast), the railway north from the port of Abidjan, on which there had been no work in the First World War, reached Bobo-Dioulasso, some 500 miles inland. Yet, the role of rail in the French colonies continued to be affected by meagre infrastructure. Investment was limited, while many tasks, it seemed, could be handled better by road or air services. In Congo, the first line was built by Belgium in 1910–18, from Sakania to Bukama. More lines followed but the mileage completed was limited given the immense scale of the country.

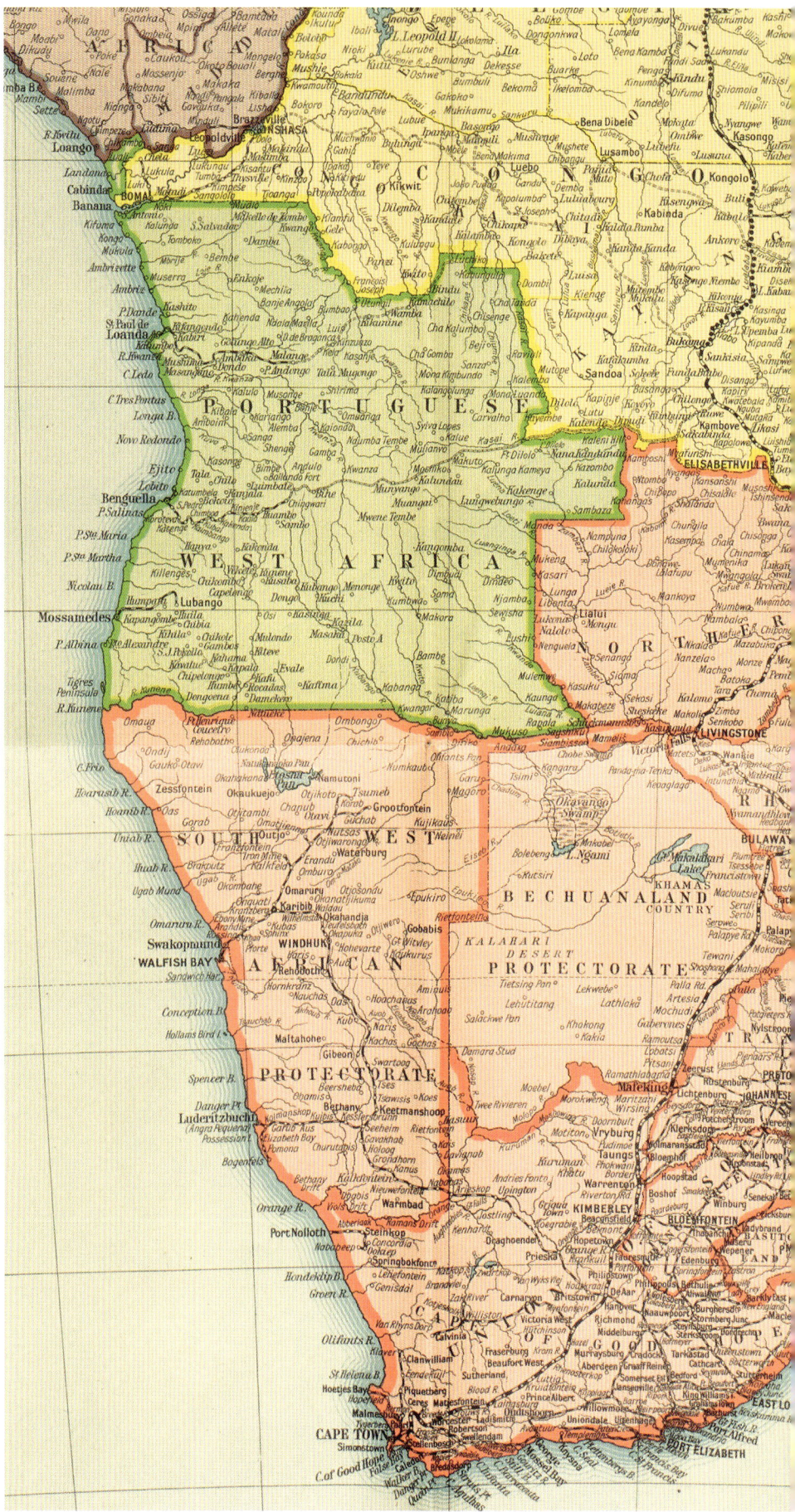

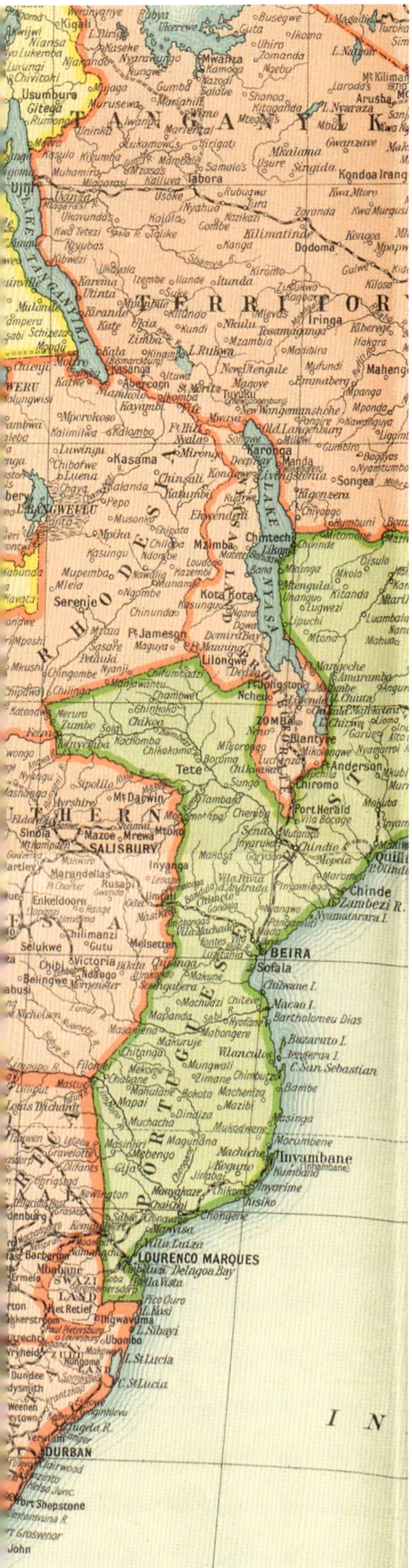

MAP No 8 Section 6.

AFRICA 1922.
EGYPT
AND THE
ANGLO-EGYPTIAN SUDAN
SHOWING THE VALLEY OF THE NILE.
Specially drawn by A.J. CLEVELY F.R.G.S. for the
STORY OF THE CAPE TO CAIRO
RAILWAY AND RIVER ROUTE.

SCALE 1 : 4,500,000 or 1 inch = 71 Miles.

RAIL ON THE PAMPAS

Buenos Aires & Pacific Railway Co, 'Map of the Argentine Railways, Presented by the Buenos Aires & Pacific Railway Company, Limited', 1925.

The map very much shows the influence of Britain. The key is in English, and the scale is shown in miles as well as kilometres, while the Buenos Aires & Pacific Railway Company's offices are given in both Buenos Aires and London. The locating of the information is impressive, with the coverage of lines extending to neighbouring Uruguay and Chile, but much of Argentina is ignored because the lines did not cover lightly populated Patagonia, where the main settlements were more readily reached by steamship. The use of colour is effective. It is a major feature of the mapping of the period, and is seen also for example in the maps of Indian railways (see pages 170–171). Colour serves here as a clear way to differentiate between lines. Interestingly, the red used in the interwar maps is brighter than the more subdued red used in the late nineteenth century.

The spread of transparent plastic in map drawing in the interwar period was part of an important shift in the compilation and production of maps that put a stress on mass production – on mechanisation, speed and quantity, rather than on craftsmanship. The changes threatened traditional standards of quality. There were problems with inconsistent line weights and coarse line elements, affecting both precision and clarity of line and general aesthetic appeal. It is not surprising that many map readers today prefer nineteenth-century maps, with their copper engraving or lithographic drawing, to the atlases of the first half of the twentieth century. Yet, if graphic quality and appeal were compromised, most atlases were still well up to their task technically. A decline in precision of line was not too serious given the scale used in most maps.

There was major expansion of the Argentine network in the 1900s and 1920s (the period of the First World War was less benign), with the cargo transported rising from 11.8 million tons in 1900 to 45.5 million in 1920, and the system length in kilometres increased from 16,500 to 47,000, a figure that was then essentially to be maintained until the end of the following world war. The network was one of the larger ones in the world. The Retiro–Tigre line of the Central Argentine Railway (visible in the Buenos Aires inset map) was the first in South America to be electrified in 1916, while from 1910 the prominent express services began: the El Rápido offered a fast service from Buenos Aires to Rosario, and from 1930 the Estrella del Norte linked Buenos Aires and Tucumán.

There was, however, already pressure on liquidity and investment as a result of the First World War and, later, the Depression: strikes had begun in 1917, and there were freight and other disputes with the government under Hipólito Yrigoyen, president from 1916 to 1922 and 1928 to 1930. Yrigoyen was opposed to the British-run companies and sought, instead, to develop a national, government-run system, not least to develop links with Bolivia, Peru and Chile, in order to help the border provinces. Under the administration of Maredo de Alvear from 1922 to 1928, the emphasis switched to cars and this was linked to the increase in American investment and activity, from which British interests suffered.

The railways were nationalised in 1948 when 20 companies were incorporated into the Ferrocarriles Argentinos (FA): ten British, seven Argentine and three French.

MAP OF THE ARGENTINE RAILWAYS.

PRESENTED BY THE BUENOS AIRES & PACIFIC RAILWAY COMPANY, LIMITED.

CALLE FLORIDA 783, BUENOS AIRES. 1925. DASHWOOD HOUSE, 69, OLD BROAD ST. LONDON, E.C.

HIVE OF INDUSTRY
Whitney-Graham Company, 'Erie Railroad: Serving the Heart of the Industrial Empire', 1927.

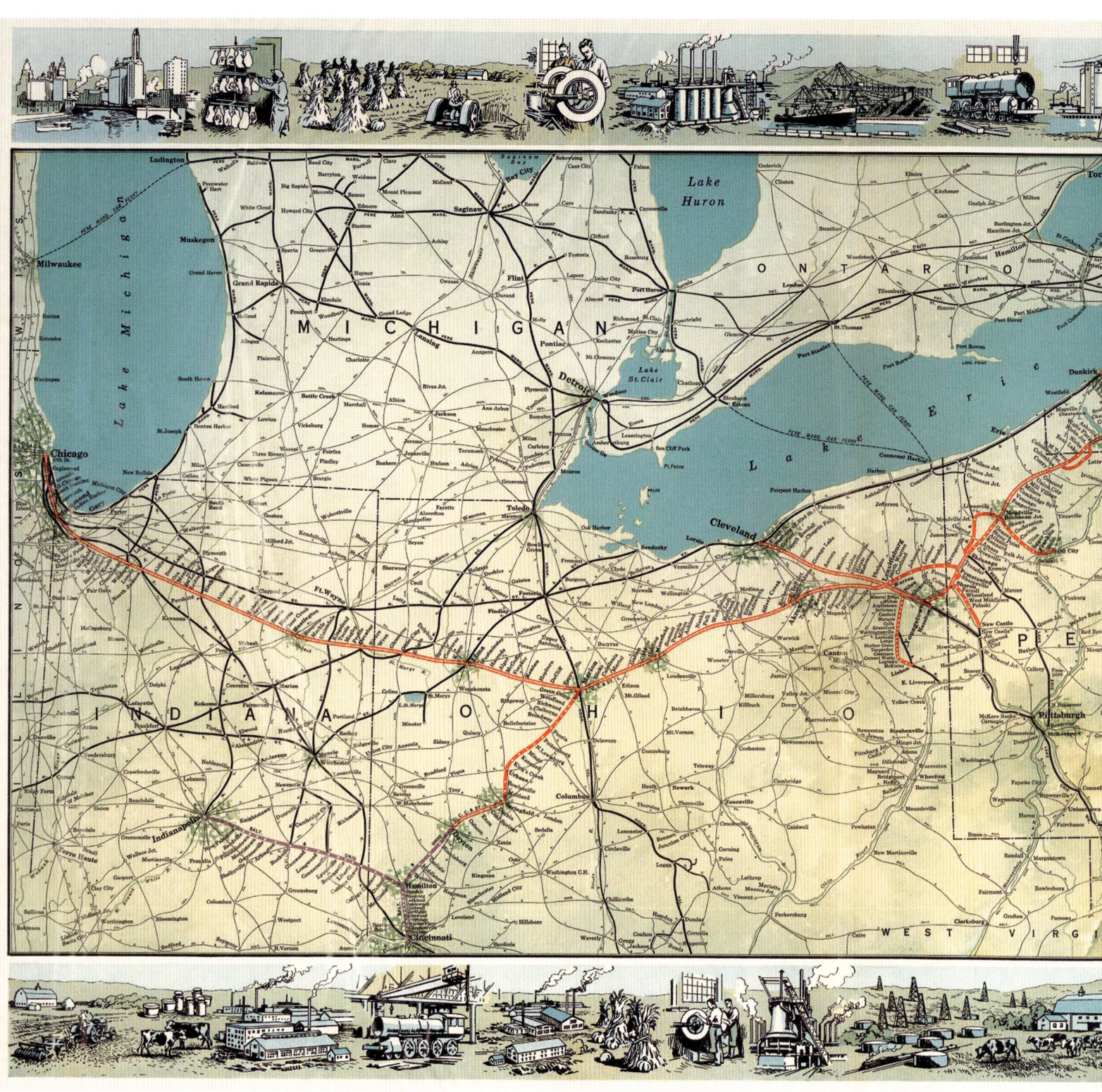

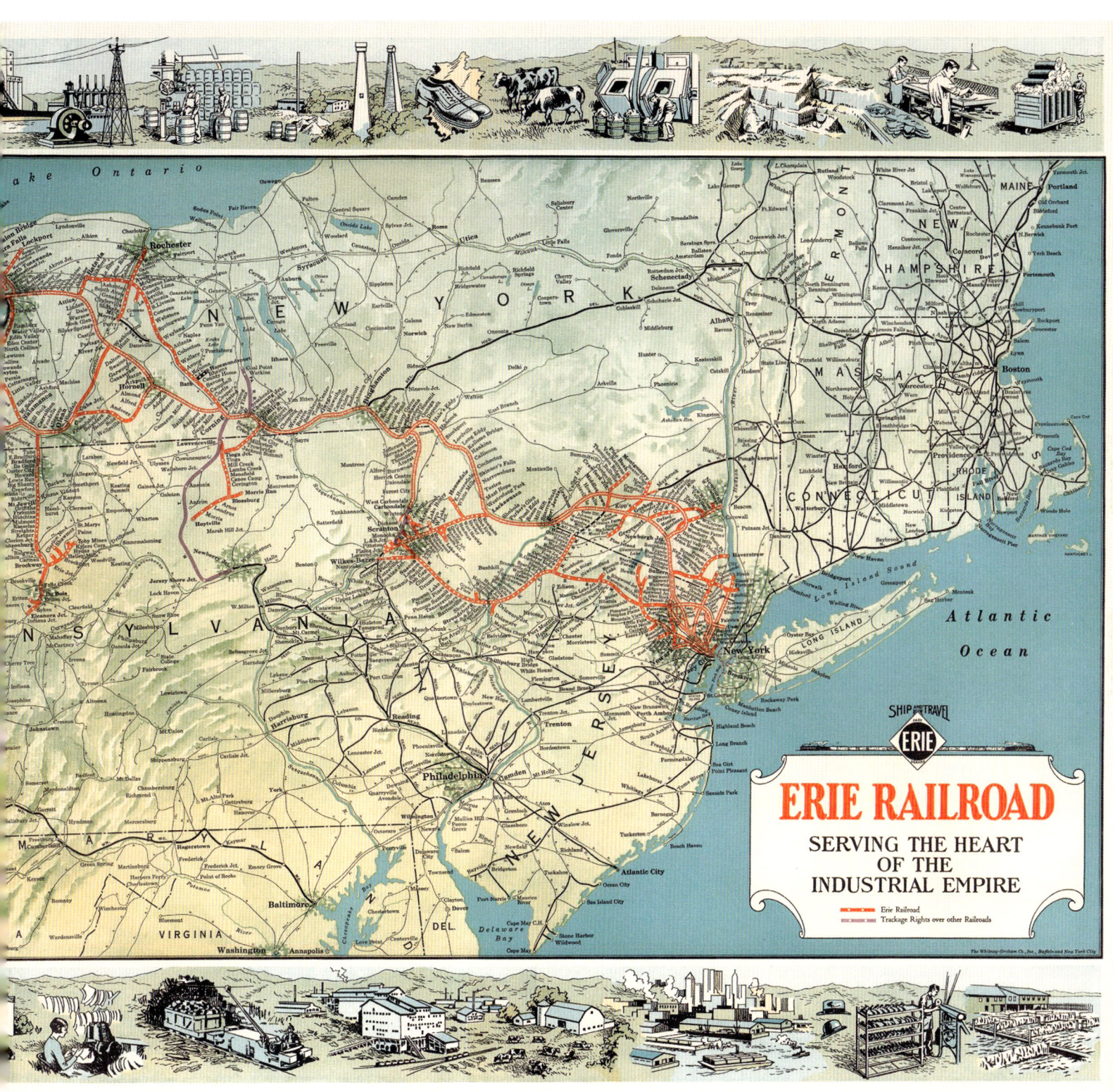

Published by the Whitney-Graham Company of Buffalo, the map depicts the railway and its trackage rights with illustrations of industry and agriculture on the upper and lower panels (previous pages). This work reflects the confidence, epitomised by the illustrations, of the United States in the 'Roaring Twenties', and the extent to which, although the headline news was of cars, Hollywood, jazz and oil, the economic fundamentals in fact remained the industrial belt shown in this map: from New York, Massachusetts and Pennsylvania to the Midwest around Chicago. The colour depiction of the lines brings out this geographical concentration and the pattern of the lines.

A longstanding element of this belt is depicted on this map, which is the linkage from New York to Lake Erie. Originally tackled by water, via the Hudson River and the Erie Canal, this route had subsequently been developed by and for rail. Chartered in 1832 and built in 1836–51, the New York and Erie Railroad originally was intended to connect the Hudson, at Piermont north of New York City, to Lake Erie at Dunkirk. However, the cost of construction led to bankruptcy in 1859 and reorganisation in 1861 as the Erie Railroad. The maintenance facilities were moved from Dunkirk to Buffalo in 1869. Fresh financial problems, due to switching from wide gauge to standard gauge, led to the company being reorganised as the New York, Lake Erie and Western Railroad in 1878. Bankruptcy in 1893 resulted in reorganisation as the revived Erie Railroad in 1895, which it remained until 1960, with Conrail then taking over in 1976.

In the period of this map, the Cleveland-based brothers Oris and Mantis Van Sweringen controlled the Erie Railroad as part of a 30,000-mile rail empire bought on credit largely financed by the J.P. Morgan bank. As throughout the history of American rail, credit from banking houses was crucial to railroad companies' activities, and notably so for the repeated takeovers that were linked to reorganisations. In 1929, headquartered in Cleveland, the Van Sweringens' holdings, which included America's largest rail network, were valued at US$3 billion, and in 1930 the dedication of the Cleveland Union Terminal testified to the brothers' prominence. Their railways included the Chesapeake and Ohio; the New York, Chicago and St. Louis; and the Missouri Pacific. They standardised the Erie Railroad's locomotives and rolling stock. As the map shows (previous pages), the railway offered one route from New York to Chicago, the most competitive service in the country.

But in 1929 and the subsequent Depression the worth of the Van Sweringens' shares was hit hard. This affected their ability to meet interest obligations and the security they could offer for their debts, while, at the same time, passenger and freight revenues fell. In the sustained liquidity and credit crisis that began in 1929, the Van Sweringens' rail interests were affected by loan foreclosures and the need to sell capital assets in order to service their debts. As a result, the brothers' wealth fell greatly. The Erie Railroad entered bankruptcy in 1938, but was then successfully reorganised.

Buffalo, where the map was published, was a major industrial centre, not least for steel production, with Bethlehem Steel, which had purchased the Lackawanna Steel Company in 1922, particularly important. This helped ensure the need to move large quantities of coke and iron ore by rail. With fresh investment, the Lackawanna site was modernised in the 1930s. The rail focus on Buffalo also arose from its significance for trans-shipment onto the Great Lakes, which was an element that is not captured by the map (but can be readily appreciated from its location in the detail, right).

The map appeared in a promotional atlas, *The Blue Book of Map Making: America's Map Makers*, issued by the publisher, which consisted of a page of introductory text followed by a total of 20 colour maps that had been pasted on. The maps depict railway or utility networks in the northeast, with an emphasis on the Midwest, and include a telegraph map of the world as well as small maps of Europe and the Adirondacks.

CANADIAN PACIFIC

Poole Bros., 'Map of the Canadian Pacific Railway: the Minneapolis, St. Paul & Sault Ste. Marie Railway; the Duluth, South Shore & Atlantic Railway; the Spokane International Railway; Northern Alberta Railways and connections', 1931.

The early decades of the twentieth century saw further expansion of the Canadian rail system. In part, this was a matter of new lines, as with that from Toronto to Sudbury (1908), which improved the situation in Ontario. There were also bridges, notably across the Red Deer River (1909) and the Oldman River (1909), and tunnels, particularly the Spiral Tunnels (1909) in British Columbia and the Connaught Tunnel (1916) under Mount Macdonald. Acquisitions were also important, notably the Quebec Central Railway, which was leased by the Canadian Pacific Railway (CPR) in 1912 for 99 years. The Quebec Central Railway and the 1912 deal reflected the significance of the links between Canada and the United States. Its longest route ran from Quebec City to Newport, Vermont, where passengers could transfer to trains of the Boston and Maine Railroad bound for Boston. The profitability of such links reflected the importance of American-Canadian rail services prior to the age of air passenger services and while road transport was limited.

Canadian National Railways (CNR) was established by the government in 1919 in order to deal with the financial problems hitting many companies. It encompassed the Canadian Government Railways (CGR), which included the Intercolonial Railway of Canada (IRC) and the National Transcontinental Railway (NTR), and those that, having become bankrupt or near bankrupt, had fallen into government hands, notably the Canadian Northern Railway (CNoR), which ran a main line from Quebec City to Vancouver, and the Grand Trunk Pacific Railway (GTP). This was very much government intervention aimed at preserving national infrastructure, and appeared necessary due to the crisis touched off by the Russian Revolution and the associated public disorder and labour unrest in North America, notably the Winnipeg General Strike in 1919.

CNR became the major competition for the CPR. As the latter was a private company, it argued that its taxes were being used to

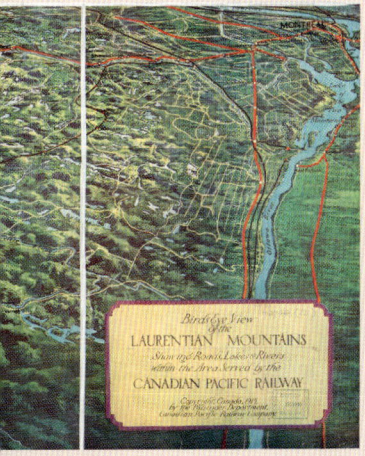

finance its competitor. However, CNR had many problems arising from its legacy, not least an inheritance from unviable rail companies as well as those serving areas of limited population and restricted economic activity. Moreover, CPR benefitted from avoiding the government direction to which its rival was subject.

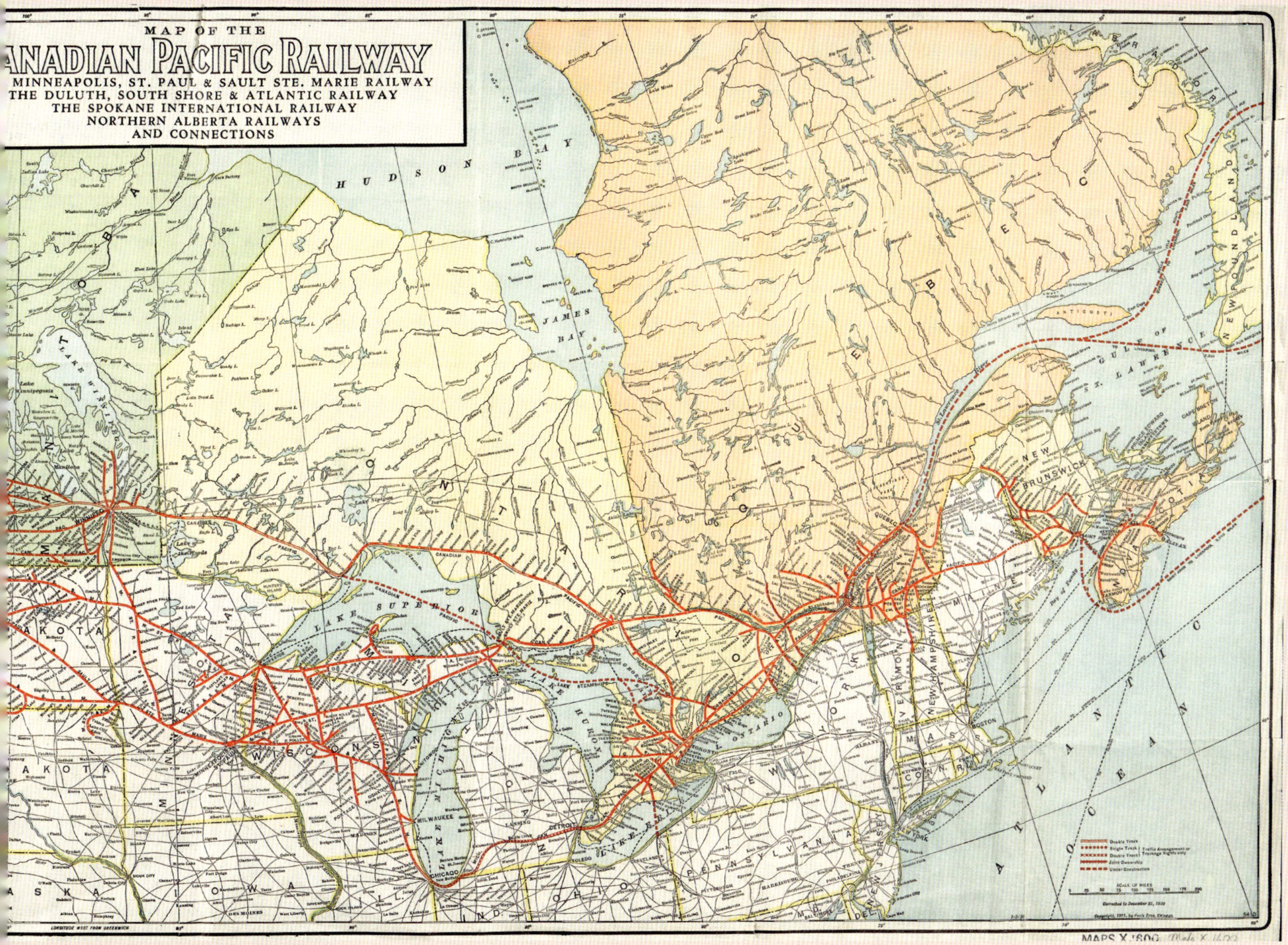

This 1915 'Bird's-Eye View of the Laurentian Mountains' (left) shows the area, around Montreal's Gatineau Valley, served by the CPR. The use of bright colour to show the route also captures the way in which the engineers had to adapt to a topography made difficult by the impact of glaciation. This was very different to the straight lines on a plain background that was typical of all too many maps of the period. However, for maps to be produced of the type depicted it was necessary to have a clear understanding of the terrain and this was not readily obtained until the use of aerial photography.

4. ALONGSIDE ROAD 1919–39 169

AN UNRIVALLED NETWORK
W. & A.K. Johnston, 'Railway Map of India', 1939.

Colour-coded by gauge and system, this sophisticated map conveys a mass of information. The numbers between stations are the distances in miles. Railways under construction are shown. Inset maps provide details on the areas of particularly dense rail services, notably the Bengal coalfields. The river and services in Bengal connected with the railways are also shown. This was the period of maximum mileage in India.

The very large and impressive Railway Staff College inaugurated in 1930 at Dehradun, India, showed the ambition of the railways as well as an emphasis on sophisticated planning. However, in 1931 a Retrenchment Advisory Committee was established by the Railway Board, and as a result the college was closed that year. The site was sold to the Indian Army and in 1932 the Indian Military Academy was opened there. A comparable rail college was not revived until 1952.

British investment in India was principally in railways. By the 1870s it was already about £100 million, with the rate of interest guaranteed by the government. The last was crucial to attract that investment. Separately, there were lines owned and run by the state from the late 1860s, as well as state-owned and privately run ones, but these categories also depended on investment. After prosperous years for rail in the mid- and late 1920s, investment became harder to obtain in the 1930s. As a result, the electrification that had begun in the mid-1920s was not continued and track, locomotives and rolling stock were not upgraded. Although low wages locally helped with undertaking track work, capital investment was inadequate. Investigation of the Bihta rail disaster of 1937, when derailment of the Punjab express resulted in 107 deaths, indicated the problems created by the instability of the locomotive on the Indian permanent way.

The establishment of the Railway Board in 1905 had brought a measure of coherence, which was necessary given the scale of the railway system, which had 437,539 permanent employees, of whom 112,710 were in the state system (excluding the princely states).

Nationalists pressed for state control of the entire system, with this control also seen as necessarily more Indian in goals and leading to a change to senior personnel. This call was a parallel to the more general situation across the world and a pushback against the liberal international system, a pushback that affected railways from Latin America to East Asia. The British government of India was not necessarily hostile: in 1921 the Acworth Committee, appointed by the Secretary of State for India, had unanimously recommended that all Indian railways should be managed in India; in short, it heralded the end of a role for private companies with British headquarters. There was also majority support in India for most railways being state-managed. It was in this context that contracts with private companies were not renewed, so that they were brought into the state railway system: the Great Indian Peninsular Railway (GIPR) in 1900; the East Indian Railway (EIR), state-owned but company operated, in 1925; the Bombay, Baroda and Central India Railway and the Assam Bengal Railway (ABR) in 1942; and the Bengal Nagpur Railway in 1944. As a result, by the time of this map the privately operated railways were on the way out.

Expansion of the network continues to the present day. In November 2023, commercial services began in Bangladesh on the 82-kilometre-long Dhaka to Bhanga line via the 6.2-kilometre-long Padma Bridge built in 2014–2022 over the Padma River, a part of the Ganges system.

MAP LOCATION	RAILWAY SYSTEMS
Mg	Ahmadpur-Katwa-Burdwan
Kf	Arrah-Sasaram Light
Of	Assam-Bengal
Lg	Bankura Rainagar
Mg	Baraset-Basirhat Light
Ei	Barsi Light
Ke	Bengal and North-Western
Me	Bengal Dooars
Lh	Bengal-Nagpur
Mg	Bengal Provincial
Ch	Bhavnagar State
Ed	Bikaner State
Dg	Bombay, Baroda, and Central India
Kf	Bukhtiarpur-Bihar Light
Qh	Burma
Hp	Ceylon Government
Bg	Cutch State
Me	Darjeeling-Himalayan
Kf	Dehri-Rohtas Light
Fe	Dholpur State
Pe	Dibru-Sadiya
Mf	Eastern Bengal State
Kf	East Indian
Kf	Futwa Islampur
Dg	Gaekwar's Dabhoi
Dg	Gaekwar's Mehsana
Ch	Gondal
Gf	Great Indian Peninsula
Ge	Gwalior Light
Mg	Howrah-Amta-Sheakhala Light
Fc	Jagadhri Light
Ee	Jaipur State
Cg	Jamnagar and Dwarka
Mg	Jessore-Jhenidah Light
De	Jodhpur
Pe	Jorhat
Ch	Junagad State
Go	Kalasakarapatnam-Tissianvillai
Mg	Kalighat-Falta
Fl	Madras and Southern Mahratta
Cg	Morvi
Lh	Mourhbanj
Fm	Mysore
Gk	Nizam's State
Ec	North Western
Ki	Parlakimedi
Bh	Porbandar State
Gd	Rohilkhand and Kumaun
Fd	Shahdara (Delhi)-Saharanpur Light
Gn	South Indian
Oe	Tezpur-Balipara Light
Df	Udaipur Chitorgarh

BROAD 5' 6"
METRE 3' 3⅜"
NARROW 2'6" & 2'0"
K.
A.S.
A.B.
B.R.
B.B.L.
B.L.
B.& N.W.
B.D.
B.N.
B.P.
B.S.
B.B.& C.I.
B.B.
B.
C.G.
C.S.
D.H.
D.R.
D.S.
E.B.
D.S.
E.I.
F.I.
G.D.
G.M.
G.
G.I.P.
G.L.
H.A.S.
J.
J.S.
J.& D.
J.J.
J.
J.
J.S.
K.T.
K.F.
M.& S.M.
M.
M.
M.
N.S.
N.W.
P.
P.S.
R.& K.
S.S.
S.I.
T.B.
U.C.

RAILWAY MAP OF
INDIA
printed and published by
W. & A.K. JOHNSTON, LIMITED, EDINBURGH & LONDON.
AGENTS IN INDIA
MACMILLAN & Co. LIMITED.
CALCUTTA, BOMBAY & MADRAS
CONTINUATION WEST
On the same Scale
BAY OF BENGAL
CHINA
BURMA
ASSAM
BENGAL
SIAM
ANDAMAN ISLANDS
NICOBAR ISLANDS
SUMATRA
Mergui Archipelago
EXPLANATION
Railway Time (Standard Time) which is 5½ hours in advance of Greenwich Time, is kept at practically all Stations on Indian Railways.
RAILWAY SYSTEMS
GAUGE
Natural Scale 1:3,150,000 or 50 miles to an inch
English Miles
Conic Projection
East 84 of Greenwich
86
88
90
92
94
96
98

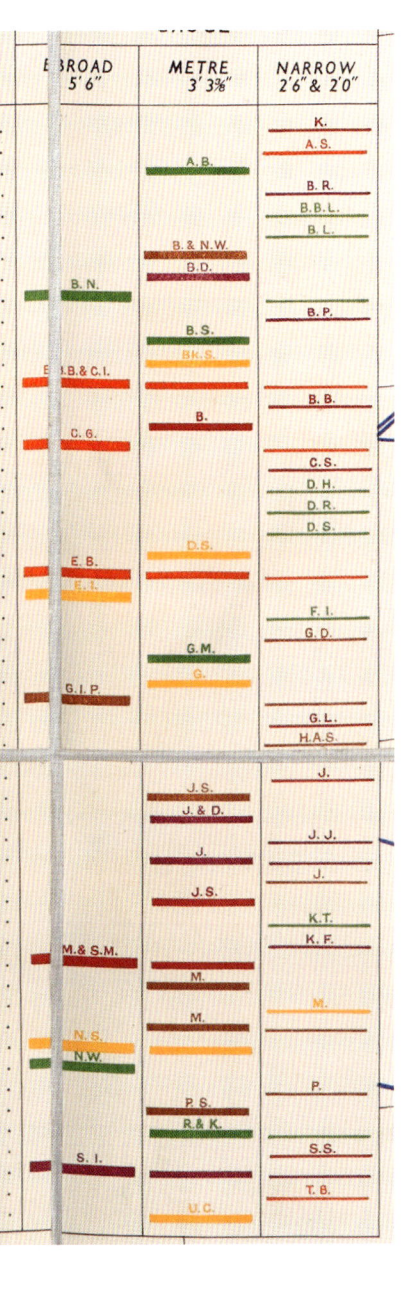

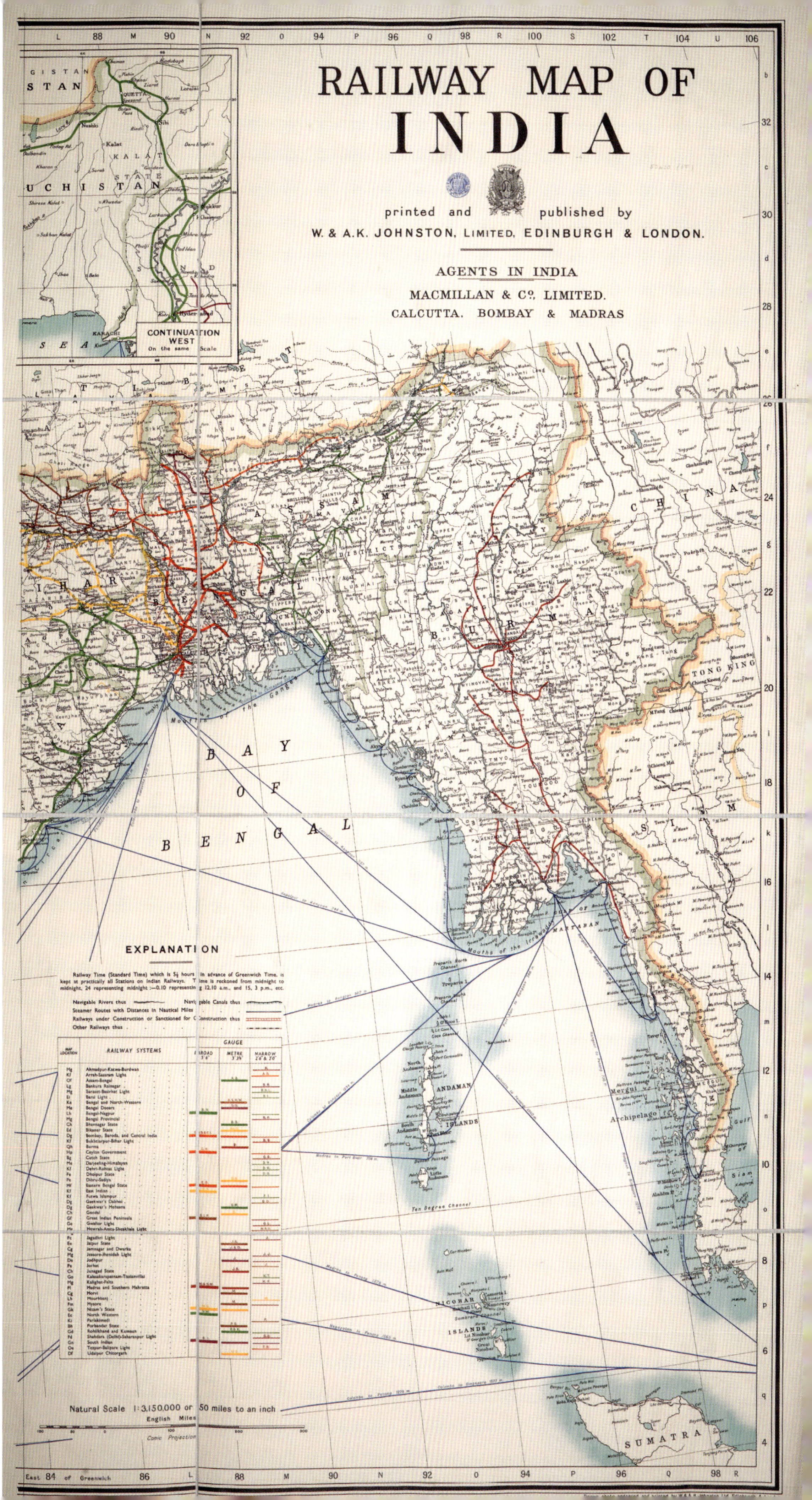

RAILWAY MAP OF CHINA

M.G. Bouillard, 'Railway Near Beijing', 1925, and British War Office, 'Canton–Kowloon Railway: Raids 14 October 1937–9 February 1938', 1938.

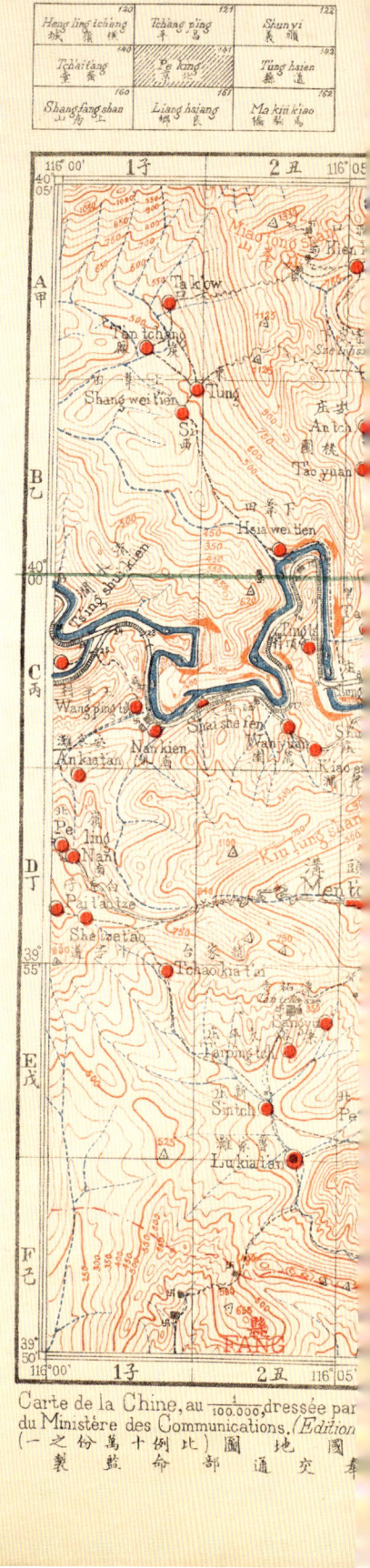

In China, where the new republican government had started to nationalise Western-owned lines after the 1911 Revolution, railway expansion and Chinese ownership were seen by nationalists, such as Sun Yat-Sen, as highly important to forging a sense of national unity, although Japan and the Soviet Union dominated the Manchurian lines. The significance of the railways was also enhanced by the scarcity of motor transport. In the 1920s, a national rail system existed in China and its steel framework provided a form of modern battlefield in the civil wars of that decade: most of the fighting was along, around or for control of the railway. The rail system connected mostly relatively new industrial cities, a situation that matched that in the West the previous century. The rail-plus-the-cities was in effect a separable realm from the countryside and its farmers.

As in the American and Russian civil wars, the ability to move troops by rail helped to shape the conflict, at least in operational terms, as well as providing clear targets in the form of junctions. Advances were described as moving along the railway, while cutting railways became a purpose of hostile military moves. *The Times* noted on 26 September 1924:

'Nearly 300 trainloads of troops and supplies have gone forward in 20 days, a considerable performance for a single-line railway. This result has been made possible by withdrawing practically the whole of the rolling stock from all the railways north of the Yangzi river, thereby completely stopping goods traffic and permitting only a precarious minimum mail and passenger traffic.
'There is much confusion at the concentration point and the empty trains are not returning. Military officers are seriously interfering with the railway administration and enforcing their demands at the point of the pistol. In one case where a number of officers were disputing for precedence the difficulty was solved by joining five trains together and sending them forward as one train, measuring one and a half miles long. Speed was reduced to a walk, but the "multiple tandem" arrived safely to the satisfaction of the 4,000 soldier passengers. The locomotive crews having objected to long hours are working double shifts, with sentries to see that the men off duty sleep soundly and do not gallivant in the darkness.'

There was further expansion in the Chinese rail system before the outbreak of war with Japan in 1937 led to an abrupt change. The railway approaches to Beijing shown in this 1925 map (right) by French engineer Georges Bouillard were those that later fell to Japanese attack. Bouillard had gone to China to build the Peking–Hankow (Beijing–Hankou) railway, the first major line built in imperial China from 1899 to 1905, and in the early years of the republic he was asked by the Chinese government to map China's railway network. The map is focused not on China as a whole, for much of Manchuria, occupied by Japan in 1937–38, is left out, as is western China. Instead, the area where the bulk of the population lived is covered, because it was there that the railways could be found, essentially from Beijing via Shanghai to Canton. Looked at alongside any information of Japanese conquests in 1937–38, the map reveals the extent to which China and its railways came under Japanese control.

The other map (following pages) was compiled more than a decade later by the British and it records Japanese air raids on the Canton (Guangzhou)–Kowloon railway in the winter of 1937–38; the first figure is the number of raids and the second the number of bombs dropped. The Japanese sought to cut China off from outside links in order to reduce the possible flow of foreign supplies. In doing so, they inflicted serious damage on the Chinese rail system. Canton fell to Japanese attack in 1938.

4. ALONGSIDE ROAD 1919–39

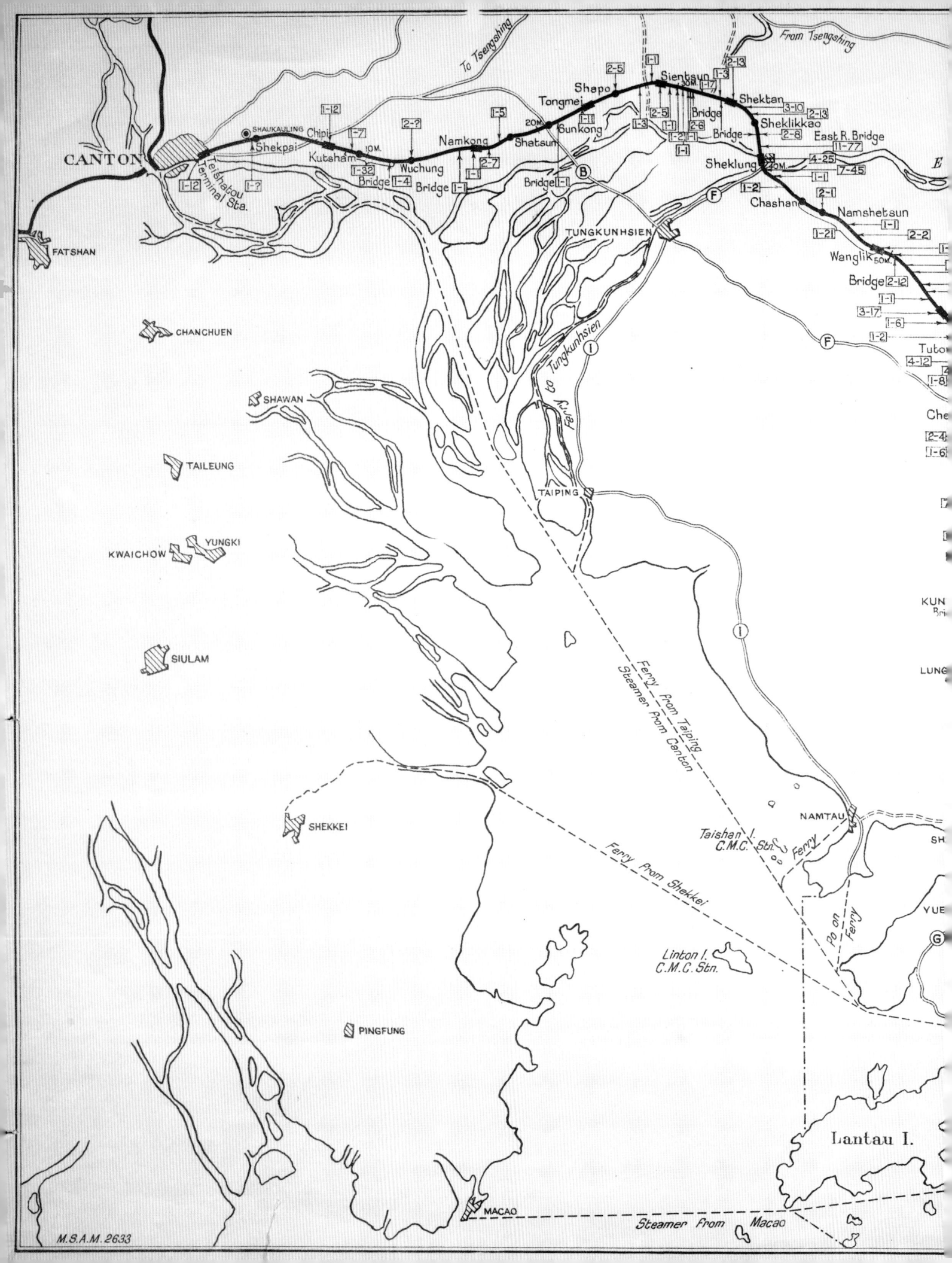

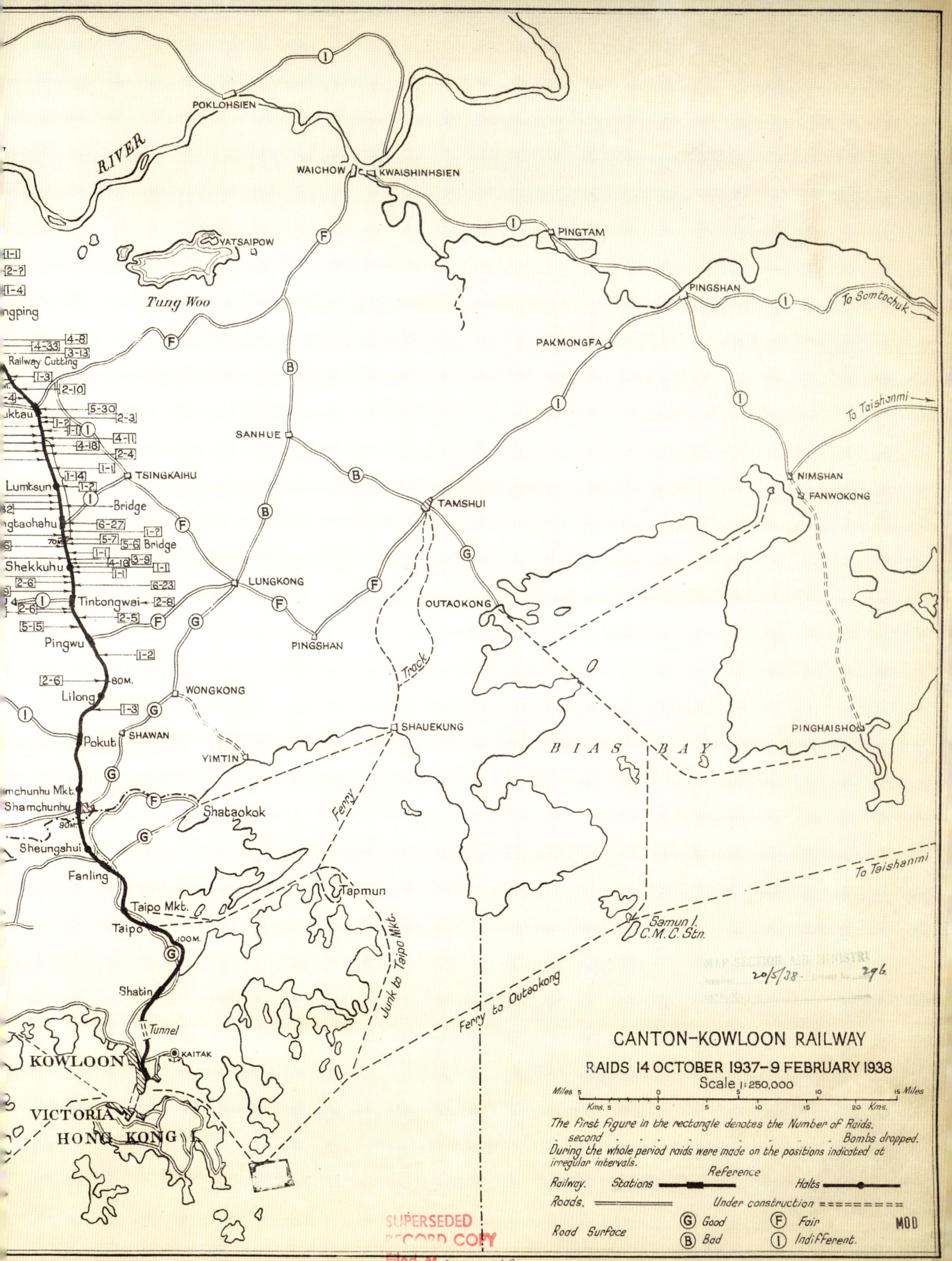

MANCHURIA

League of Nations, 'Military Situation in Manchuria before 18 September 1931', 1932.

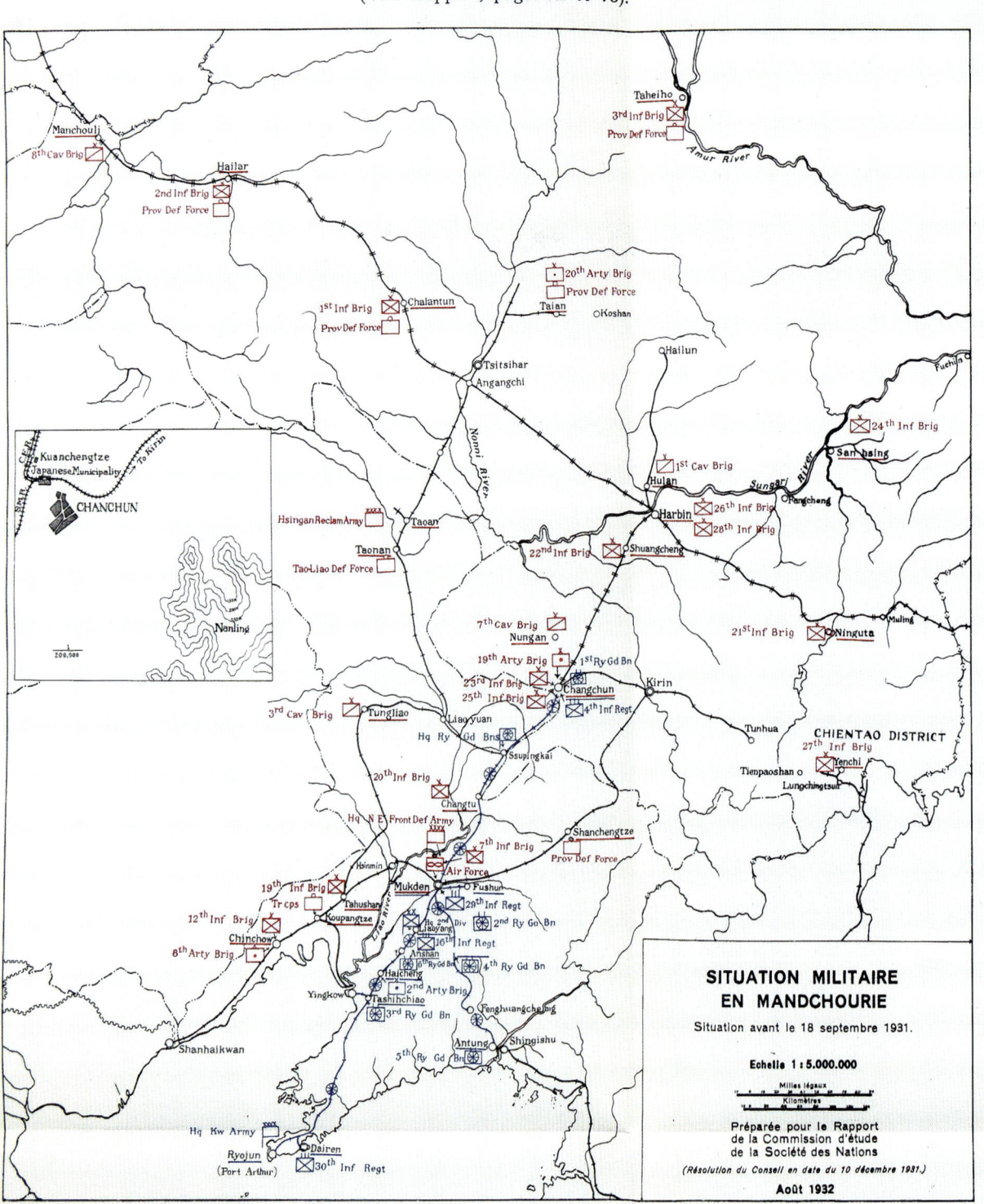

ÉGENDE

ÉVIATIONS.

Hq	Quartier général
Inf	Infanterie
Kw	Kwantung
MG	Mitrailleuses
Mxd	Mixte
NE	Nord-Est
Pres	Abri
Prov	Provisoire
Reclam	Réclamation
Ry Gd	Gardes de chemins de fer indépendants
Regt	Régiment
Serv	Services
Tr	A l'entraînement

COULEURS.

SYMBOLES.

Lignes (indiquant les emplacements occupés, les zones assignées aux unités, etc.).

——— Brigade
——— Division
×——— Armée
——— Arme ou unité mixte
——— Troupes japonaises
——— Anciennes troupes chinoises pro-japonaises
——— Troupes régulières chinoises (y compris les troupes provinciales avant le 18 septembre)
——— Troupes provinciales chinoises après le 18 septembre
·········· Volontaires, bandits, etc.
⇒ Direction de l'attaque

The Manchurian Crisis of 1931–34 led to much international attention, not least a 1932 report for the League of Nations prepared by British statesman V.A.G.R. Bulwer-Lytton for the League of Nations from which this map of the military situation pre-mid-September 1931 is taken. The crisis revolved in large part around control over the Manchurian railways, with the Japanese expanding their earlier position in southern Manchuria, the South Manchuria Railway Zone established for Russia in 1896 and transferred to Japan in 1905, by advancing along them. The pretext for the crisis was an explosion near a railway line belonging to Japan's South Manchuria Railway near Mukden (now Shenyang), an explosion set off by a Japanese officer but blamed on the Chinese. This was used as an excuse for invasion that rapidly became successful, in part due to only limited resistance.

The jurisdictional and military significance of railways helps to account for their importance on this map. The two rail lines from Korea marked in the southeast corner of the map were significant because in 1910 the country had been annexed by Japan. Korea contained substantial Japanese forces, and from there more troops could be readily deployed into Manchuria. In 1911 the completion of a bridge across the Yalu River established a direct link with the South Manchuria Railway, which from 1917 to 1925 was also responsible for the management of the railways in Korea, expanding them greatly. Although there were private holdings in the company, generals controlled its leadership and most of the staff were Japanese rather than Chinese. Although in 1925 management was returned by the South Manchuria Railway to the Railway Bureau of the Government-General of Korea, the South Manchuria Railway continued to control the North Chōsen West Line in northern Korea, which was designed to provide a rapid route from Japan to eastern Manchuria.

The highly profitable South Manchuria Railway provided a large amount of Japanese tax revenue, much of the money deriving from soybean exports for which Manchuria was then the leading source in the world. In 1936 the company owned 466 locomotives. Unlike the army strategists, the company was interested in business.

The sketch map dating from 1939 (below) shows the Japanese advance into central China in 1937–38, emphasising the extent to which railways were important to the axes of Japanese operations in China. In the upper right corner is a photograph of Lieutenant General Isogai Rensuke, the commander of the 10th Division, and (in the upper left corner) his successor in June 1938, Lieutenant General Shinoduka Yoshio. In late 1937, in what was called the Beijing–Hankou Railway Operation, the division had advanced south along the Beijing–Hankou railway as far as the Yellow River. The contemporaneous Tianjin–Pukou Railway Operation was an advance further east to the Yellow River.

4. ALONGSIDE ROAD 1919–39 177

LONG-DISTANCE LUXURY LINE
J. Barreau & Cie, 'Simplon-Orient Express Taurus-Express', 1930.

Operated by the Compagnie internationale des wagons-lits (International Sleeping Car Company), the Taurus Express, running from the Asiatic side of the Bosphorus, was introduced in 1930 as an extension to the Simplon Orient Express, the southerly rail route to Istanbul, with connections provided in Syria to Palestine, Egypt, Iran and Iraq. The poster emphasises that over a distance of 7,226 miles it links three continents and 14 countries, which was more than any other rail service.

As so often, the brochure map is misleading, and in this case in terms of both routes and carriage. The map suggests that the main route terminated at Cairo, with connections to Iran and Iraq. Instead, the main service was to Baghdad, the capital of Iraq, and a centre of British influence after Britain had gained control of the region following the First World War. There were also connections from Syria to Cairo, and from Iraq to Tehran. As an additional attraction, the connection from Baghdad to Basra, the port for Iraq, offered a route to India. This would be a means to avoid the long sea passage around the Cape, and it also represented the way in which rail meant that overland routes could be reconceptualised and pursued. As another instance of this process in the case of routes between Europe and India, rail links between Alexandria, Cairo and Suez were a way to speed passage to the Red Sea, thus complementing the Suez Canal, and notably so for passengers.

As far as carriages were concerned, there was no train at that stage between Nusaybin in Turkey and Kirkuk in Iraq and it was necessary to transfer to a coach for that stage, a service introduced in 1930 and one that was not appropriate for freight. The Baghdad Railway was not completed until 1940, and only then did the Taurus Express make its first complete journey between Istanbul and Baghdad. The following year, Iraqi State Railways took delivery of British-made Pacific steam locomotives to haul the Taurus Express to Baghdad. The first Taurus Express had a locomotive, two coaches and two baggage cars, and the train ran weekly.

Agatha Christie travelled on the Taurus Express, which is also referred to in her *Murder on the Orient Express* (1934). Poirot takes the train from Aleppo to Istanbul on which there is also the governess Mary Debenham, who had travelled from Baghdad: '...the train grandly designated in railway guides as the Taurus Express. It consisted of a kitchen and dining-car, a sleeping-car and two local coaches.'

The railway network that the line followed reflected the creation of new states. In particular, Yugoslavia brought together sections of Austria-Hungary, leading in 1918 to the National Railways of the Kingdom of the Serbs, Croats and Slovenes, renamed Yugoslav State Railways in 1929, only to be divided by the Axis conquest in 1941.

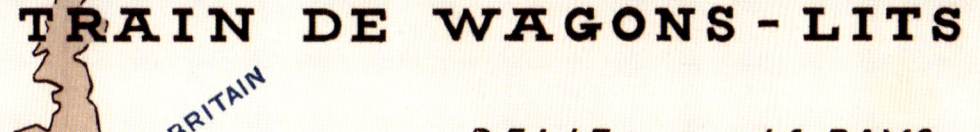

'GOD'S WONDERFUL RAILWAY'
Geographia, 'Great Western Railway: Map and Index of Goods Stations', 1933.

The 'Map and Index of Goods Stations' produced by the London map publishers Geographia for the Great Western Railway (GWR) in 1933 captures a role for that railway that has been underplayed due to the emphasis, instead, on passenger services, notably into London from the west as well as to the West Country holiday resorts, and for general passenger services across a wide tranche of England and Wales. The map shows both the network and related 'Foreign Lines', which are clarified by initial, as in 'SR' for Southern Railway. For freight, there are country lorry centres at goods stations, goods stations, company docks such as Plymouth, Weymouth and Brentford, railway distribution centres and railhead distribution centres. Inset maps capture the most complex areas: South Wales, the West Midlands and London, the latter including lines over which GWR had running powers, so that its freight system reached to depots such as Chelsea and South Lambeth. The post-war amalgamation had led to the gain of Cambrian Railway and Taff Vale Railway lines in Wales, and the Midland and South Western Junction Railway (M&SWJR), which provided a route from Cheltenham via Swindon to Southampton. The map does not capture the intensity in usage nor changes, such as the impact of the Depression, which hit coal production and transport.

Overambitious expansion in the nineteenth century compounded by a lack of investment in the First World War left the railways in a difficult position. These difficulties were exacerbated by interwar economic problems, notably the Depression, especially in their traditional freight markets of coal, metallurgy and heavy engineering. The General Strike of 1926 proved that the nation could live without railways, and much of the lost high-value traffic never returned after the strike. In the interwar period, 240 miles of track and 350 stations were closed completely and another 1,000 miles and 380 stations were closed to passenger traffic.

Among the 'Big Four' railway companies compulsorily created in 1923 by the amalgamation of more than 120 companies, the Southern Railway (SR) was alone in paying a regular dividend to shareholders before 1948. It was the only company that electrified on a large scale and that did not serve depressed industrial areas; instead, it catered to London's massive expansion south of the Thames. The other three were the London, Midland and Scottish Railway (LMS); the Great Western Railway (GWR, which was heavily involved in the South Wales coal industry); and the London and North Eastern Railway (LNER). The lack of profit reduced investment.

Underinvestment, overcapacity and a massive wages bill were major interwar issues, as was economic transformation: the newer consumer industries tended to have lower freight needs, many of which were met by road transport. Whereas access to rail needed special facilities (notably passenger stations, freight sidings or marshalling yards), road transport was different: it offered access at every point along a road. In addition, the vehicles were far less expensive than those required for the rail system, the majority were owned by individuals, and they did not require comparable training in their use. Lorries could readily move housing materials from central sites. Buses competed with railways to take people to the new holiday sites such as the Butlin's camp opened at Skegness in 1937. The 1933 Road and Rail Traffic Act provided inadequate assistance for the railways, while price increases in 1937 hit rail freight.

More positively, the reorganisation of 1923 led to a measure of rationalisation. Locomotive and rolling stock standardisation cut maintenance costs and helped efficiency. Freight handling improved thanks to the introduction of containers and mechanised marshalling yards, such as Whitemoor. Furthermore, an attempt was made to project a modern image. LNER used Gill Sans lettering, which was invented for it, on publicity. Much building on the rail network used modern architectural techniques – for example, modernist (Grade II listed) Surbiton station and SR 'glasshouse'-style signal boxes. More generally, design-wise there was a lot of streamlining.

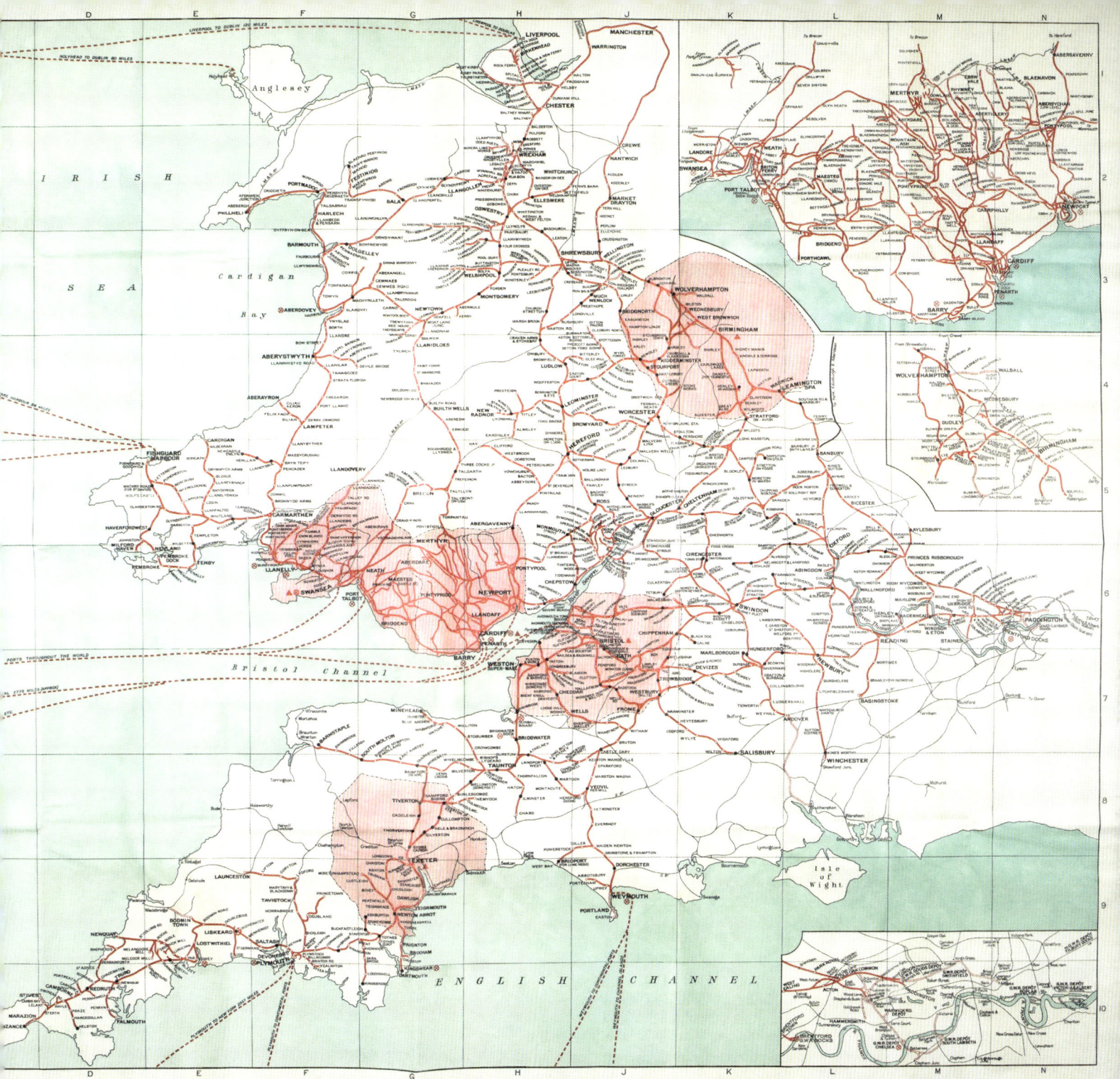

During the General Strike of 1926, the pro-government press emphasised that the rail workers were not united, with *The Times* of 10 May 1926 reporting:

'The TUC continues to urge that statements concerning the number of trains running are much exaggerated, but the public can judge of the success of the companies in establishing passenger services by the daily schedules published in *The Times*.... 106,232 regular railwaymen loyal on the four main railways groups alone.'

The *Daily Mirror* of that day also noted improving train services, including: 'Practically the whole of the Southern Railway's suburban electrified system is running a regular service.'

MOUNTAIN TOURISM
Otto M. Müller, 'Exploring Switzerland', 1939, and Heinrich Caesar Berann, 'Dolomites and European Rail Access', 1936.

This upside-down perspective of Switzerland produced by Otto Müller, a talented German artist and graphic designer, for the Swiss Central Office of Tourism in 1939 is an instance of imaginative interwar mapping, in which railway lines have been used to help fix the imagination. The map's whimsical style became increasingly prominent as part of the cartographic imagination, but this approach was less commonly employed by public institutions. The map almost seems like an attempt to suggest that Switzerland is not dull. It became characteristic of Müller's work, as in his 1956 'Les correspondences ferroviaires internationals en Suisse' ('International Train Connections in Switzerland'), which he designed for Swiss Federal Railways.

Although it appeared novel, a similar form of pictorial map was produced in Germany in 1938 by Leo Faller after the forcible union with Austria, 'Die Schöne Eisenbahnreise' ('The Beautiful Railway Journey'), and earlier still by the Austrian cartographer Heinrich Caesar Berann whose map (see following pages) of rail access to the Dolomites was published in 1936 by the Italian State Tourist Board. Berann's impressive map focuses attention on a developing tourist destination. A painter, Berann found his forte in producing panoramic maps, beginning in 1934 with that of the Grossglockner High Alpine Road, which was built in Austria over a difficult pass in 1930–35, with the first car crossing in 1934. His panoramic maps included works for the Olympics.

Skiing was becoming a major leisure activity for the wealthy, with rail the main means of access to the mountain resorts. This owed much to the extent to which rail was seen as a luxury means of travel – one, moreover, that was well suited to transporting bulky skis. The modern pattern of travel by air to ski resorts was not yet a realistic option. It was possible to drive to the Dolomites, but there was no road network prefiguring the modern motorways that were extensively developed throughout Europe from the 1960s.

In *Crossed Skis: An Alpine Mystery* (1952), 'Carol Carnac' [Edith Caroline Rivett], who had been Alpine skiing, described a train journey from Victoria via a Channel crossing, and the Calais-Basle express, and thence the Arlberg express, to Langen:

'It's a perishing long journey to Austria, and those continental thirds are simply grim ... Crossing Switzerland in daylight was a very different matter from rumbling across France at night. Once they were beyond Zurich the mountains seemed to close in on them, and in the latter hours of the journey they had the thrill of seeing others doing what they had come to do themselves, as they watched the tiny figures of distant skiers come flying down the slopes with the precision and speed of animated toys.'

POLAND'S INHERITANCE

Wladyslaw Groszek, 'The Railway Network of the Republic of Poland', 1938.

Poland's modern railways were only fixed in 1945, as part of the post-war territorial settlement. When previously independent in 1918–39, Poland, having re-emerged from the Austrian, German and Russian empires, had sought to adapt to its new identity (independence had ended with partition in 1795) and to the new borders of the period. It did so with new lines and locomotives – from 1923 made in Poland – and, from 1927, some electrification. In 1936 the first electric line was completed. Whereas the 1920s had seen some new lines, there was far less new construction in the 1930s and following the Depression of the 1930s Poland's railways remained in serious difficulties until 1937 when profitability improved. The map from 1937 (published in 1938) provides details of the network as well as an inset of the situation at the capital, Warsaw. The thickness of the lines provides clarity as to the hierarchy of the system, with much of Poland, particularly its more agrarian east, singularly poorly served by rapid service.

In Poland, as with many countries previously ruled by empires, there was the interplay between the legacy infrastructure inherited from earlier powers and new needs. In the case of Poland, there were new national boundaries and different economic priorities. Thus, for the German Empire the railways in Silesia were designed to help the economic development of this industrial region. This continued to be a priority during the twentieth century, but after 1945 Silesia formed part of Poland.

Post-war repairs were followed in the 1950s by a greater commitment to electrification, and by new rail lines. The most significant, built in 1971–77, was a direct rail connection between Warsaw and Upper Silesia, which became the Central Rail Line. It had a more ambitious plan, for a new line as far as Gdansk (Danzig), moving coal for export and thus to earn revenue. The economic difficulties that followed the 1973 economic crisis led instead to a terminus at Grodzisk Mazowiecki near Warsaw, where it joined the Warsaw–Vienna railway. As was normal in the communist years, the emphasis was on freight, so that in 1980 the line carried 73 freight trains and only four passenger trains daily, although the latter became more important in 1984. The 245-mile long Broad-Gauge Metallurgy Line, built in 1976–79, was solely for freight, taking iron ore from the Soviet Union to the Katowice Steelworks. Coal was moved in the other direction.

There was also a key strategic dimension in the strategic planning and operation of Poland's rail system, with the Polish railways being the crucial means of transit for Soviet forces in East Germany, and also part of their operational rear.

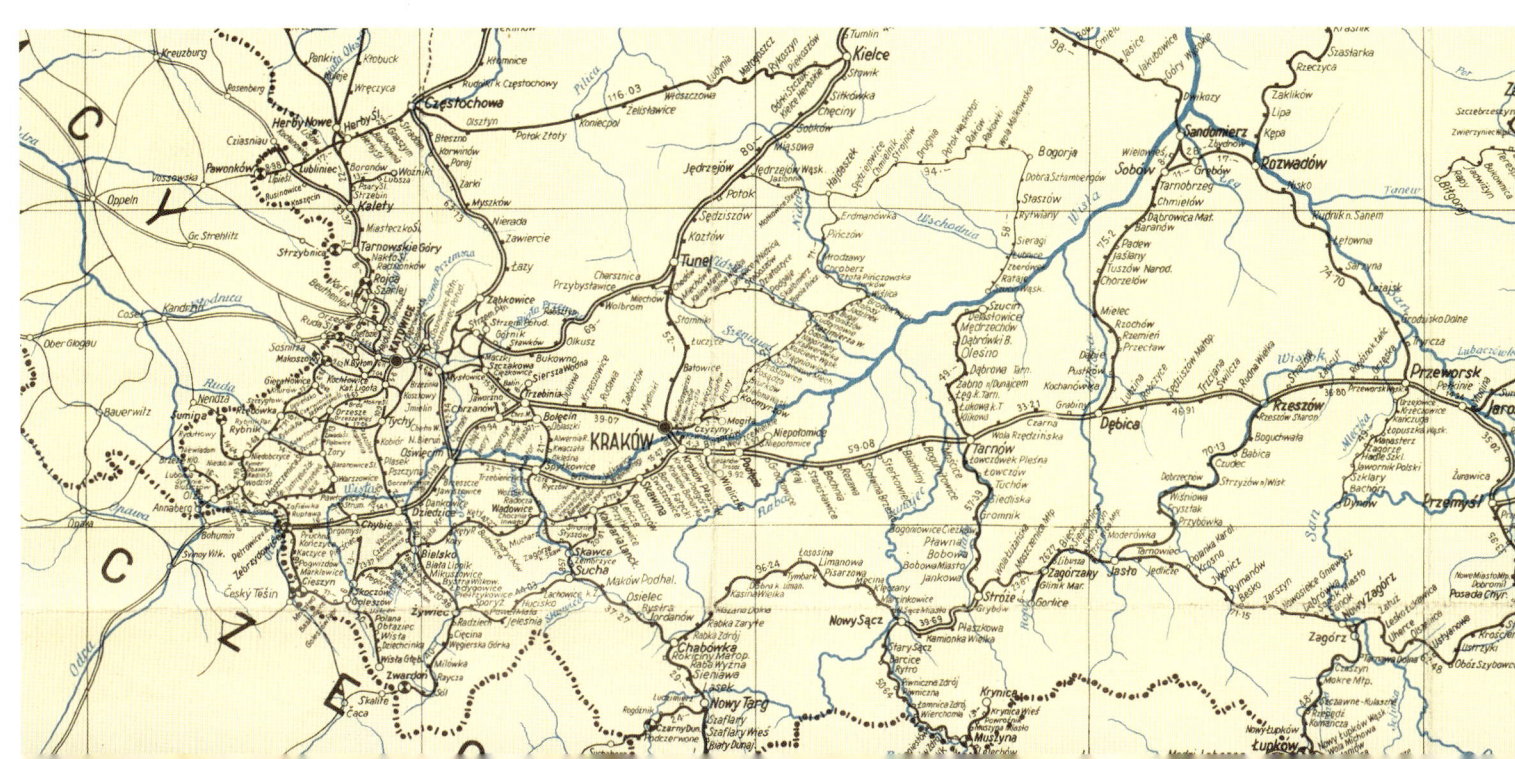

EURASIAN RAIL
'Soviet Rail Routes', 1938.

Politics was to the fore in the determination of governments to develop rail systems. This was the case with the Soviet Union and its communist ethos of heavy industry, which was intended to contribute to the desired 'dictatorship of the proletariat' among whom the rail workers were regarded as a key element.

This map addresses the immense scale of the Soviet Union by being double-sided. This makes the map unsuitable as a wall-map, but provides an effective way to convey both the relative density of railways in European Russia and the far sparser network in 'Asiatic Russia'. The insert reflects the significance of industrial activity, being devoted not to Moscow or Leningrad but to the Donbass economic zone. Rivers provide background location and also the significance of crossing places over rivers, such as Kyiv. The map distinguishes between double- and single-tracked lines and includes rail-linked ferries, which were particularly significant across the Caspian. The loss of the Russia/Soviet presence in Manchuria is evident because the railways there are marked as if they were in non-Russian Europe. Also, west of the Dniester River and the Black Sea the area of Bessarabia is marked in a distinctive fashion, reflecting Soviet opposition to Romanian rule.

In the Russian Civil War (1917–21), the difficulty of sustaining operations in this vast space had encouraged the emphasis on movement and hence the significance of the railways. This helped explain the control of key junction cities, such as Moscow and Tashkent, which helped provide interior lines and the ability to move troops to areas of opportunity and need. Armed trains provided a form of mobile artillery, combining firepower with the rapid movement of troops, thus offering manoeuvre warfare and potential at the tactical and operational levels.

The civil war greatly hit rail use, but by 1926 usage had passed the level of 1913. State force played a major role in the development of the Soviet rail system, with a major source of readily controlled manpower for the new construction provided by forced labour gangs from *gulags*.

This map is typical of the type of information, cartographic and otherwise, that was subsequently to be deemed confidential. The German invasion of the Soviet Union in 1941 led to the withdrawal of many maps from libraries and other sources, and thus espionage in part involved the acquisition of pre-war Soviet material.

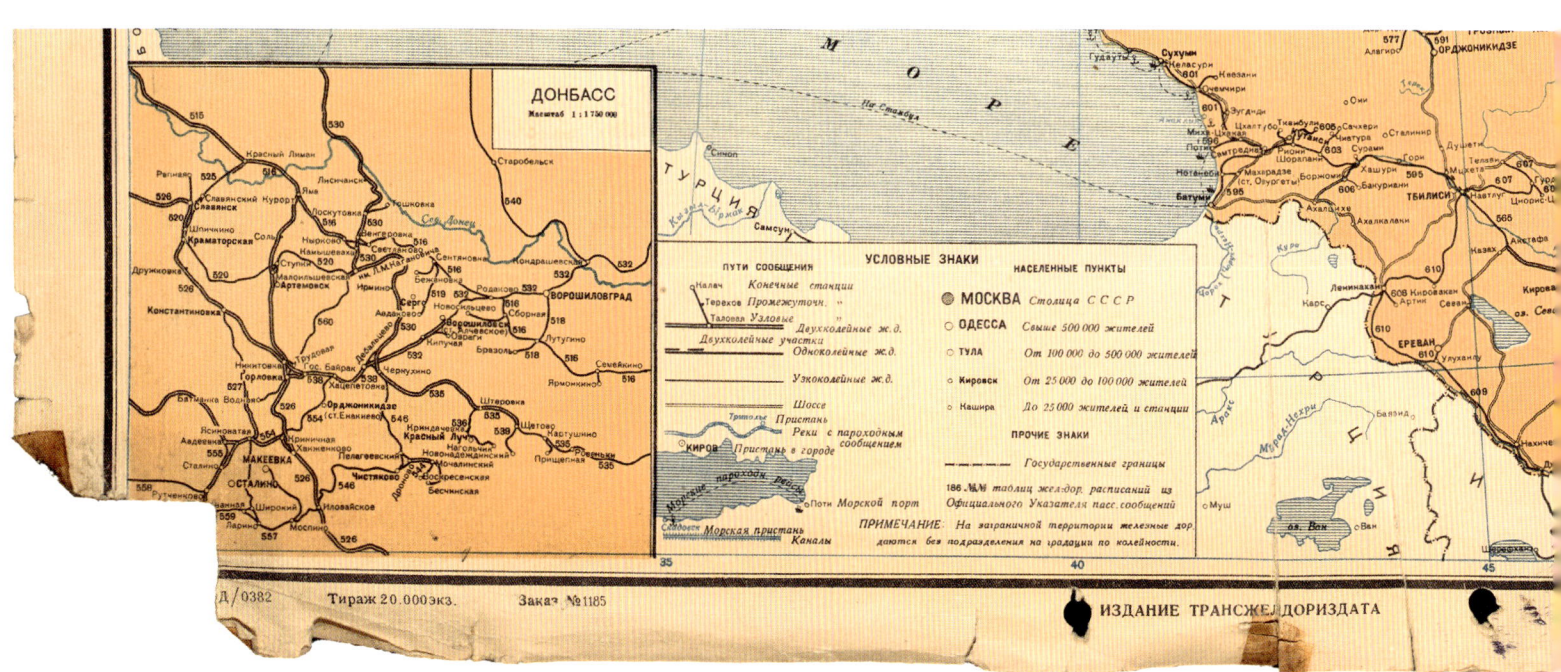

5

WAR AND THE AIR CHALLENGE 1939–70

War again prevailed as a theme in this period. This was so over three timescales. First, the heavy use of rail during the war itself, as well as the unprecedented devastation visited upon it. Second, the post-war period of repair. In the most obvious of senses, this was a matter of destroyed or damaged track, stations, bridges and marshalling yards (see pages 192–193).

Less obviously came the replacement of locomotives and rolling stock, which reflected wartime destruction, the inroads of heavy wartime use, and a lack of wartime investment in new capacity, notably carriages. There were also the immediate tasks, notably demobilisation and the movement of refugees and prisoners. Never before in European history had such numbers of people moved across borders.

Third, the longer-term damage from the conflict, notably in terms of reduced profit, lessened investment and the habit of regulation. The latter was frequently sustained post-war; thus, in 1948 rail nationalisation in Britain owed much to the control exercised by government in wartime and the corporatism to which it gave rise. Not all states saw such a continuation from war to peace, and notably not the United States, but such government control became normative in many countries whatever the formal structures of ownership.

Regulation acted as a depressant on price increases, for freight and passengers, as return on capital ceased to be a theme, major or otherwise, for most systems. In addition, the post-war territorial situation and successor conflicts introduced communist systems in Eastern Europe, China, North Korea, North Vietnam and Cuba (see pages 198–199), while anti-Western populism and nationalism had an impact in newly independent states, such as India and Indonesia, as well as in Latin American states breaking free of British influence over rail, notably Argentina. A time of political flux was not the best basis for discussing planning, attracting investment and earning returns.

And yet investment was required, not only to compensate for the damage of the past but also to embrace the opportunities of the new. In the 1950s this very much focused on the replacement of steam by diesel, which was largely complete by 1960 in North America, Western Europe and Africa. The diesel electric (or simply the diesel) had far lower operating costs. Its thermal efficiency was about four times as much as a steam locomotive and there was no need for the coaling, watering and maintenance. This made the working time of both locomotive and crew greater, and encouraged speedier runs. Diesels could accelerate more rapidly and smoothly – and achieve higher sustained speeds. Moreover, they proved more efficient in cold weather and could be completely shut down when not in use. Labour requirements were less expensive, not least because two or more diesel units could be combined and operated from one control cab. The changes required investment in new equipment and labour flexibility, but both were eased by the economic boom years of the 1950s (see pages 212–215).

As with rail, map production techniques changed. In particular, in the late 1940s and early 1950s, there was a rapid spread in the use of transparent overlays of plastic material for scribing. Maps were compiled using a multi-level system of overlays, with information registered on a typeset base map. Photomechanical laboratories then produced colour negatives from which to print plates. This complex procedure started with the preparation of the set of 'peel coats' (photographically opaque films) on which the information for each colour on the map was 'scribed' (cut) away. The reproduction house exposed

The cover of 'War Atlas', a section of *The Philadelphia Inquirer*, dated February 13, 1942, features a train as part of its depiction of an America fully committed to war production.

each peel coat to make appropriately coloured printing films, which were then combined for proofing. With one peel coat for each tint on a map – a single map might require ten or 12 – this was not only expensive, in terms of materials, but also meant that the effect of the combined colours could not be evaluated until the proof was made. Any correction to lines or colour might involve changing several peel coats – unwanted lines were covered, new ones drawn on – and this created serious problems in matching up the original and the corrected versions.

A less complex method for more simple maps involved a coloured (painted or airbrushed) baseboard, with the line work either painted on or supplied as overlays, and with each piece of type stuck down by hand on a further overlay. Corrections were made to the base by repainting it, a clumsy procedure that led to inferior quality and the risk that type might fall off during handling. The disadvantage of both these printing methods resulted in strong pressure not to make changes.

The challenge of aircraft to rail freight was restricted to high-value, low-bulk goods, notably mail – and particularly valuable parcels and express letters. For passengers, the competition was more acute, especially in large countries, notably the United States, Australia and Canada. This competition initially was limited because aircraft were expensive, slow and unreliable, but it became more significant from the 1950s with the introduction of jet aircraft. Air travel became more predictable, less expensive and quicker, while larger aircraft were able to carry more passengers.

WARTIME INTELLIGENCE
Supreme Headquarters of the Allied Expeditionary Force (SHAEF), 'State of Damage to Rail Centres in Western Germany', 1945.

The preparation for Allied attack as the 1945 campaigning season approached included air attacks on the Rhine crossings, as at Worms (circled, towards the centre), and on places near the front line – for example, Mönchengladbach (northeast of Aachen) and Trier (across the border from Luxembourg). The latter places were important both to disrupting any German response to Allied ground attacks and to preventing any resumption of the German Bulge offensive of the previous December, made possible by the train-transported movement of tanks from the Eastern Front. Railways were required to carry larger, heavier and different loads than in campaigns a half-century earlier. Not only were there large numbers of tanks but also considerably more stores per soldiers.

The map, endorsed 'Secret' (top left), shows the rapid integration of intelligence material and the use of evaluated aerial reconnaissance to ascertain the differing rates of devastation. This permitted the distinction in the key between light, medium and heavy damage, with these categories given a quantitative weight of damage inflicted – up to 15%, 15–50% and over 50% – and those results which had changed over the previous week. There was also the value qualification of no reconnaissance since the most recent attack, indicated by '?'.

As always with maps, what is not covered or highlighted is significant. In this case, the lack of bombing of rail targets in the Netherlands, which were exposed to attack from the main concentration of Allied air bases in East Anglia. However, the failure of the Arnhem offensive in the autumn was followed by a lack of interest in reviving such an axis of advance or having rail as a matter of targets of opportunity. So also with the comparative lack of Allied interest in targets in the Black Forest, although Freiburg (near the south end of the 'line at 17 Feb') was hit. Damage inflicted was the key element in the air war on rail, but also what it was possible to map – for, unlike fixed artillery, aircraft posed an inherently dynamic character whose location could not readily be depicted.

Maps were also important in strategic planning. The German High Command, in its military evaluation of Britain, produced a map of British railways and an index of stations. The Geographical Section of the General Staff in 1942 produced maps of the main railway lines in France, Belgium and Holland, differentiated between double-track, single-track and electric lines.

Railways provided a key logistical context and means in the war. Thus, when the Germans invaded the Soviet Union in 1941 they suffered not only from a shortage of lorries but also the extent to which the rail system was damaged in the fighting – notably the bridges over the wide River Dnieper destroyed by the retreating Soviets. In addition, the Soviet railway system posed major difficulties due to the change from one track gauge to another. The destruction of rail infrastructure by bombing and by withdrawing forces also affected the Allied advance across France in 1944. As a result, it was necessary to rely heavily upon lorries, although in part that was also an appropriate response to the rapid nature of the advance in the late summer.

Wartime intelligence on rail took a number of forms. A major element was logistical capability, as with the maps in the schoolhouse in Reims (Rheims) used by Eisenhower as the Supreme Headquarters of the Allied Expeditionary Force (SHAEF). Very different to those was the map on road and rail construction in the Balkans, 1940–42, produced by the Topographical Section of the General Staff, in which railways were differentiated as normal or narrow gauge. This map included supporting notes: for example, 'Maritsa bridges reported to be repaired. June 1942', 'Sofia-Karlova-Kazanlik line complete to Karlovo' and (for Rustchuk) 'Train ferry over Danube working since July 1941'.

In the cartographic rhetoric of war, that of maps in propaganda, railways were frequently identified as targets. Thus, the British poster 'Allied Air Offensive against Germany up to January 1st 1941' was designed to show that not only Britain was being bombed; the key included 'Railways', which were indicated by a pictogram of a steam locomotive, and that was provided for many targets.

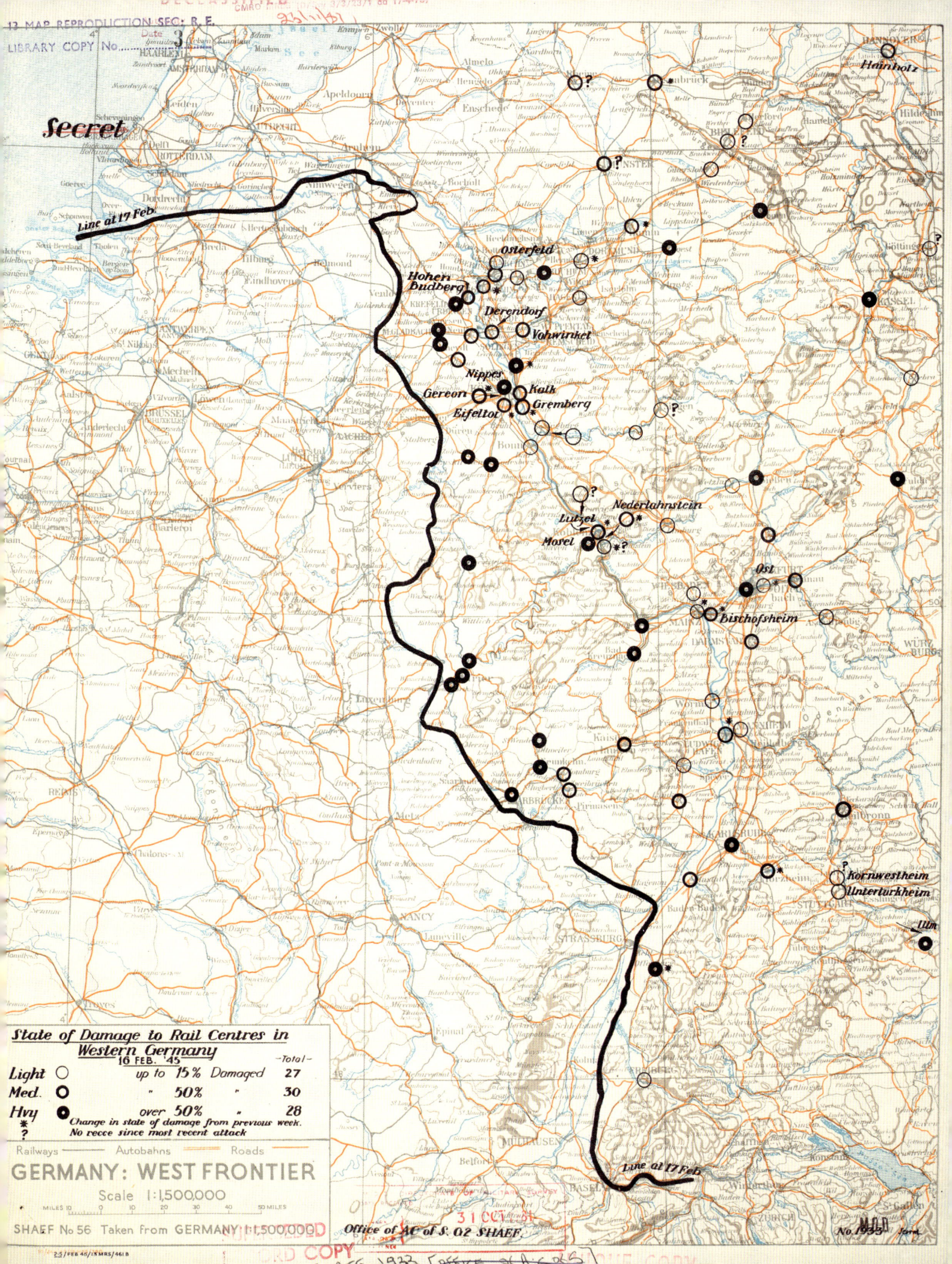

RAILROAD OF DEATH
OSS, 'Burma - Siam Railroad Installations', 1944.

This declassified map was compiled and drawn in late 1944 in the Branch of Research and Analysis of the OSS (Office of Strategic Services), the forerunner of the CIA. The map provides a high level of accuracy, drawing on aerial reconnaissance, about the railroad installations between Thanbyuzayat and Kanchanaburi. It records the railway itself with a key that lists locomotive shelters, stores depots (including rail and tie depots), sidings, spurs, passing loops, reversing triangles, bridges over 100 feet long, labour or military camps (distinguishing between those with floor space of under or over 200,000 square feet), the distance in miles from Thanbyuzayat, the international boundary, rivers, spot heights and fair weather motor transport road.

To avoid a long maritime route via Singapore, a route that would put pressure on Japan's limited shipping resources and expose them to Allied attack by submarines, surface shipping and air attack, with all the risks and protections thereby incurred, the Japanese decided to open up a rail one from the Gulf of Siam to the Bay of Bengal, from Bang Pong in Thailand (south of Kanchanaburi) to Thanbyuzayat in Burma, most of the line being in Thailand. Although Thailand had briefly resisted the Japanese landings on 8 December 1941, it ceased to do so the following day and declared war on Britain and the United States on 25 January 1942, providing a base area for the invasion of Burma. Thailand was rewarded by Japan with territory in Malaya (1942) and Burma (1943), and although it brought scant benefit to Japan in terms of industrial capacity or military assistance it offered strategic depth as well as links between the parts of the Japanese world in Southeast Asia. The 258-mile-long railway was designed to supply the Japanese forces in Burma and thus prepare for the invasion of northeast India that was launched, without success, in 1944. The railway was a physical demonstration of Thailand's alliance.

Short of construction equipment but with plentiful forced labour, the Japanese relied heavily on both Asian and Allied prisoners of war. Begun in October 1942, the work was driven even harder from April 1943, with the 'speedo' period also seeing cholera. The railway was opened in October 1943. Large numbers of the inadequately cared for and brutally, indeed murderously, treated workers died. Alongside cholera, there were many deaths from beriberi, dysentery, dengue fever and malaria. There were also many non-fatal diseases and injuries that left horrible conditions and wounds; and malnutrition and weakness made it harder to cope with them. As part of the improvised nature of Japanese rail construction, track and sleepers were reused from dismantled railways in Malaya and the Dutch East Indies, now Indonesia.

The map reflects the growing interest in exploiting Japanese weakness, both militarily, in order to drive the Japanese back in Southeast Asia, and politically, by getting Thailand to abandon its Japanese alliance. Although not a priority comparable to the Pacific advance on Japan, it was a sphere in which the OSS had a role.

In 1944 there were tensions between Britain, which treated Thailand as an enemy from which territory could be gained in the Kra Isthmus (a goal sought by Churchill in order to provide a continuous land route between Burma and Malaya), and the United States, which saw Thailand as occupied by Japan. Successful American pressure ensured the peace terms for Thailand were lenient ones.

The intelligence on the railway reflects both the identification of targets for possible attack and interest in using the route as a potential axis of advance.

The railway carried much freight during the war. A British air raid on 13 February 1945 bombed and damaged two bridges, including the one at Kanchanaburi later known as 'the Bridge on the River Kwai'. The Japanese had the bridges repaired. But problems were caused by the nature of the construction, with a post-war Allied report of October 1945 noting that the railway:

'...was not constructed in a permanent form ... some of the 688 bridges on this railway are too weak to carry locomotives, wagons having to be moved over them singly by hand; many of these bridges are liable to be washed out during the monsoon season. Under present conditions, therefore, its capacity is negligible.'

The railway remained in operation until 1946 when the Burma section was removed. Part of the Thai section remains in use.

THE HOLOCAUST
I.G. Farben Planners, 'Site Plan for Auschwitz-Monowitz', 1944.

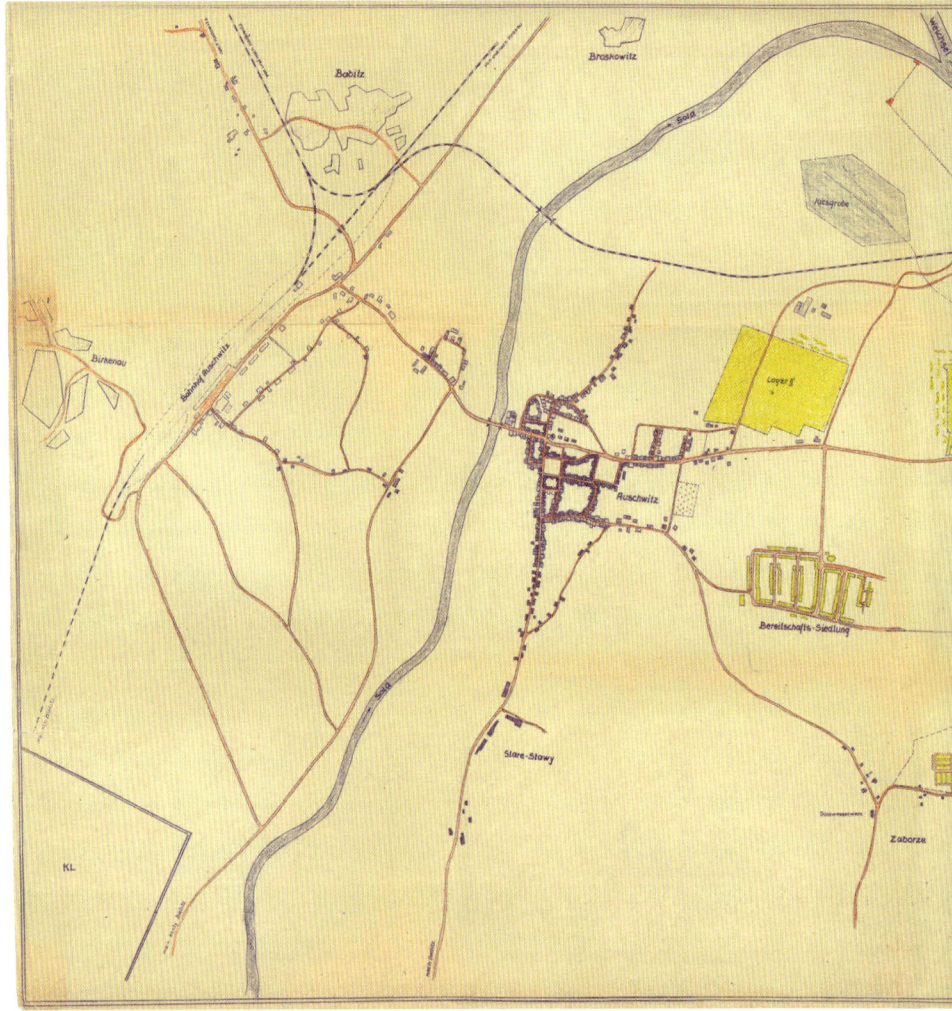

Auschwitz III, also known as Monowitz and Monowitz-Buna, was opened in October 1942, at first to supply forced labour for nearby industrial facilities run by the company I.G. Farben. Unlike the other extermination camps, Auschwitz was located in a key economic region, that of coal-rich Upper Silesia, a well-connected industrial zone. With the slave labour force from Auschwitz, the nearby I.G. Farben plant, or Buna Werke (marked Dwory I in the map), acted as a nexus of the co-operation between the SS and the Ministry of Armaments under Albert Speer. The I.G. plant for the manufacture of synthetic rubber and oil was the largest German industrial project for these products and, located in Silesia, it was beyond the then reach of Allied bombers. Synthetic rubber and oil were crucial to the German war economy, the latter especially so after Operation Barbarossa brought about the end of the delivery of Soviet oil that had been so significant in 1939–41. At Auschwitz III, the SS guarded what in effect was a private concentration camp.

In contrast, Auschwitz II, which was also the eventual destination of many of the workers at Auschwitz III, was an extermination camp, with the first gassing of Jews there in early 1942. Jews were taken there for slaughter by train, crammed into the extremely crowded freight wagons. In the rail journeys, which were lengthy, Jews were denied food, water, light, warmth, sanitation, space and bedding. Those who died on the journey were left en route in the shut freight cars.

Auschwitz II was fed by train with Jews from throughout occupied Europe, the first from Slovakia arriving in March 1942. In August and September 1942, more than 2,000 foreign Jews held in the concentration camp of Les Milles near Marseille were herded by police into railway convoys and deported to Auschwitz via the Parisian holding camp at Drancy, from which about 10,000 Jews were sent by train to Auschwitz between 17 July and 31 August. In total, about 2.5 million Jews were taken to death camps by train in 1942–43. On 19 April 1943, unusually, a train carrying deportees en route for Auschwitz was attacked in Belgium near the holding centre at Mechelen, from which almost 25,000 Jews were sent to Auschwitz in 27 trains between 1942 and 1944. Three young men used a hurricane lamp covered with red paper to bring the train to a standstill before opening one of the goods wagons, releasing 17 prisoners. Prominent groups killed in Auschwitz included, after a long train journey, the large community of over 40,000 from Salonika (Thessaloniki), a key step in German anti-Semitic violence in Greece, and, in 1944, 437,000 Hungarian Jews deported by the Germans, about three-quarters of whom were killed on arrival.

The centrality of rail to the Axis powers was shown with the rail spur constructed so that trains carrying Jews for slaughter could go directly to the extermination camp at Auschwitz II. There has been subsequent

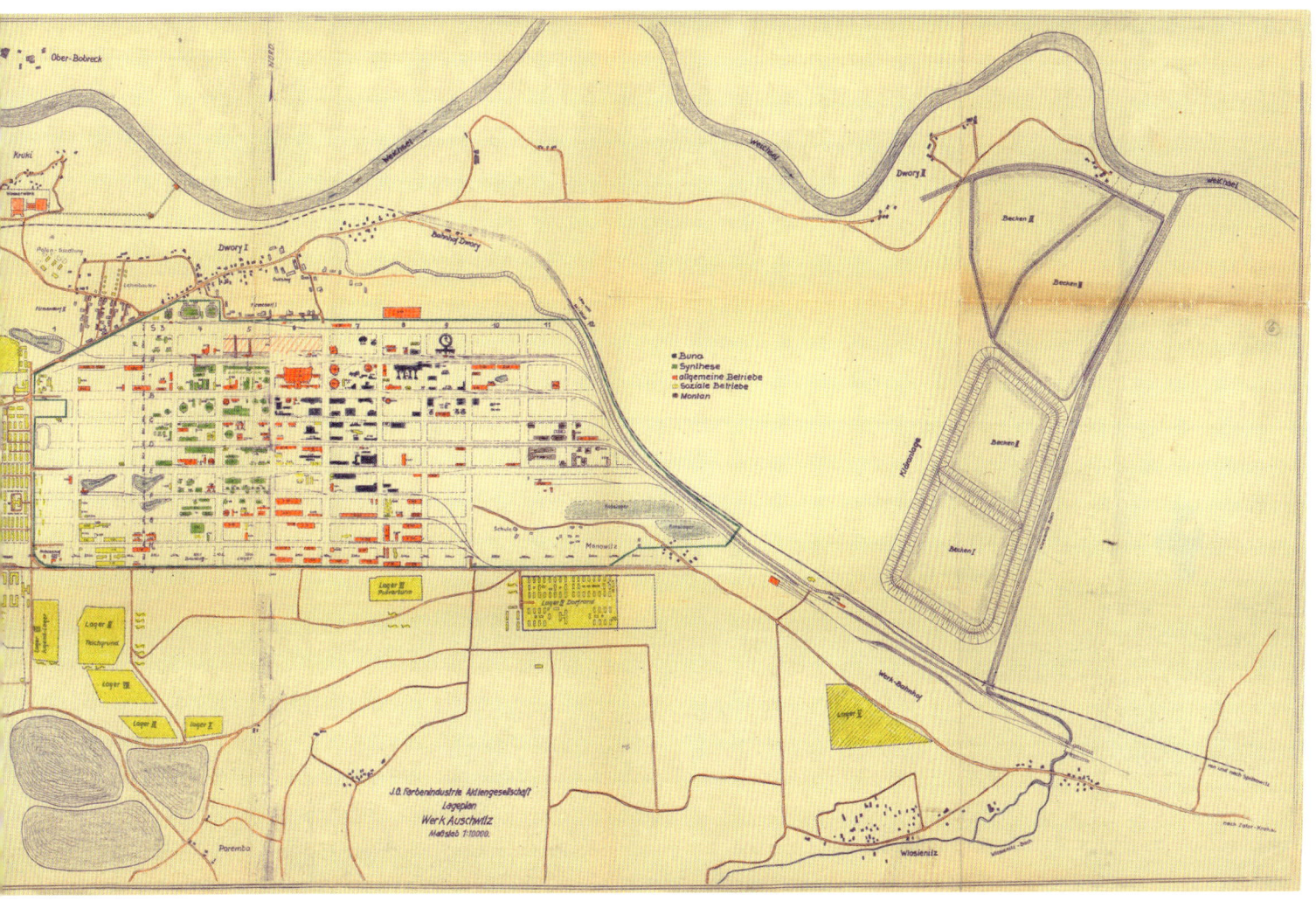

controversy as to whether the Allies should have bombed the rail links.

In 2003 Kurt Schaechter took legal action against the SNCF, the state-owned French rail company, for deporting his parents to the extermination camps at Auschwitz and Sobibor, a symbolic case designed to highlight the range of responsibility. In 2010 the SNCF expressed remorse for transporting Jews to German extermination and concentration camps. In part, this step was taken to ensure a better chance at obtaining well-paying contracts for high-speed rail lines in the United States. In 2011 a formal public apology was made directly to Holocaust victims, while the company handed the station of Bobigny over to local authorities in order to create a memorial to the Jews transported from there to the camps. In 2015 France approved an agreement with the United States to pay US$60 million in compensation to foreign nationals who had been deported to extermination and concentration camps on French trains. Previously, French citizens who were victims had been paid US$60 million under a 1946 scheme. Other transit camps, notably Mechelen in Belgium and Westerbork in the Netherlands, were closely linked to the rail system.

Trains were used for deportation in other states, for example from the Baltic states occupied by the Soviet Union in 1940–41 and from 1944. The extent to which recent history is politically contentious is shown by the degree to which such movements are noted in public memorialisation. Thus, the end of Soviet rule was followed by the display of wagons used for such deportations, as in the Latvian Museum of the Occupation.

5. WAR AND THE AIR CHALLENGE 1939–70 **197**

UNION OF SOVIET SOCIALIST REPUBLICS
British War Office, 'Transport Engineering', 'Principal Land Communications' and 'Principal Rail Traffic Flows', 1951.

With intelligence-provided overprints to maps compiled by the Ordnance Survey, these three maps were produced for the Geographical Section of the General Staff in Britain in 1951. The use of colour in order to distinguish between types of factories, transport routes or goods was effective; as was the scaling of quantity of annual freight tonnage in 'Principal Rail Traffic Flows' by the width of arrow. This was a standard cartographic device but one that was successful both in visual terms and because it provided a sense of relative significance. This was material that the Soviet Union sought to keep secret. The movement of coal, oil, ores and timber captured the Soviet focus on heavy industry. The stress in the data was on European Russia, the Urals and western Siberia. In contrast, British knowledge of the situation in eastern Siberia was limited. The key rail centres are apparent, including Volgoda and Sverdlovsk. 'Principal Land Communications' shows how industrial and transportation development was disproportionately concentrated in the westernmost part of the Soviet Union.

The unspoken agenda is significant. During this period there was an escalation of Cold War capability with the deployment to East Anglia from 1949 of American nuclear-capable B-50 Superfortress bombers, followed by the British acquisition of similar aircraft and weapons. Britain's first atomic bomb was detonated in 1952, followed by the introduction of British Blue Danube atomic bombs in 1953; although the Vickers Valiant bombers to deliver them were not available until 1955.

Moreover, the Korean War (1950–53) threatened to broaden into a third world war, as indeed it came close to doing in 1953 when the Soviet Union considered an invasion of Western Europe, while the Americans made nuclear threats as a way to bring the conflict in Korea to an end.

In such a context, targeting for air attacks became highly significant. This was a response to the conventional land strength of the Red Army and in part to the lessons 'learned' from the Second World War. Any new conflict, it was believed, would be quicker and go nuclear from the outset, and thus it was necessary to have accurate targeting information. The emphasis here was on the Soviet economy, including railways, as the basis of a threat, and not on the population as a nuclear target. In 'Transport Engineering' the map records the railways in relation to factories making locomotives, wagons, passenger coaches, motor vehicles and tractors. As a literary parallel, in Ian Fleming's opening novel *Casino Royale* (1953), James Bond is instructed to compromise 'Le Chiffre', the paymaster of the communist-controlled trade union in the rail and heavy industries of Alsace, the most vulnerable part of France to Soviet attack and one that was crucial to the NATO response to any such attack.

Railways as well as railways under construction were marked on other Intelligence maps – for example, that of 1965 on China, North Korea and North Vietnam produced by the General Staff Map Service of the British Ministry of Defence. In turn, that was used for the Joint Intelligence Committee in 1968 to produce a map of airfields.

Map 26

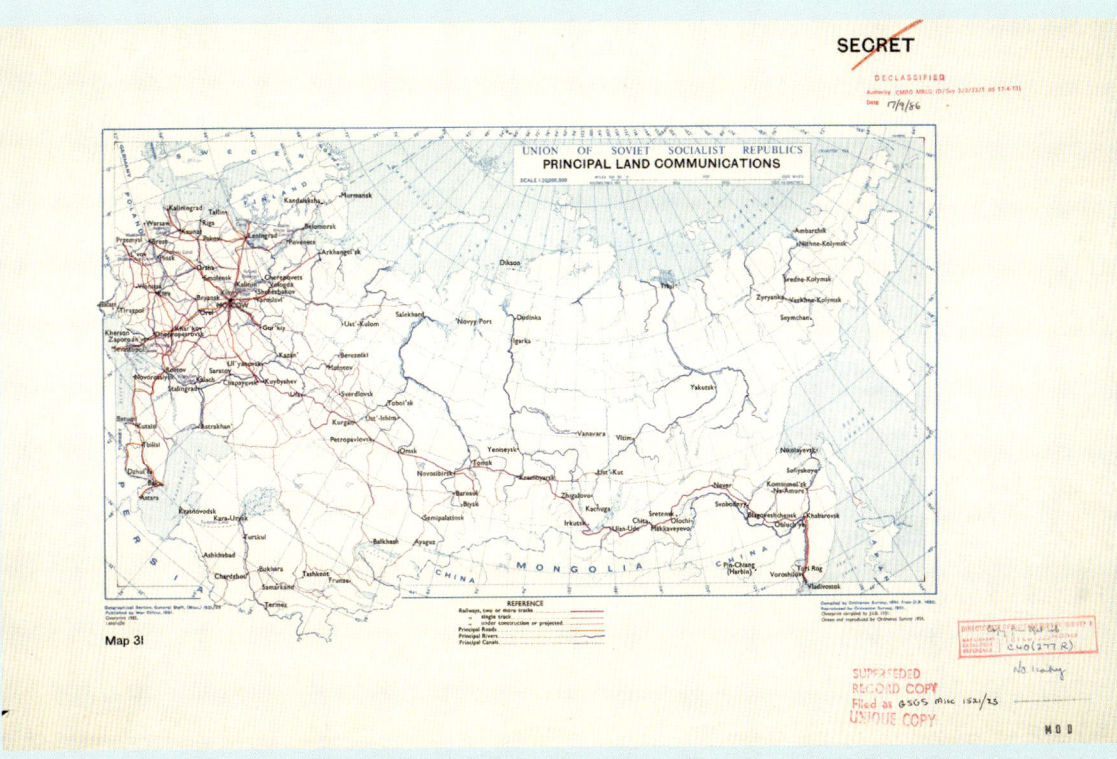

Map 31

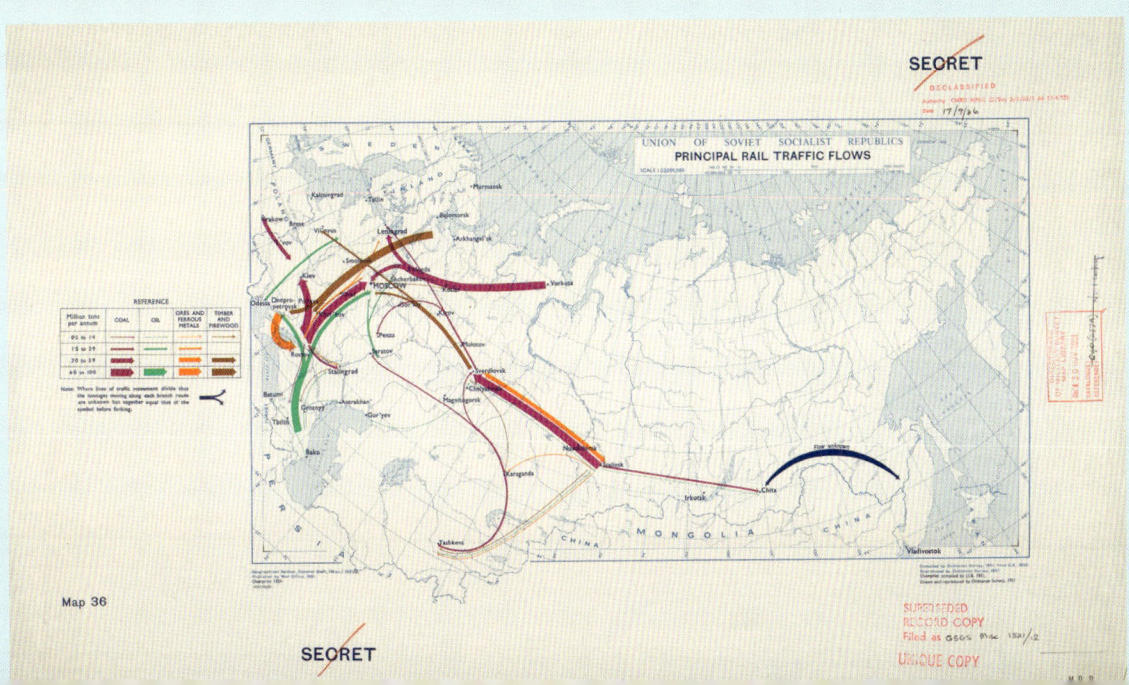

Map 36

RAILWAYS AT THE END OF EMPIRE
Saben and Company, 'Railway Map of East Africa', 1956.

This British map was published in 1956 by Saben and Company in Kampala, the capital of the colony of Uganda. Saben published an annual commercial directory and handbooks on the British model, and this map was a supplement to the 1955–56 directory. The map uses red to direct attention to its subject, the railways in British East Africa. After the First World War this included Tanganyika, the former German East Africa, as well as the already established British colonies of Kenya and Uganda. On the general pattern, routes focus on ports, although in Kenya there is the additional major hub of Nairobi, the inland capital. The railway between northern Tanganyika and Mombasa provides a link between the two soon-to-be-former colonial systems although it is indirect: there is no direct line between Mombasa and Tanga, but rather a line via a junction southeast of Moshi, although from Arusha a road continued southwest to join the main Tanganyikan railway.

The map indicates the linkage of motor, lake and rail routes, as well as only limited plans to expand the rail system, principally in Uganda from Kampala westward to the border with 'Belgian Congo' (where a projected extension from the existing system would link Mityana, Kabagole and Kasese by 1956 to Kisindi). There was also a modest western extension of the railway in southern Tanganyika. The Kenya and Uganda Railways and Harbours had merged in 1948 with the Tanganyika Railways and Ports Services to form the East African Railways and Harbours Corporation.

This map is an instructive commentary on earlier bold hopes for a Cape to Cairo railway, notably via Sudan, Uganda and Congo, or indeed one from West Africa to Mombasa.

There was expansion in other overseas empires. In French West Africa the line north from Abidjan in Ivory Coast reached Ouagadougou, 1,145 kilometres away, in 1954 and the discovery of manganese in Tambao in 1959 led to work on an extension. That was very typical of the resource-linked nature of imperial rail expansion.

The availability of fuel was an important issue for the railway history of Africa, as the distribution of coal in Africa did not match requirements, while the cost of its movement was high. There was the additional problem that lorries, which relied on oil, benefitted greatly from the relatively low oil prices of the period up to 1973. The 'fuel mix' for transport, that of availability and cost, varied greatly by empire and, indeed, colony.

The speed of the abandonment of empire was far faster than had been envisaged, and notably so in the case of Britain, which had expected a slow process of decolonisation. In the Portuguese Empire, where a sense of mission remained strong, there was no intention to end empire and the railway system continued to be expanded. Indeed, in the leading colony, Angola, the most southerly railway, from the port of Mocamedes, reached Menongue in 1961, having been re-gauged to a broader gauge in 1950. The most northerly, the Benguela Railway to Congo, completed in 1929, proved profitable in the early 1970s after Zambia closed its border with Southern Rhodesia (Zimbabwe). This line then became the key means for moving Zambian exports, notably copper, to the coast for trans-shipment. The commitment of the imperial government was seen in 1941 in the Portuguese colony of Mozambique when the last of the companies that had been given a concession to operate railways was nationalised.

Aerial surveying was used extensively after the war, with the expertise developed in wartime put to use mapping large areas previously mapped only poorly. This was particularly valuable in inaccessible terrain. Aerial photography became central to the surveys carried out by Britain's Directorate of Overseas Surveys (DOS), for example in Kenya and Gambia, not least because it could achieve more rapid results. The costs and time taken by ground surveying engendered greater support in the 1950s for aerial photography, leading to the taking of many photographs in parallel traverses. Height could be measured by the use of overlapping images but the unpredictable weather was a problem, while ground-based observations remained necessary for the most accurate interpretation of aerial photography.

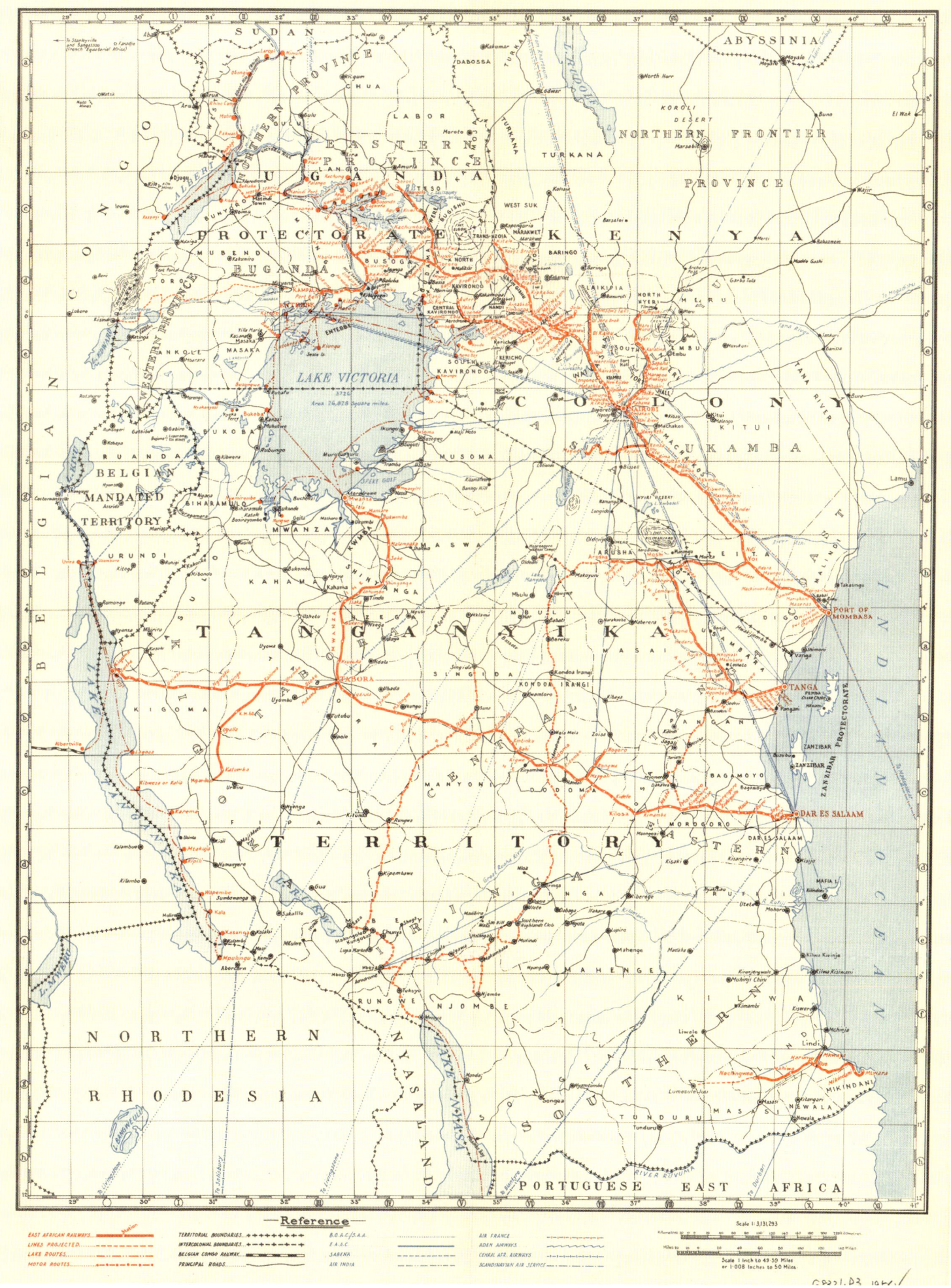

TURKISH-IRANIAN RAILWAY
T. Filipeyan, 'Plan and profile – Qureh Tapeh to Tabriz', 1959.

This plan and profile for a rail link in northwest Iran from Tabriz to Qureh Tapeh (Qareh Tapeh) is a copy held by Britain's Directorate of Military Intelligence (DMI). The scheme for improved rail connection between Turkey and Iran was supported and largely funded by the Central Treaty Organisation (CENTO), a mutual security organisation, designed to resist Soviet influence and expansion, composed of Britain, Iran, Pakistan and Turkey, which until 1959 also included Iraq.

As the map survey notes by showing at the bottom the relief with spot heights, beyond Qureh Tapeh, on the northern edge of the highly saline Lake Rezaiyeh (now Urmia), in particular it was challenging terrain. The railway opened in stages in 1971 and 1977, and was a welcome release from isolation for Iran during the 1980–88 war with Iraq. However, the railway was suspended in 2015 due to opposition in Kurdistan.

The west–east railway line has recently been revived. The first train from Pakistan arrived in Turkey via Iran in January 2022, a 3,666-mile route that took 12 days and 21 hours. This journey is clearly an aspect of the geopolitics of Asian rail links, but in the case of Turkey, Iran and Pakistan the geopolitics is not that of anti-communism, as had been the case in the 1950s, but instead an Islamic alignment aimed in part at providing unity against both the West and Russia. Moreover, the three powers had shared interests in opposing separatism, respectively by Kurds (Turkey and Iran) and Baluchis (Iran and Pakistan). The railway system could also be expanded into Iraq.

The interwar Trans-Iranian Railway ran north to south – and was extended in 1961 to Gorgan, near the Soviet frontier east of the Caspian Sea. In the context of the period prior to 1947, the revived railway would have linked British India to the British zone of influence in Iraq, and for this reason some British commentators had wanted the Trans-Iranian Railway to be west–east.

In Persia/Iran there had been tentative attempts at railway operation in the 1880s with a horse-driven suburban railway south of Tehran from 1886 that was later converted to steam, and a railway from Mahmudabad and Amol established in 1887 but not brought successfully into use. The first significant railways came under the new, self-consciously modernising Pahlevi dynasty, and were built from Tabriz to Jolfa in 1914 and Sufiyan to Sharafhaneh in 1916. The Mirjaveh–Zahedan line followed in 1920, but the most important was the 865-mile Trans-Iranian Railway from Bandar Shah on the Caspian via Tehran to Bandar Shahpur on the Persian Gulf. A pre-war project that had seen a significant jockeying for advantage involving, in particular, Britain and Russia, the railway was begun in 1927 and finished in 1938. It was a difficult project that involved the building of many tunnels. This line was taken over by Britain and the Soviet Union in 1941 and its capacity raised, a process furthered when the Americans replaced the British in 1942.

Tabriz, the major town in northwest Persia/Iran, was of strategic significance in controlling the links between Persia and the Soviet Union. As such, it was of longstanding interest to the British, keen to limit Russian expansion. The city had been the centre of the constitutional revolution movements of 1905–11 against the ruling dynasty and was occupied by the Russians in 1909 and 1911–17, before being seized by the Turks. However, in August–September 1941 the two wartime allies jointly invaded Iran in order to prevent it from turning towards Germany. The advance made much of railway links. Soviet troops occupied the Tabriz area from 1941 until 1946 and once they had withdrawn the Iranian army then crushed the Soviet client state, the Azerbaijan People's Government.

As a further instance of the geopolitics of rail, in 2014 a new line connected Gorgan to Etrek in Turkmenistan. This connected Iran to Turkmenistan and Kazakhstan, as intended by the International North-South Transport Corridor (INSTC) agreed in 2002 and designed to move goods, by sea and rail,

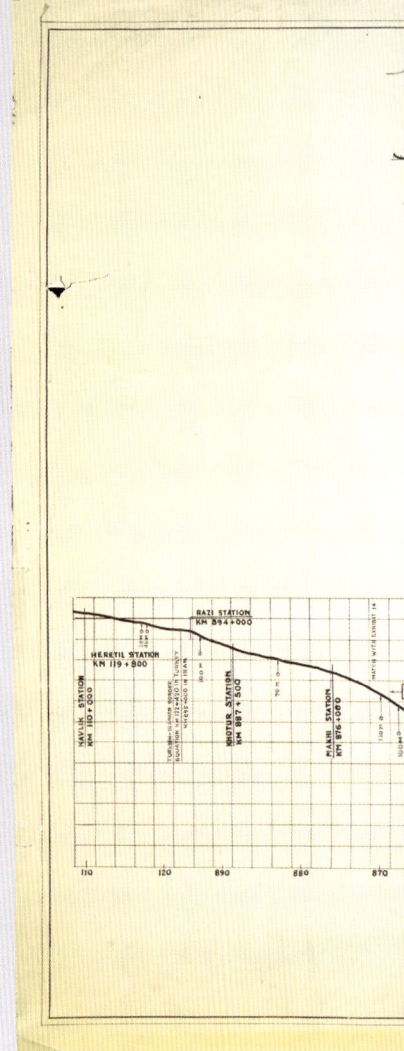

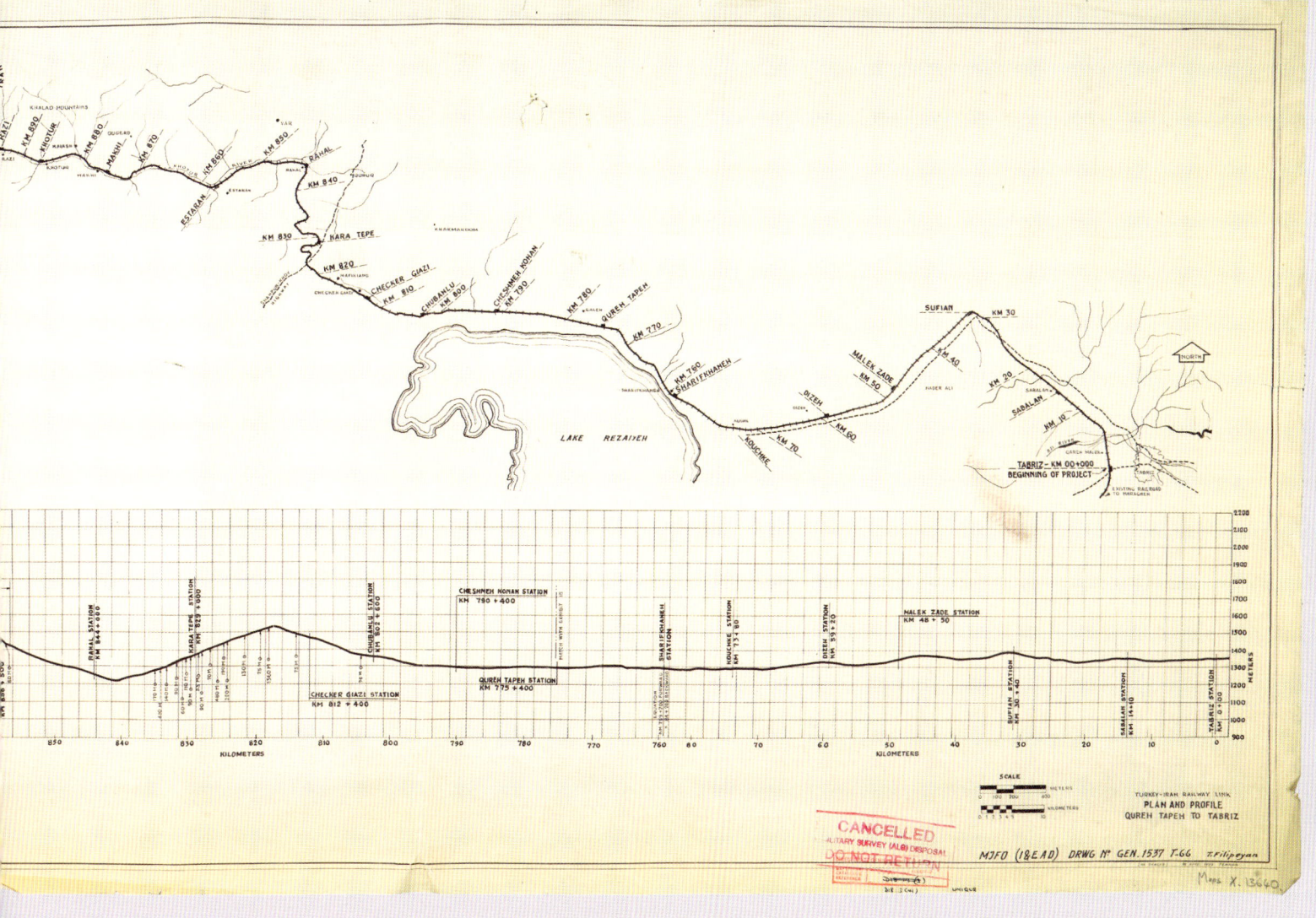

from India via Iran and Russia to Europe. In 2016 the completion of the Rasht–Astara section greatly improved rail links between Iran and Azerbaijan. The INSTC returns attention to the area covered in this map. Indeed, the two show the continuity of imperial interests and geopolitical concerns in railways in particular areas.

The construction of a cross-border rail link between Iraq and Iran was formally inaugurated in September 2023 with a foundation stone-laying ceremony at the Shalamcheh border crossing near Basra.

ACROSS THE NULLARBOR PLAIN
Department of National Development, 'Commonwealth Railways: Map of Australia…', 1960.

Commonwealth Railways was created by the Australian government in 1917 partly to administer Trans-Australian Railway. This map was compiled and drawn by the Division of National Mapping of the Department of National Development, and printed by A.J. Arthur, the Commonwealth Government Printer in Canberra. The map includes what is in effect an advertisement for the Trans-Australian Railway, with a picture of the train above the line, the only one to be picked out by name and boldly so. Moreover, above the picture is the motto 'Travel in Comfort by Fast Diesel Electric Air Conditioned Trains on the Trans-Australian Railway'.

Opened in 1917, the service operated between Port Augusta and Kalgoorlie in Western Australia as part of a longer service that was extended to run as far as Port Pirie to the east from 1937, and then on to Adelaide in 1982; while, from 1969, it reached Perth to the west. This was possible as a result of an end to gauge changes. The route to Kalgoorlie across the arid, desolate Nullarbor Plain includes the longest straight stretch of track in the world – 297 miles without a curve. The Trans-Australian Railway relied on sleeping accommodation until the 1960s, and in 1975 it was absorbed into Australian National (AN). Due to competition from air services, the service was cut to three weekly in 1977 and was cancelled in 1991.

The map uses a four-colour key to present the different gauges employed, and differentiated between railways built (solid line) and those under construction (broken or dotted line). The inset route mileage table broke down the total of government railways (as opposed to private lines) by state.

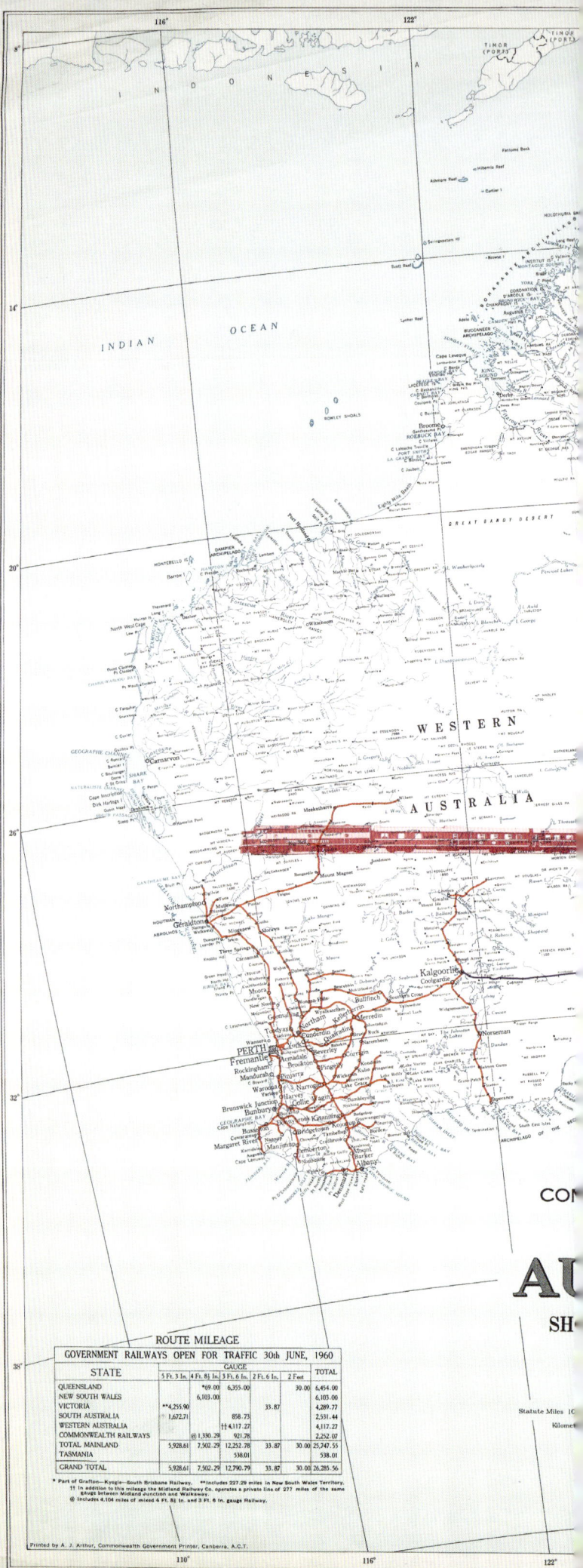

A FADING SYSTEM
Pennsylvania Railroad, 'Regional Map', 1955, and
W. James Shaw, 'Major Rail Network: 1966', 1970.

Founded in 1846, the Pennsylvania Railroad (PRR) was the largest railway in terms of traffic. Developed in part by the acquisition, merger and part ownership of at least 800 other railways, PRR was in difficulties in the 1950s due to serious loss of revenue and its reliance on costly steam locomotives, whose support facilities, notably for coaling, watering and maintenance, were expensive. As a whole in the United States, revenue freight traffic in millions of passenger miles and net ton miles was lower in 1960 than in 1925 or 1944. This encouraged recourse to a number of remedies, including electrification, containerisation and, for the PRR, a new regional organisation, for which this 1955 map (right) provides the clear guidance of a simple colour palette. Containerisation had the virtue of greatly easing the problems caused by the managing of freight. So also with the use of largely automatic processes of freight wagon classification.

The adventure and romance of American train travel was captured by Ian Fleming in *Live and Let Die* (1954) with a description of The Silver Phantom service from New York to St. Petersburg, Florida. Fleming based this on a train known as the Silver Meteor, begun in 1939 as the first diesel-powered streamliner to Florida and operated by Amtrak from New York to Miami, with the west coast Florida, Silver Phantom, section lost in 1968:

> 'It lay, a quarter of a mile of silver carriages ... up front, the auxiliary generators of the 4000 horsepower twin Diesel electric units ... the usual American train-smell of old cigar-smoke ... The crack train thundered on ... streaked past ... the steady gallop of the wheels ... pounding out the miles.'

The romance of train was also to the fore in Alfred Hitchcock's highly successful film *North by Northwest* (1959), with a romantic journey on the 20th Century Limited express, which ran from New York to Chicago from 1902 to 1967, a train that had had streamlined train sets from 1938.

Long-range passenger services moved between major termini, such as the New Orleans Union Passenger terminal completed in 1954, but services were greatly hit by aircraft competition. The onset of air travel contributed to a crisis in the rail passenger system in America, where it ceased to be viable across much of the country. In 1952 an American Airlines advertisement had contrasted a trainbound Don, who 'made the trip – too late again', and his younger colleague John, who 'made the sale – he took the plane'. The development of bus companies, such as Greyhound, operating on interstate and regional highways, also proved important.

Individual rail companies, big and small, reflected these pressures. For example, the requirements of the Acme Plaster Company's plant in Quanah, Texas, had led to the founding in 1902 of the Acme, Red River and Northern Railway. Having done well in every decade except the Depression-hit 1930s, it declined rapidly in the 1960s in the face of road freight competition, and was abandoned in the early 1970s.

The crisis of rail at the national level was seen in the cutting back on the maintenance of track and equipment, which led to a slowing down of services. The bankruptcy of major companies included the Penn Central in 1970, the Rock Island in 1975 and the Milwaukee Road in 1977. As a result, there was the establishment of Amtrak (a government-funded body that took over inter-city passenger services in 1971) and Conrail (a public body that from 1976 ran the lines of many bankrupt companies). This was followed by the deregulation provided by the Staggers Act in 1980, which was to help make the freight system profitable, indeed highly profitable in the case of many lines. Conrail cut its network and was successfully privatised in 1987.

This major transportation map (following pages) of primary and secondary railroads was produced by the Geological Survey of the Department of the Interior in 1966, and was published in 1970 as part of *The National Atlas of the United States of America*. It depicts the network as being dense in the

east, and more particularly the Midwest and the eastern Great Plains, but far more limited further west. The map also displays political boundaries, time zones, topography, bodies of water, drainage, coastlines and islands. Relief is shown with shading. Two inset maps show Alaska and the principal islands of Hawaii.

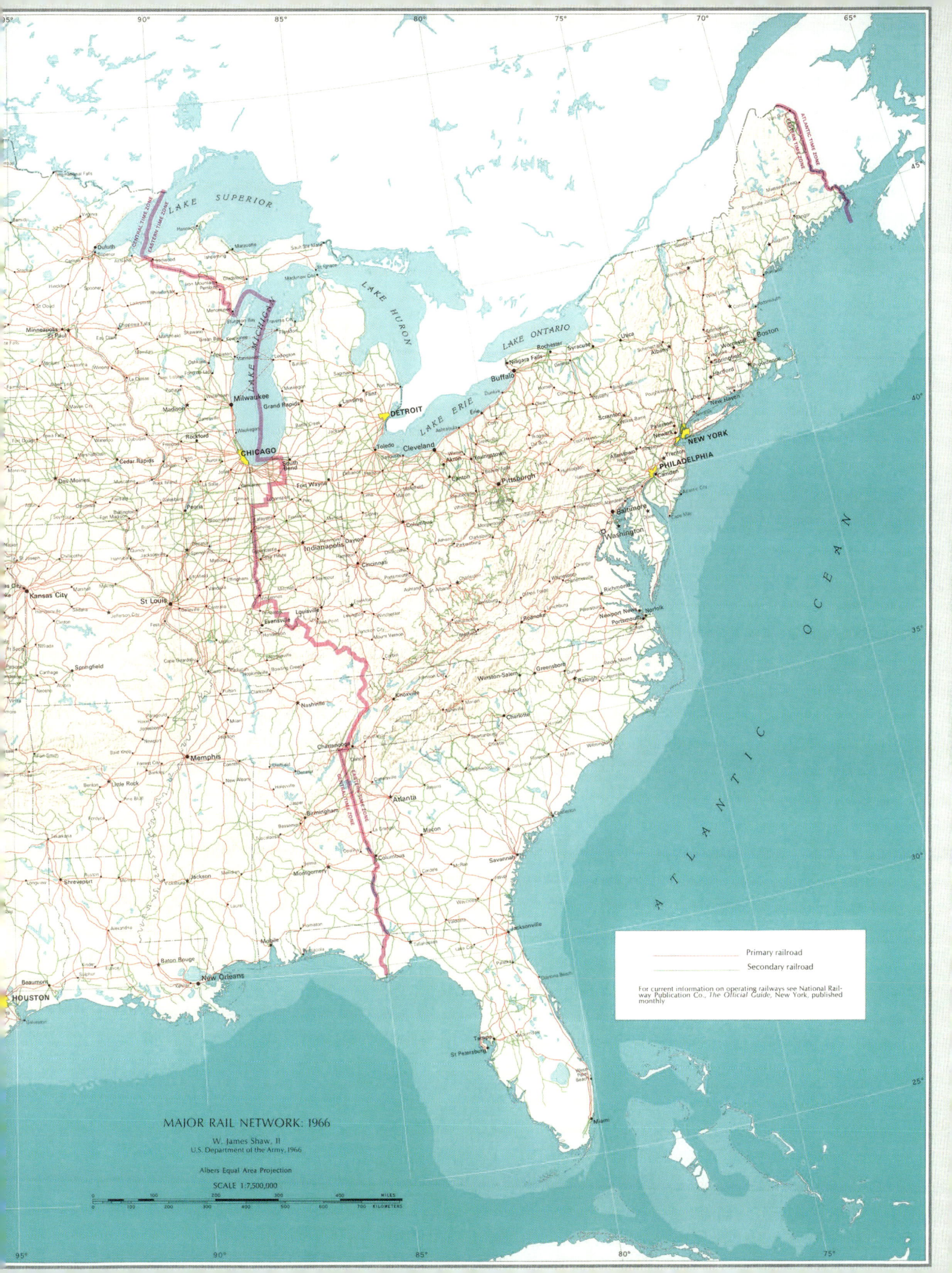

CHANGING THE TRAIN SYSTEM
British Railways, 'Liner Train Routes and Terminals Under Consideration', 1963.

The Reshaping of British Railways (1963), a report better known as the Beeching Report after Dr Richard Beeching, the Chairman of the British Transport Commission from 1961 to 1965, and the first Chairman of the British Railways Board, proposed major changes to the network. Maps were employed to depict current usage for freight and passengers, and maps were used to suggest remedies, not only in the shape of closures and service modifications but also in terms of proposed new facilities aimed at helping bulk freight flows. Map No. 11, shown here ('Liner Train Routes and Terminals Under Consideration), tends not to attract attention, but it is a modernising attempt to shape an efficient and productive network linked to other aspects of the economy. It anticipated the container revolution. Moreover, the Treasury was pressing for clear financial targets and economic charges by nationalised industries in order to limit their debt and investment requirements, and the pressure thereby put on the rest of the economy through higher public expenditure.

A government-imposed ban on fare and freight increases led to a serious fall in net earnings in 1945–47. Nationalised in 1948, the railways were placed under the British Transport Commission, which was also made responsible both for London Transport and for a large amount of road transport. The British Transport Commission was supposed to integrate the transport system and rebuild it after wartime damage and a serious lack of investment due to wartime and post-war financial difficulties. However, such integration was not pursued, not least because road haulage was denationalised in 1953. Moreover, British Rail lost money from 1953, with the losses rising from £15.6 million in 1956 to £42 million in 1960; although how much was lost depends on which accounting convention is used and there have been significant disagreements on this issue. There was also concern that the £1,200 million Railway Modernisation Plan launched in 1955, under which steam was being replaced by diesel and electricity, was being seriously mismanaged. The British Transport Commission was abolished under the Transport Act of 1962 and replaced, for rail, by the British Railways Board, which operated from 1963 to 2001.

The Beeching Report showed the role of rail in the different but complementary strategies of modernisation. It was commissioned under the Conservative government of Harold Macmillan, which had a clear corporatist stance – as shown by the National Economic Development Council, established in 1962 and instructed to create a national economic plan. In turn, the report was carried forward from 1964 by the Labour government of Harold Wilson, which focused on technological modernity and had scant time for steam trains, rural services and those that were not used intensely. Instead, its new Department of Economic Affairs and Ministry of Technology embraced the new. Under Labour, the Beeching-inspired closures continued, while expenditure on rail modernisation decreased. Labour, however, had rejected Beeching's second report, issued in February 1965, in which he recommended that of the 7,500 miles of trunk railway, only 3,000 miles 'should be selected for future development', with traffic concentrated on nine lines, and the East Coast Main Line closed north of Newcastle.

Beeching represented a more general atomisation of a system, because assessing the profitability of each section – line, station or rolling stock item – independently meant failing to understand the value of each part to the network as a whole. *The Daily Telegraph* of 14 August 1961 had recorded other signs of change, with British Railways revealing that it was proposing to redevelop the main stations at Edinburgh and Glasgow. Thus stations were to be part of bigger mixed-use complexes.

Although public investment remained focused on roads, the national train system

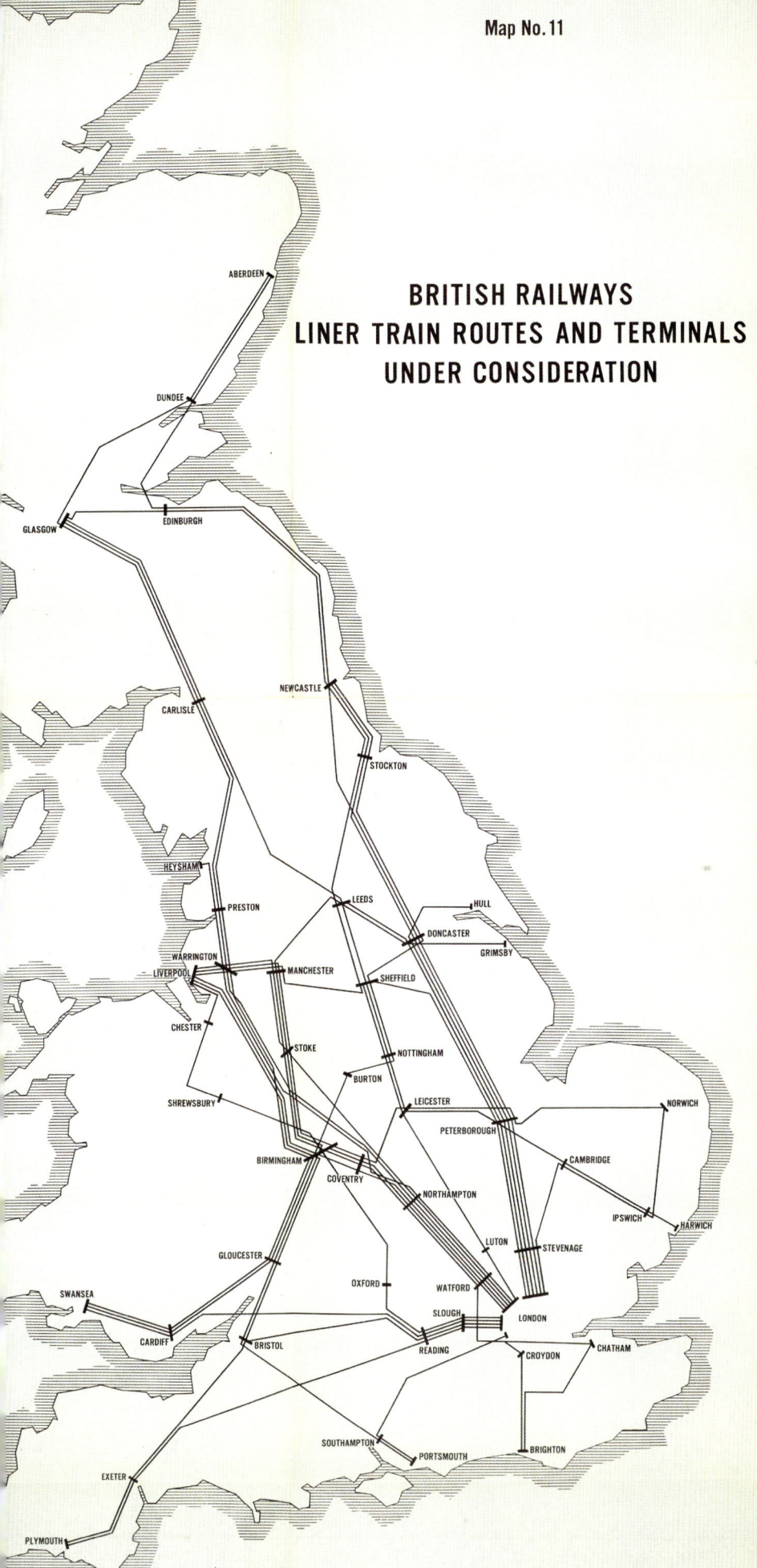

Map No. 11

BRITISH RAILWAYS
LINER TRAIN ROUTES AND TERMINALS
UNDER CONSIDERATION

did not experience a revival of the core system after the closure of branches, but in the following years it was spared a repetition of Beeching, notably the dramatic cuts proposed as one option in the unpopular Serpell Report of 1982. These would have reduced the system to 1,600 miles, essentially a few major routes, notably from London to Cardiff, Newcastle and Scotland via Manchester. Instead, the system, with passenger numbers reviving from the mid-1980s, stabilised at just over half the interwar mileage (10,261 miles).

Furthermore, there were important improvements. The InterCity 125, a high-speed diesel train capable of travelling at 125mph, was introduced in 1976 on the Paddington to Bristol and South Wales routes and then spread to other non-electrified main lines. The tilting Advanced Passenger Train did not prove a success. In contrast, on the East Coast Main Line, from London to Edinburgh, fully electrified 225s, capable of travelling at 140mph were introduced in 1991.

5. WAR AND THE AIR CHALLENGE 1939-70   211

SIGHTSEEING

British Rail, 'Get Out and About on the Southern', 1966.

The use of pictorial techniques to advertise rail services frequently involved new versions of old devices, as in this 1966 map's employment of pictograms. The stress is on what could be seen from the train, with the result that the train is seen as a means for leisure and speed therefore is not at a premium. Such advertisements built on earlier publications, such as the guides published by the Great Western Railway, for example the *Cathedral Line of England* (1908). The emphasis in the map was not on travel for work (the basis, in practice, of the Southern Region), which would have placed a premium on speed and reliability. The sybaritic MP Henry 'Chips' Channon had been atypical in writing in 1950: 'I came up [to London] from Chichester by train. I hardly ever travel by train and hate them; one gets claustrophobia.'

Any map recording the use of the network would have shown it was concentrated on commuter services, and that times of travel are a key factor. Instead, 'Get Out and About' drew on the Londoner's idea and phrase of getting 'out' of town. The pictograms did not emphasise trains (only one, a two-carriage train, is depicted) or stations. Instead the map tries to capture the essence of a location through an image of a building or leisure activity: the beach huts at Bognor Regis, cathedral in Salisbury, zoo at Chessington or Royal Pavilion at Brighton. Places of interest were also to the fore in the comparable British Rail map of Essex, Suffolk and Hertfordshire.

The Southern Region had not done too badly out of the Beeching cuts that proved a background to this map. Having electrified early, the Southern Region was challenged less than the other regions by the technological and organisational changes of the 1960s. The first Beeching report, *The Reshaping of British Railways* (1963), had recommended the closure of all lines on the Isle of Wight, but the Ryde–Shanklin line was reprieved. However, the Cuckoo, Cranleigh, Steyning, New Romney and Bexhill West lines were all cut.

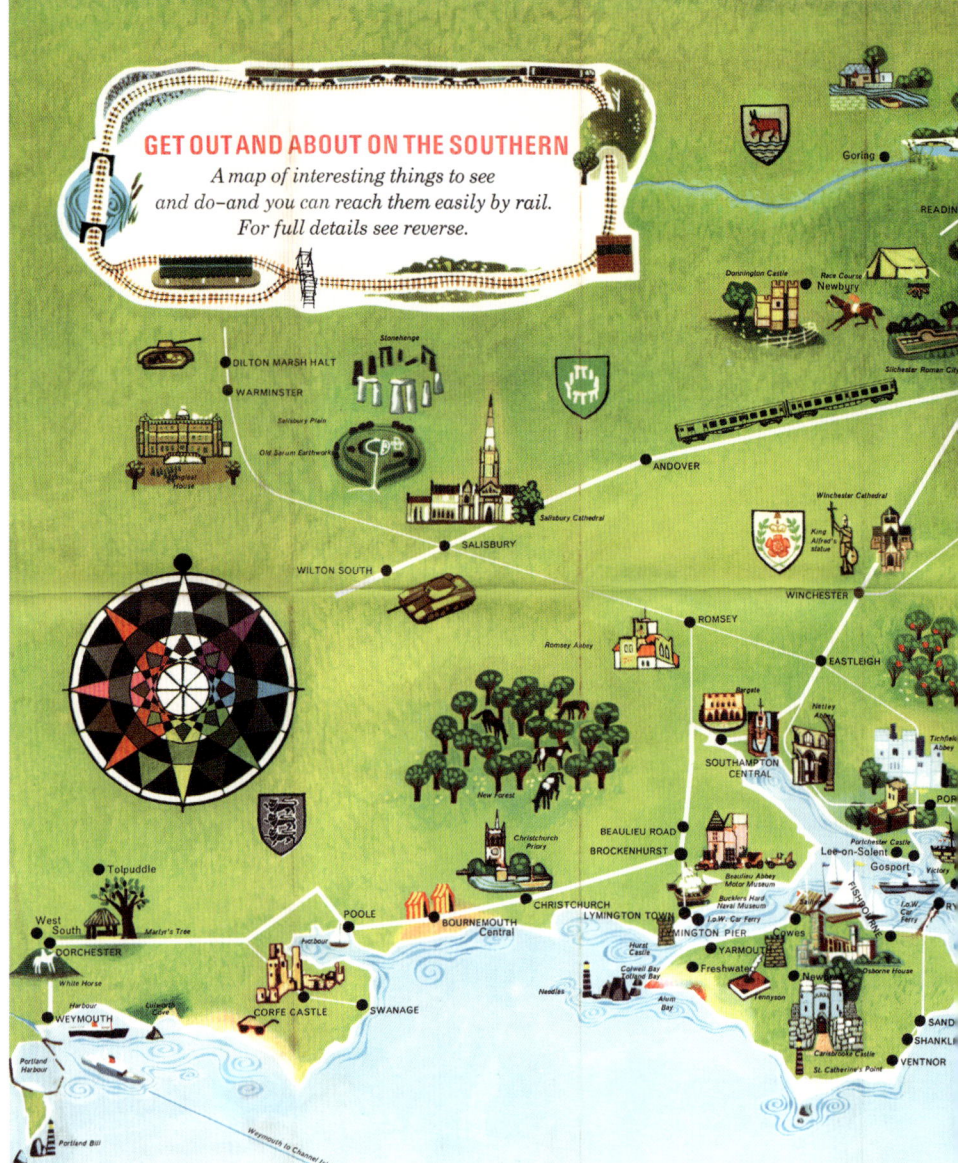

Beeching's second report, that of February 1965, identified major trunk routes, which included the London to Brighton, London to Portsmouth, and London to Bournemouth via Southampton routes, as well as a line from Reading, via Guildford and Ashford, to Folkestone and Dover. Fortunately for Southern Region commuters, the further cuts in that report were not pursued.

Very differently, the emphasis was on terrain and places of interest in *A Map of Yorkshire* (1949) produced by Estra Clark for the newly established British Railways. The railways were shown and, for Darlington, the *Rocket*. Doncaster was described in terms of its locomotive and carriage works, and the map carried the arms of the London and North Eastern Railway. A modern steam train was shown.

ELECTRIFYING ITALY
Ferrovie dello Stato (FS), 'State Railway Network', 1959.

The impact of electrification is captured in this 1959 map of Italian railways from the annual report of that year by Ferrovie dello Stato (FS). The effort reflected by this map can be understood only if the extent of heavy wartime destruction is appreciated: aside from bombing, there was bitter ground fighting from 1943 to 1945 and much of the track and many of the bridges had to be extensively repaired or replaced.

The fundamental distinction is between (blue or red) electrified (*trifase* means three-phase and *continua* indicates electrification 'ongoing') and (black) non-electrified lines, because there are few double-tracked non-electrified lines. In contrast, the major routes have not only been electrified but also double-tracked. This is true of the Milan–Bologna–Florence–Rome–Naples axis in its entirety. Milan, the economic centre, rather than Rome, the political capital, emerges as the prime node of the system. The ports are still major rail centres, notably Genoa. Italy's industrial triumvirate – of Genoa, Milan and Turn – is quite visible, albeit it lacks an electrified transportation triangle because of the non-electrified Milan–Turin line, two cities connected by motorway in 1932. In contrast, rural centres in the poor south such as Potenza, Cosenza and Catanzaro, are not the beneficiaries of electrification, which is true more generally of southern Italian society. Similar maps could be produced for other systems, but in general there is an emphasis on the network, rather than on the qualitative nature of the system. This can be highly misleading.

In Italy, significant breaks in the network continued, notably between Sicily and mainland Italy, a route greatly affected by earthquake risks. Plans for the replacement of the ferry route between Messina and Reggio di Calabria by a bridge across the Strait of Messina were advanced in detail from the 1990s and rejected by Parliament in 2006, despite being strongly pushed from 2008 by Silvio Berlusconi, who had Sicilian political interests. The plan, which has so far proved fruitless, is for a road-rail bridge, the latter to be linked to a proposed high-speed railway from Naples to Reggio di Calabria. In 2013 a lack of funds led to the abandonment of the project, but it has been discussed since, notably in 2020 by then prime minister Giuseppe Conte, and again in 2023 by the Meloni government.

Italy's main rail corridors are north–south, both west and east of the Apennine Mountains, and east–west, the latter from Turin via Milan and Venice to Trieste in the north and from Naples to Bari in the south. These corridors have been important from the early days of planning Italian railways. Links to and between ports, such as Genoa, Trieste, Naples and Bari, have always been significant.

In 1977 Italy opened its high-speed Treno Alta Velocità (TVA) network with the Direttisima line between Florence and Rome. The TVA network was completed in 1992 and also designed to increase freight capacity by night. In 2018 the first high-speed freight rail service in the world began operating, between Bologna and Caserta near Naples.

STATION BUFFETS
SNCF, 'Regional Cuisine is One of the Wonders of France', 1954.

This 1954 map by SNCF (Société nationale des chemins de fer français) captures an unfamiliar aspect of rail travel, notably for British and American travellers. Arguing, with reason, that regional cuisine is one of the marvels of France, the map provides illustrations of key dishes, such as the duck of the southwest, apples and dairy for Normandy, sauerkraut and sausages for Alsace, and seafood and oysters for the Atlantic coast. For each station listed, the name of the restaurant concessionary is given, as is the price of the set *menu touristique* and that of a standard dish. The central place in the system is Paris, and the assumption is that the poster informs Parisians of rural delights. The map captures the French idea of the *pays*, the fixing of regional identity in part in terms of food, and the degree to which from the 1950s leisure was of increasing importance for rail travellers, with passengers given further choice by the ongoing expansion of road transport. Drawing on the strong French tradition of poster art, SNCF proved effective as an advertiser.

After large-scale closures of narrow-gauge railways from the 1930s and Depression-era financial difficulties for many private rail companies, the Popular Front government pushed through nationalisation in 1938, establishing the SNCF to run the railway, with the government then owning 51% of the company. The railway was initially organised on a regional basis. Heavy wartime damage contributed to the post-war difficulties, but a more positive note was pushed by the transition away from steam. The train shown in this 1950s' map (left) is electric.

Another instance of French rail modernity was provided by Rudi Meyer's 1976 map (see page 13) of the mainline national rail network in which no boundaries by land or sea were provided, while space was effectively shaped by the rail hierarchy. Meyer was an imaginative Swiss graphic designer who produced effective poster designs and logos, and in 1964 established a School of Design in Paris. His commitment to graphic design and infographics was important to what became the general character of more and more railway maps. This change – to a new type of diagrammatic map – poses a challenge to collectors, readers and libraries familiar with more conventional methods and forms; and, indeed, as a result to those planning books such as this one. However, the tendency to focus on the standard earlier format can be highly misleading.

France has sought to present its railways as modern, a process that was accelerated with the commitment to high-speed TGV trains. Nevertheless, there was also a reality of an expensive, heavily unionised service that was prone to strike action. France has had a rail strike every year since 1947, with particularly major ones in 1995, 2019–20 and 2013 over opposition to pension reforms. Fortunately, there has been no repetition of an episode in 1961 not recorded in train maps: the Organisation armée secrete (OAS), an extremist terrorist group of French settlers in Algeria opposed to independence for the latter, carried out a bomb attack on the Paris–Strasbourg express that killed 28 people and injured over 100. In 2023 there were no casualties when, in protest against pension reforms, a puppet of President Macron was laid on rail tracks in Nice to block a TGV service.

BUFFETS	Concessionnaires	PRIX	
		menu touristique	Autour d'un plat
Agen	M. Charbonneau	750	400
Aix-les-Bains	M. Hybord	750	400
Angoulême	MM. Baby Père et Fils	750	350
Annecy	M. Rassat	750	400
Avignon	M. Silvestre	800	400
Aulnoye	M. Janssoone		400
Bar-le-Duc	M. Chebret		400
Bayonne*	M. Arnaud	750	350
Belfort	M. Magat	850	400
Béziers*	M. Colson	800	400
Bordeaux-St-Jean*	M. Mégret	850	450
Bourg	M. Gonon	800	450
Bourges	M. Maillard		350
Brest	M. Picot	750	400
Brive	M. Dykczyk		350
Caen	M. Ecuvillon	800	400
Calais-Maritime	Cⁱᵉ Intˡᵉ Wagons-Lits		350
Cerbère*	MM. Deléon et Fonters	750	350
Chagny*	Mme Laclède	800	450
Chalon-sur-Saône	M. Philippon	750	350
Châlons-sur-Marne	M. Le Bris		400
Chambéry	M. Paupardin	750	400
Chantilly	M. Berrux		350
Chartres	M. Boyer		350
Châteauroux	M. Pounel		350
Châtellerault	M. Dupin		400
Chaumont	M. Huguier		450
Clermont-Ferrand	M. Bru	800	400
Colmar	M. Antzemberger	850	450
Compiègne	M. Hautcœur		350
Dieppe	M. Lorite	750	350
Dijon	M. Parizot	800	450
Dinan	M. Guizlen		350
Dreux	M. Dupuy	800	400
Dunkerque	M. Meurette		400
Épernay	M. Tollet	850	400
Grenoble	M. Colard	800	450
Hendaye	Mme Courrèges	750	400
Hirson*	M. Sol		350
Jeumont*	M. Berné		400
Langres	M. Tamisier		350
Laon*	M. Supply	800	400
La Rochelle*	M. Kuhn	750	400
Le Havre	M. Bouchard	850	450
Le Mans	M. Bouveret	750	400
Lens	Mlle Desauty		350
Lille*	M. Thésio	950	500
Limoges	M. Jonvaux	800	450
Longuyon*	M. Bauzon		350
Longwy*	M. André		350
Lons-le-Saunier	M. Cleyet		350

* Buffet-hôtel.

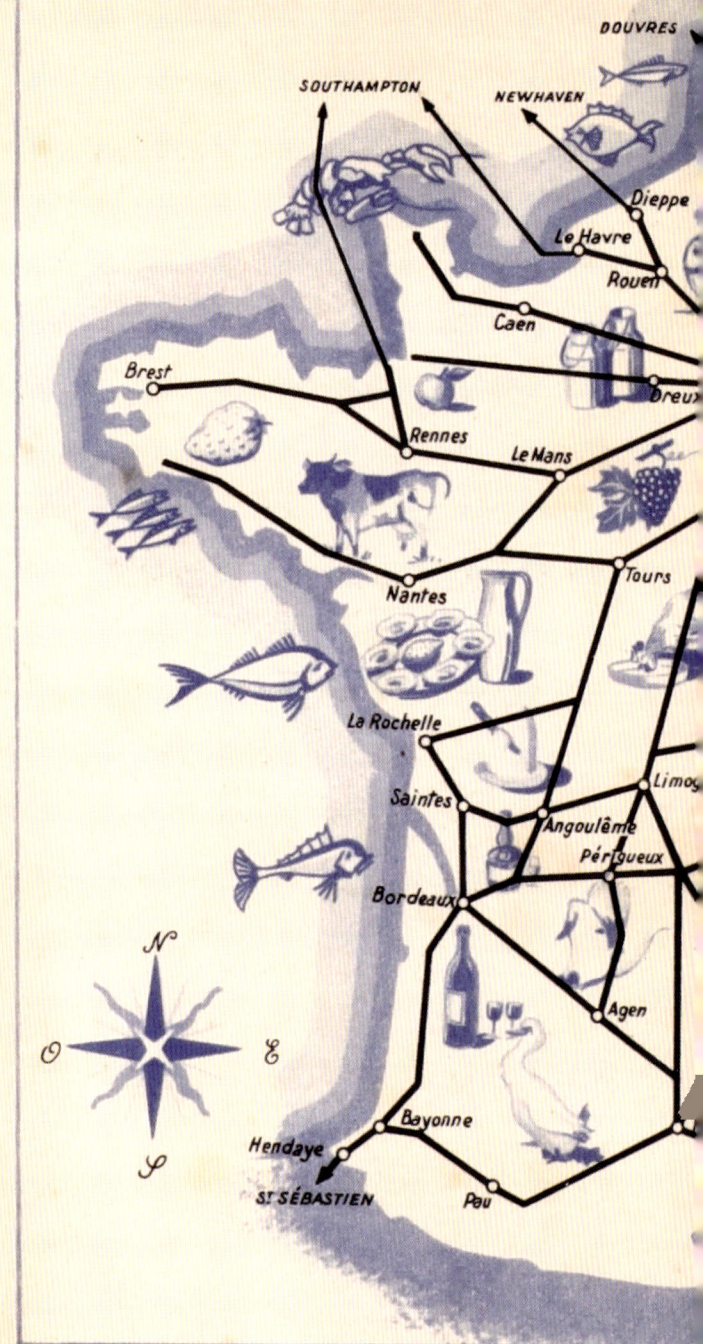

LA CUISINE RÉGIONALE EST UN

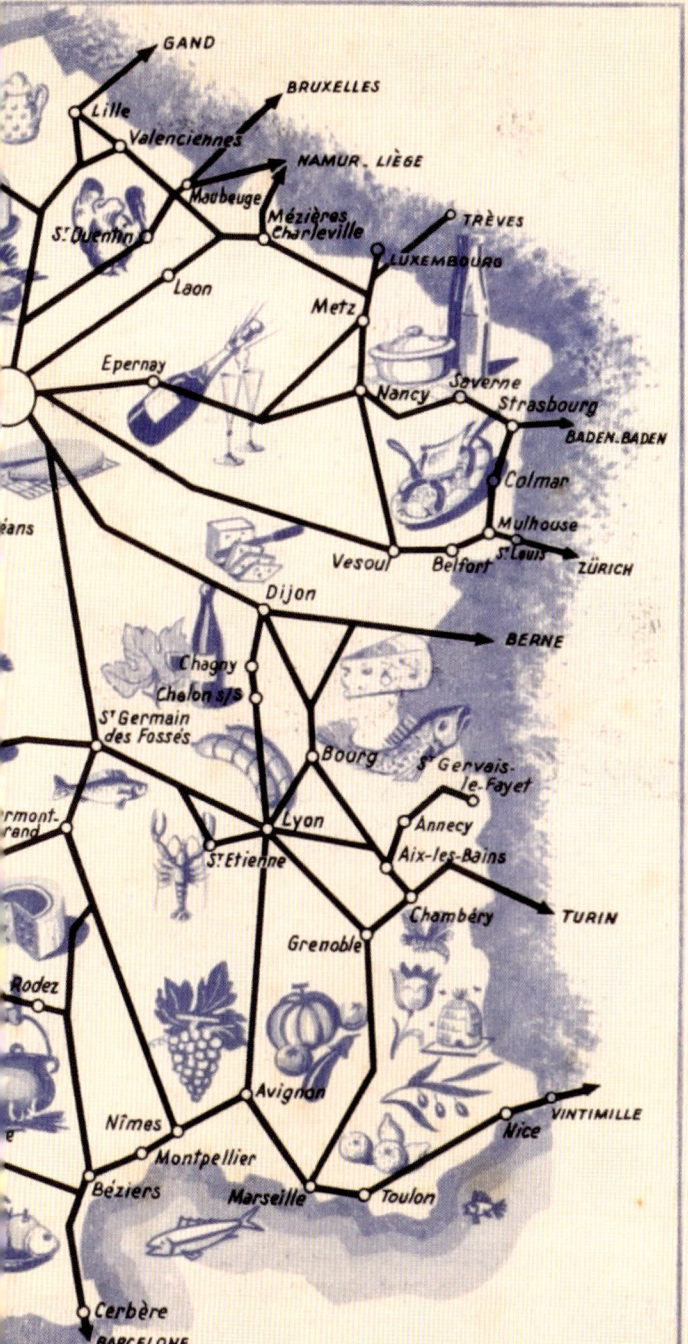

DES MERVEILLES DE LA FRANCE

BUFFETS	Concessionnaires	PRIX	
		menu touristique	Autour d'un plat
Lyon-Brotteaux	S.N.C.F.	800	400
Lyon-Perrache*	S.H.T. (Hôtel)	915	450
Lyon-Perrache	S.H.T. (Buffet Quais)		400
Marseille-St-Charles*	S.H.T.	915	400
Maubeuge*	Mme Merciant	800	400
Metz		850	450
Mézières-Charleville*	M. Auffret	850	400
Modane	Mme Chabert		350
Montargis	M. Avandet		400
Montluçon	M. Daureyre		350
Montpellier	M. Ribes	750	350
Mulhouse	M. Greth	850	450
Nancy	M. Friedel	800	400
Nantes	Mme Suisse	750	400
Nice	M. Donarel	800	400
Nîmes	Mme Silvestre	750	350
Orléans	Mme Paliès	750	350
Paris-Austerlitz	M. Rouillon		400
Paris-Montparnasse	M. Pringuet		400
Paris-Nord	M. Hazard		450
Paris-St-Lazare	M. Auset		450
Pau	M. Blois	800	400
Périgueux	M. Bonnet	750	350
Perpignan	M. Mouche		350
Poitiers	M. Souchaud		350
Quimper	M. Vilain		350
Reims	M. Bouteillez		350
Rennes	Mme Bouveret	800	400
Rodez	Mme Fraux	800	400
Rouen (rive droite)	M. Le Jan	800	400
Saintes	M. Daveux	750	350
Sarrebourg	M. Klintz		350
Saverne	M. Billé	750	400
Sélestat	M. Mermet		400
Sète	M. Mauran		350
Strasbourg	Mme Schmitt	850	500
St-Brieuc	M. Gros		400
St-Dié	M. Boyon		350
St-Étienne	M. Douziech	800	400
St-Germain-des-Fossés	M. Moine	750	350
St-Gervais-le-Fayet	M. Grosset	750	350
St-Louis (Haut-Rhin)	M. Stiegler	800	400
St-Quentin*	M. Bouteville	915	450
Thionville	Brasserie Basse-Yutz		500
Toulon	M. Cormenier	800	400
Toulouse*	C¹ᵉ Int¹ᵉ Wagons-Lits	915	450
Tourcoing*			400
Tours	M. Benoit	750	450
Valenciennes*	M. Drumont	850	450
Vesoul	M. Hadey	750	350
Vierzon	M. Cellier		350

MODERNISING SPAIN

RENFE, 'Spanish Railways', 'Density of Circulation', 1968 and 1971.

These two maps, for the same network, provide a vivid contrast in terms of usage as opposed to extent. The same was the case in other countries. This highlights the mistake in assuming that the network provides guidance to the usage of particular parts of it and to the relevant managerial issues.

Heavily damaged during the Spanish Civil War of 1936–39, the rail network had been nationalised by the Franco government to form RENFE (Red Nacional de los Ferrocarriles Españoles) in 1941 (now Renfe-Operadora). But, in a recurrence of the situation in the 1920s and early 1930s, there was a shortage of investment in the 1950s, a period in which Spain was poor, and improvements had to wait until the more prosperous 1960s, not least with the introduction of Talgo coaches and engines in 1964.

In 1963 RENFE had begun to produce a series of booklets, which were published annually until 1989. The booklet included various quantitative maps to illustrate the density of rail traffic through a combination of different colours and line thicknesses. This 1968 map (right), distributed and marketed by the publishing house Paraninfo, largely follows the design used by previous ones of Spain's national rail network: a white background overlaid with the red-coloured standard-gauge tracks and the black-coloured narrow-gauge tracks. The main urban centres are also shown as inset maps. The 1971 map (following pages) shows the density of circulation based on average daily numbers of *mercancías*, goods or freight (yellow), and *viajeros*, passengers (green). From the seven insets provided it is noticeable that there is a higher proportion of goods traffic flowing between Madrid and the major cities and ports of northwest Spain.

In 1971 RENFE introduced its double-arrowed logo. Usage that year was greatest in Catalonia, the most prosperous region, and also from the Basque Country to Madrid and then on, via Córdoba, to Seville. These pressures were to influence the long-distance, high-speed AVE (Alta Velocidad Española) rail system, for which construction began in 1988.

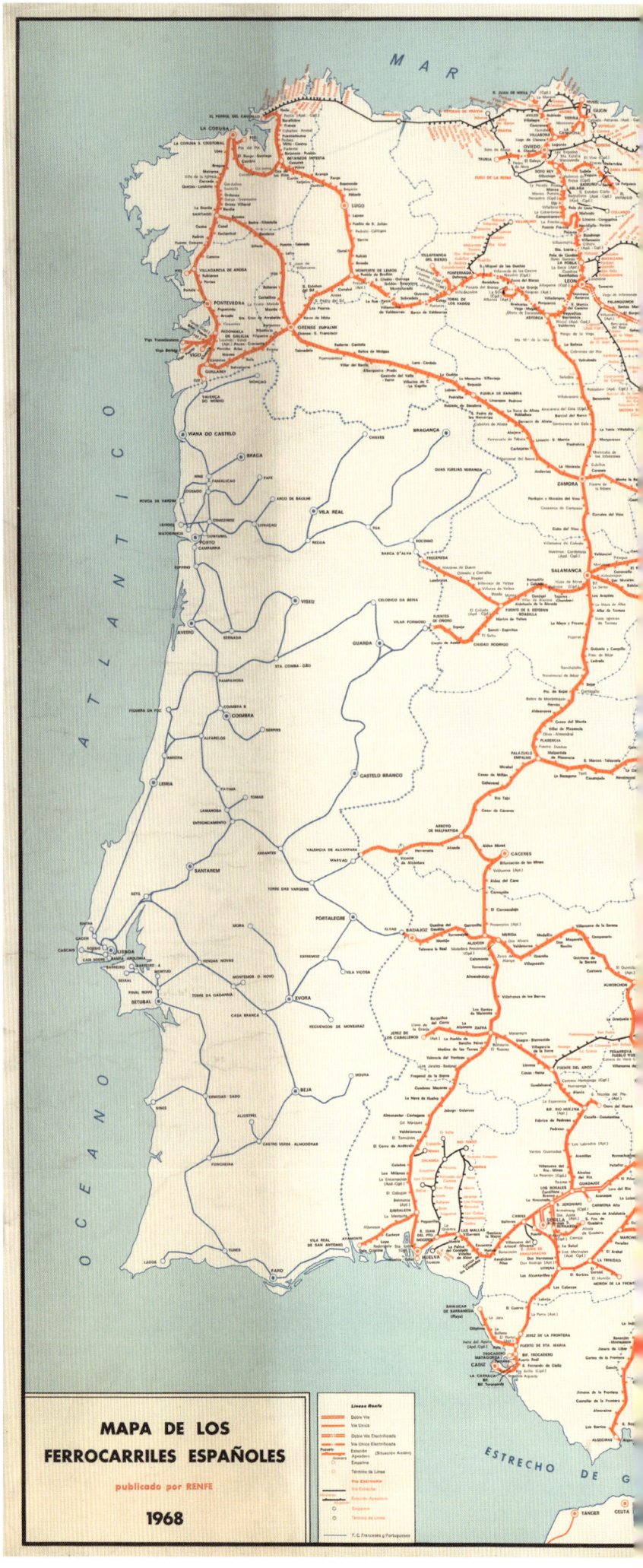

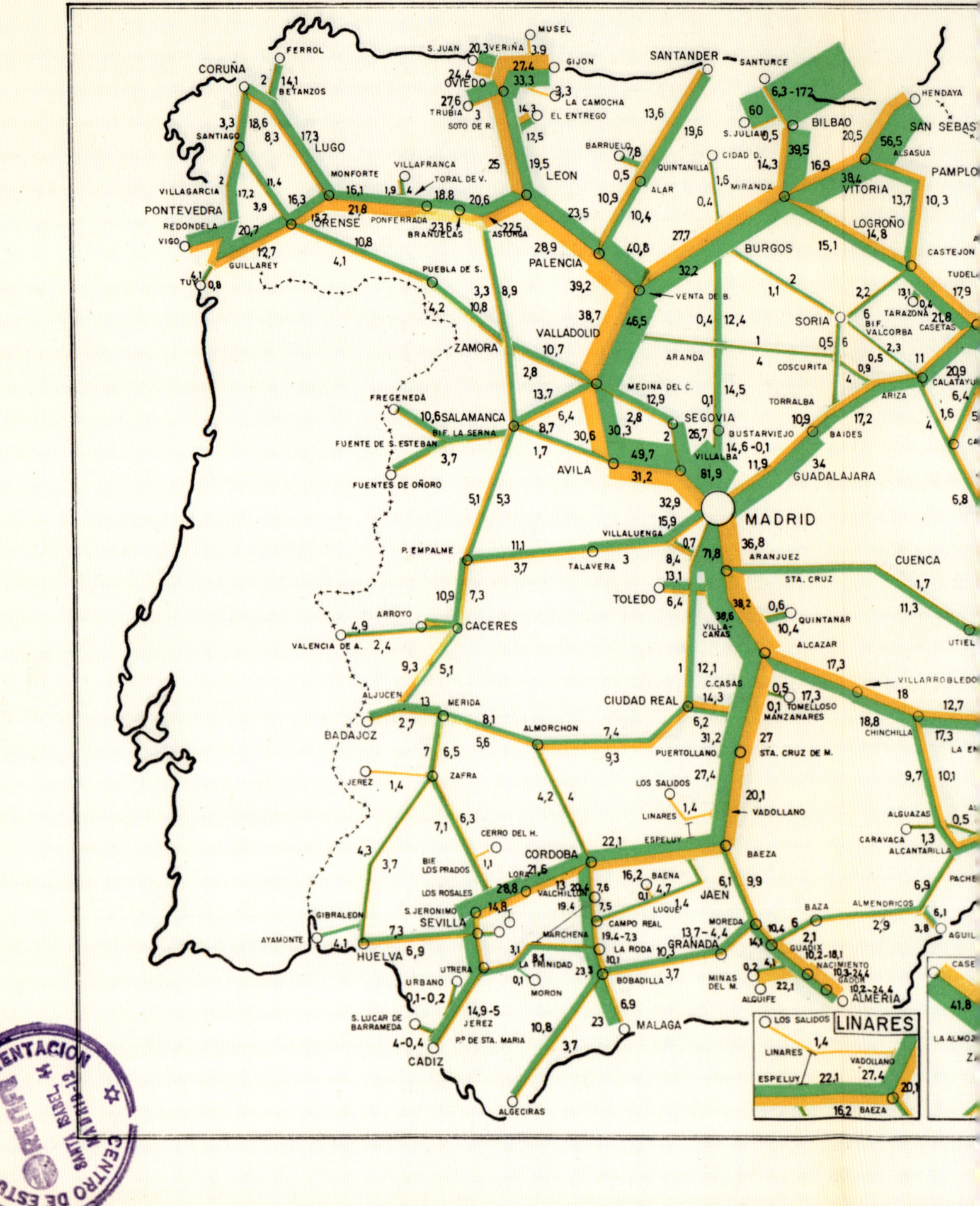

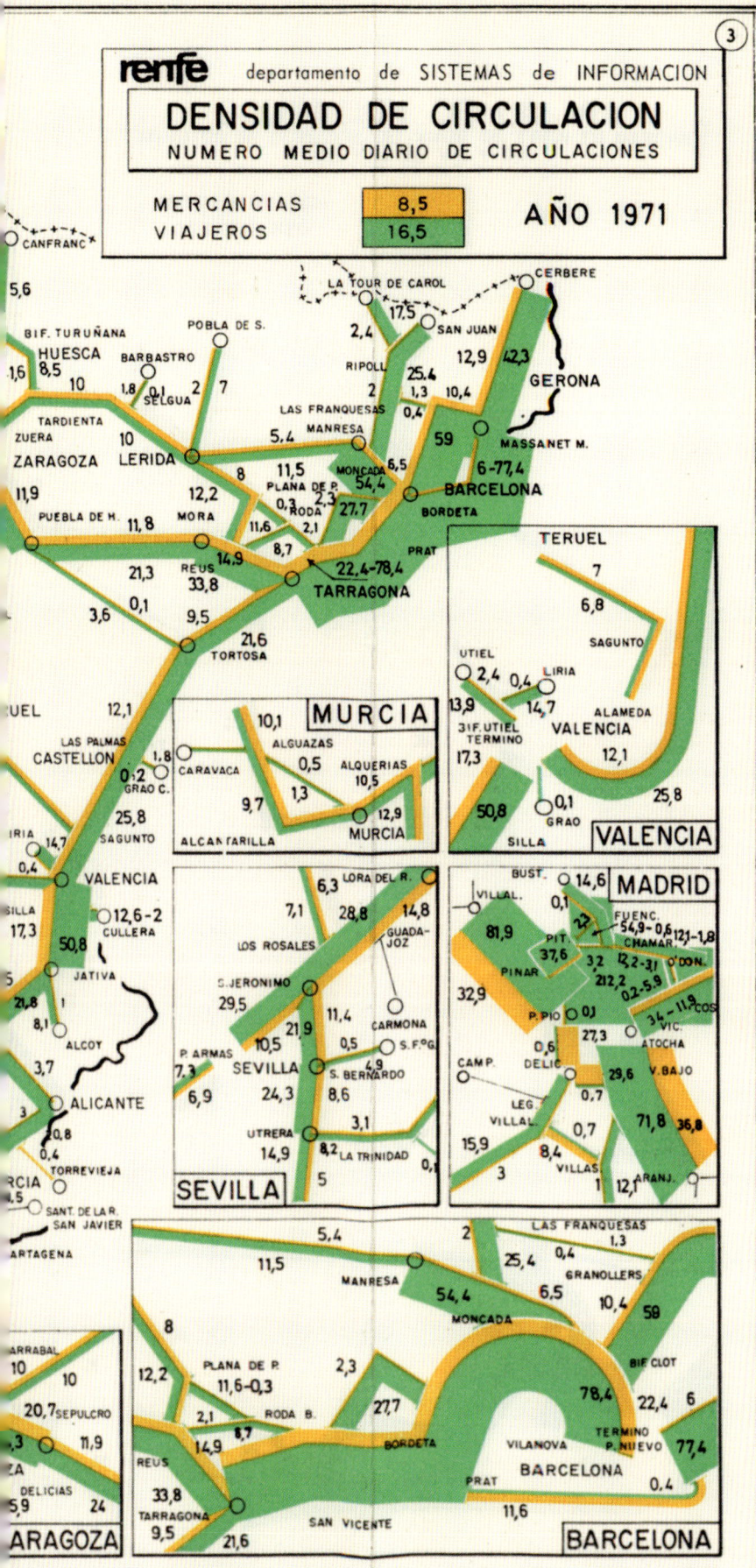

Part of the underlying political motivation for the major upgrading of Spanish railways with the AVE was opposition to regional separatism. The governing Socialist Party was willing to spend very heavily, and notably on transport. In 1992, the same year as the Seville Expo, the AVE line opened linking Madrid to Seville. In 2015 the line was extended to Cádiz. Crucially, the line from Madrid to Barcelona via Zaragoza opened in 2008 and the 386 miles are covered in two and a half hours. The third line (Madrid to Toledo) opened in 2005, and others followed to Valladolid (2007) and Valencia (2010). There are also AVE connections with France. The opening in 2007 of an AVE line from Málaga to Córdoba makes it possible to travel from the Costa del Sol to Madrid, then on to Barcelona. The daily number of AVE passengers rose from 4,878 in 2006 to 22,370 in 2019, before being hit by Covid-19.

By December 2021 Spain's high-speed system, at 2,251 miles, was the longest in Europe. Like those elsewhere, it reflected a process of concentration. Thus, the AVE trains, of which there were 96 in the fleet, called at only 30 stations. Essentially, they offered high-speed city-to-city travel, and indeed the AVE operated on a different gauge to that of the standard Iberian broad gauge. Lines included a northern corridor, with services from Madrid to Asturias and the Basque Country, with work due to start in 2023 on a line to Santander. In 2005 a second expansion plan was designed to ensure that a 10,000-kilometre (6,200-mile) network was constructed. This plan has been indefinitely postponed, but it would have included a Trans-Pyrenean Central Corridor from Zaragoza to Toulouse via Huesca and a new tunnel through the Pyrenees.

INTERNATIONAL RAIL TRAVEL
Information and Publicity Centre of the European Railways, 'Trans Europ Express', 1965.

Developed in the late nineteenth century with impressive through services and an enviable reputation for luxury, international rail travel had been hit hard by both world wars and by post-war restrictions on travel in each case. In turn, there were significant attempts to reinstate international rail travel in the 1920s, and then again in the 1950s. In 1922 new trains colloquially known as Train Bleu (Blue Trains) were put on the run from Calais to the Côte d'Azur on the French Riviera. These Blue Trains were an important part of the post-war programme of the Compagnie internationale des wagons-lits (International Sleeping Car Company). Hitherto, sleeping cars had been constructed with a framework of wood resting upon a chassis of metal. In the new cars, wood was only used for interior decoration, and the framework, partitions, ceiling and floor were all made of steel. The first sleeping cars, built in 1876, had had four berths per compartment, and later ones had two. The new 1922 trains had, as formerly, single-berth compartment coaches each taking 16 passengers but with the total length increased from 66 feet to 77 feet.

The 1950s' attempt to reinstate international rail travel focused on the Trans Europ Express (TEE), a network founded in 1957 by the rail systems of France, Italy, the Netherlands, Switzerland and West Germany, with Belgium and Luxembourg joining in 1964 (which helps to provide a background for this 1965 map), and Spain, Denmark and Austria also joining. First class only, and charging an additional supplement, the express trains were aimed at business and wealthy travellers.

The map lists various routes, notably between Italy and France, Switzerland and Germany, but also captures the greatest concentration of provision, that between Paris and Brussels. The TEE network reached a peak in 1974, but ended in 1995 as other services replaced it, not least the growing role of domestic high-speed trains that also carried second-class passengers and were controlled by national carriers, thus being politically more acceptable in terms of expenditure and subsidy.

International train travel was also affected by the changing nature of movement for business and wealthy travellers. Air travel became easier within Europe, with business-class seats and airport lounges providing a sense of exclusivity that trains could not always, or even generally, match. Jet aircraft made air services quicker, and also helped aircraft appear yet more modern. So did the development of rapid transit public transport to take passengers into city centres.

It was no accident that luxury travel by rail declined in importance at the same time as such travel did by ship. Indeed, airports became more important than ports as the point of entry into Britain, the United States, Australia and other countries with an oceanic coastline. With airports increasingly significant for transcontinental movements, it is not surprising that many travellers then moved onto regional air routes – as in an itinerary of John F. Kennedy International Airport (New York)–Aéroport Paris-Charles de Gaulle–Aeropuerto Josep Tarradellas

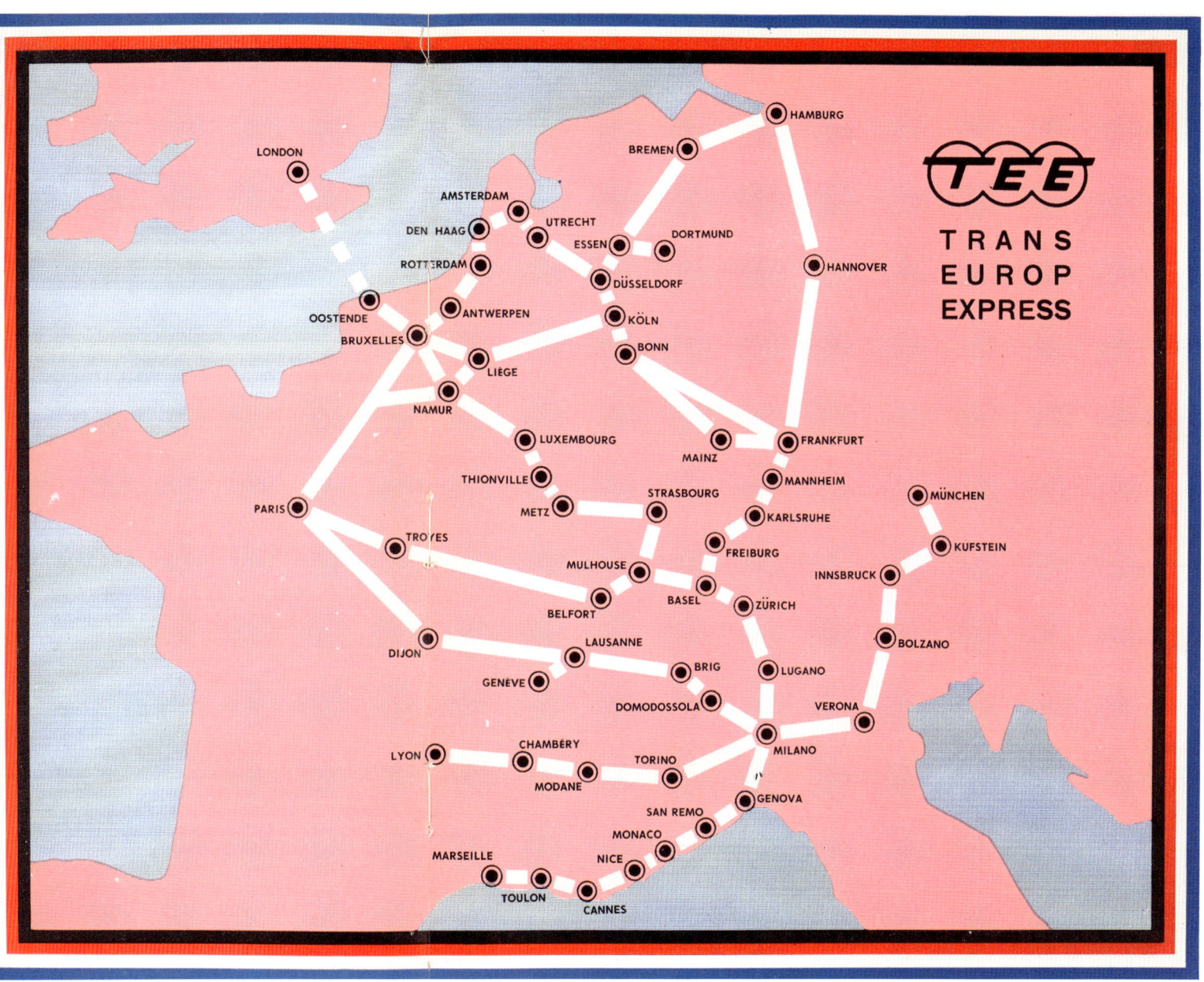

Barcelona-El Prat —rather than transferring at the transit point to the rail network. Indeed, the importance of the train system in part became a matter of its ability to help travellers going to airports or coming from them. Accordingly, rail system maps were to devote more attention to such links, which was not the case with the users of the Trans Europ Express.

In turn, environmental concerns and the accompanying opposition to short-haul flights led to a revival from the 2010s of plans for business-class international train travel, with a sleeper train introduced between Brussels and Berlin in 2022, and others to follow.

6

RAIL DEVELOPMENTS 1970–PRESENT

The mapping of railways continued to face the specific issues of rail and the more general questions of mapping. A classic instance remained that of topology. Maps that were correct in depicting a system but not in showing exact location – for example, rapid transit urban systems (see pages 236–237), or the London Underground – continued to vie with those displaying exact location type, such as those in Britain drawing on Ordnance Survey data (see pages 234–235). There was also the tension between maps of networks that provided no hints on the usage of the network and maps that offered some guidance to usage. The latter, furthermore, could be approached from two different angles: one was usage in terms of frequency of services and the other was usage in terms of demand (see pages 230–231). Both were generally depicted in terms of line width.

The mapping of rail in this period benefitted from changes in cartographic method. This was a matter of organisation and technology. Maps came to be generated more rapidly and in greater quantity. Their central point could be altered readily, different projections and perspectives could be adopted easily (see pages 228–229) and complex data sets, including those of railways, could be mapped rapidly. The standard coordinate-based mapping was integrated with the computer processing and depiction of data to create an effective mapping technology.

Computer technology transformed the printing of maps, notably by giving greater flexibility. Computers can store the source material for maps in a single database and produce maps as a vertical scan made up of many minute dots, which records symbols as dots, lines or surfaces. Design and cartographic processes could be integrated more easily for specific rail-linked purposes. Bases could be modified for particular design requirements and it became easier to incorporate maps with other material such as text, artwork and photographs. The results did not always appeal to those whose imagination of the map was based on nineteenth-century models, which, however, became unrealistic as a guide to mapmaking. The latter increasingly adopted a systems-based approach, one that was topological in approach and primarily concerned with the internal dynamics of the system.

Maps, meanwhile, provided only limited guidance to the ability of railways to cope with crisis, let alone their more general problems. For example, competition from air passenger services was supplemented by that from air freight, notably a major rise in overnight delivery services. These took a lot of high-value goods from rail and road, although rail remained particularly important for bulk goods for which there was no real time constraint, such as coal, iron ore and stone. There was a degree of going back to the original uses of rail, and this focus proved profitable.

Separately, there was a hidden government subsidy in the case of government-owned air traffic control systems, but also for rail networks.

In America, the crisis of rail led to intervention in the shape of support for Amtrak (see pages 238–241), but, as in Europe (see pages 248–249), cuts were made to the system.

Most countries that have railways lack a complex system that can handle major demands, for example those of war, or a breakdown due to major difficulties for a particular line or junction. Maps, moreover, do not necessarily provide insight on the alternatives to rail, as with the North Vietnamese ability to cope with American air attacks during the Vietnam War, notably during the Linebacker II campaign in 1972 when there were extensive air strikes on rail facilities, including the Kinh No, Duc Noi, Trung Quang, Hanoi and Haiphong Railroads, and the Yên Viên and Lang Dang railyards, in or near Hanoi. The last were important to the supply route from China. The US Air Force claimed that the bombing destroyed 372 pieces of rolling stock and caused 500 rail interdictions, but aside from the degree to which the damage was rapidly repaired, in part due to the deployment of large numbers of workers, the Vietnamese simply made extensive use of human porterage.

At the same time, war often leads to a focus on rail. In Ukraine since the invasion in 2022, the railway has made a 'comeback' – or, rather, was forced back because the Russians lack sufficient lorries and are developing a system to the north of the Sea of Azov in order to provide a link to Crimea through conquered territory.

The city of Kramatorsk, the headquarters to the Ukrainian army in the Donbass since 2014, has a fast rail link from Kyiv (an aspect of the impressive nature of the Ukrainian rail system), which helps in the deployment of troops. This capability led the Russians to bomb Kramatorsk station in April 2022. Rail is also used by Ukraine to move goods overland to foreign markets, as there is no security for Black Sea maritime routes. In 2022, as part of the rail war, rail links between Russia and Belarus were damaged, probably by disaffected Belarusian rail workers in order to make it difficult to transport Russian forces.

A very different political context for modern rail is provided by environmental challenges. New lines often require environmental impact studies and engineers strive to create less-polluting locomotives. At the same time, the degree to which states have responded to these factors varies greatly, and this has become another context that is framed and handled very differently.

STRATEGIC CONSTRUCTION

Main Directorate of Geodesy and Cartography, 'Soviet Rail System', 1978.

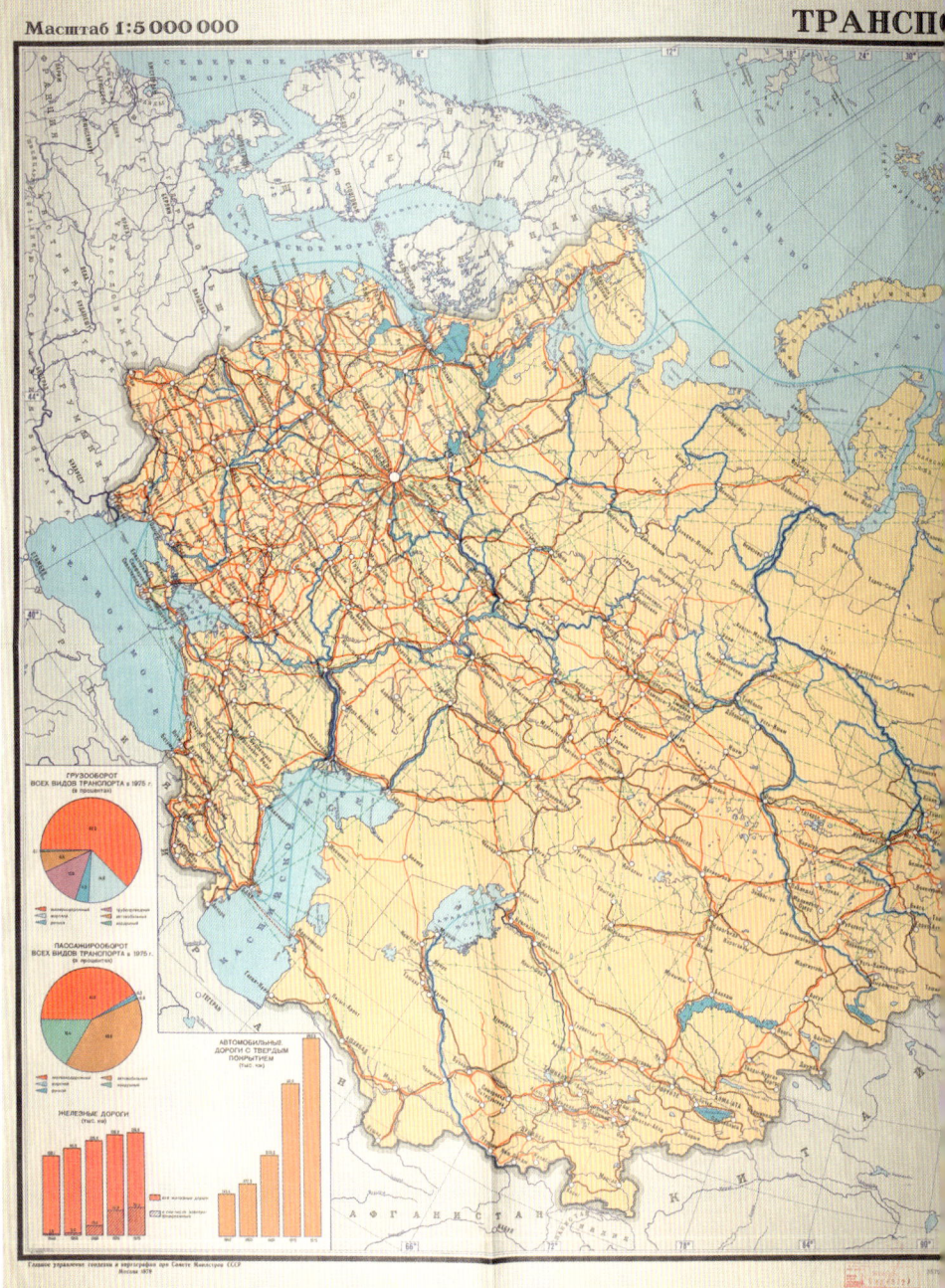

While all data of economic performance from the Soviet Union was problematic, and remains so for the Russian Federation, maps tend to be more accurate than quantitative data. Nevertheless, as with the latter, there was a tendency to draw lines and chart the route rather than to assess working capability and map data.

Because the economic and political fundamentals remained largely constant, the Soviet rail system did not change greatly in the last decades of the Soviet Union, which ended in 1991. The state-owned system was, at 91,600 miles, the largest in the world, although the size of the Soviet Union meant that the network's density was low, certainly compared to the remainder of Europe. This, however, looked somewhat different if attention was devoted solely to European Russia, where the density was far higher. This is particularly the case on an axis from Leningrad (now St. Petersburg) via Moscow to the Donbass mining and industrial belt.

Soviet railways were crucial to an economy that did not rely heavily on road transport despite the ready availability of oil. In large part, this role reflected legacy factors from earlier Soviet history, as well as the ideology of communism. Just over one-third (33,500 miles) of the network was electrified. Postwar expansions were at the annual rate of around 400 miles from 1965 to 1980, and notably so in Siberia where the existing capacity was limited. The wish to bring new rail capacity to areas with developing oil and mineral extraction was crucial. Driven by strategic considerations, not least concern about possible hostility with China (which in part replicated the earlier British position in Canada vis-à-vis the United States), the Baikal Amur Mainline (BAM) was later a key addition parallel to the Trans-Siberian

Railway to the south. However, that new line suffered greatly from poor management and congestion, and thereby highlighted more general issues for the Soviet network as a whole. In 1974, Leonid Brezhnev, the head of the government, presented this railway as 'the construction project of the century'. It certainly confronted a harsh terrain, notably permafrost, which was difficult to build on and raised the question of climate change causing damage through the melting of the permafrost layer. The railway has numerous bridges and tunnels, all of which increased construction and maintenance costs. Much of the labour was provided by volunteers from the Young Communist League. The railway was declared complete in 1984, when that was not the case, and again in 1991. Many mining and industrial projects were linked to the railway, but it did not match expectations in terms of revenue.

This experience does not augur well for the ambitious Chinese plans of large-scale expansion (see pages 232–233 and 256–257), which appear to be strategic – in the sense of the presentation of state power – rather than economic, in pursuit of financial profitability. Looked at in this way, in both the Soviet Union and China the underlying prime logic was and is strategic.

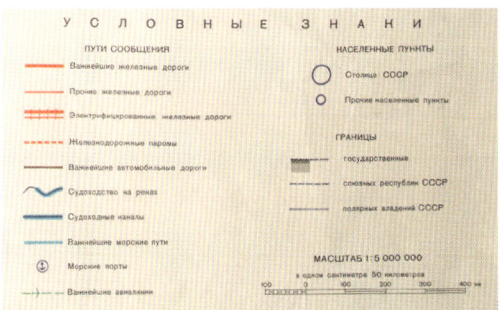

Rail and Lake Transport

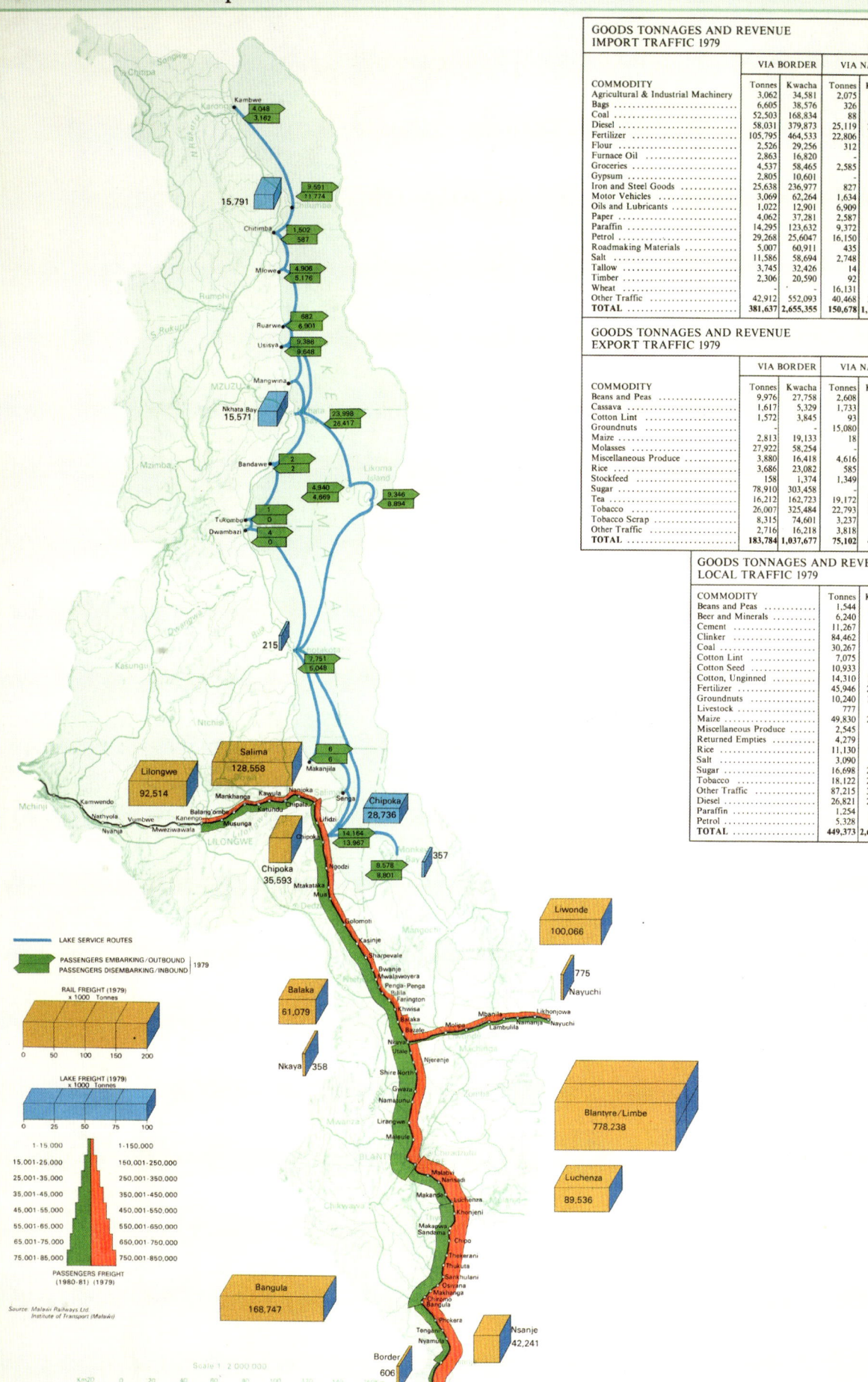

GOODS TONNAGES AND REVENUE IMPORT TRAFFIC 1979

COMMODITY	VIA BORDER		VIA NAYUCI	
	Tonnes	Kwacha	Tonnes	Kwacha
Agricultural & Industrial Machinery	3,062	34,581	2,075	15,976
Bags	6,605	38,576	326	1,632
Coal	52,503	168,834	88	288
Diesel	58,031	379,873	25,119	287,099
Fertilizer	105,795	464,533	22,806	61,496
Flour	2,526	29,256	312	1,727
Furnace Oil	2,863	16,820	—	—
Groceries	4,537	58,465	2,585	19,749
Gypsum	2,805	10,601	—	—
Iron and Steel Goods	25,638	236,977	827	5,200
Motor Vehicles	3,069	62,264	1,634	20,087
Oils and Lubricants	1,022	12,901	6,909	58,442
Paper	4,062	37,281	2,587	17,430
Paraffin	14,295	123,632	9,372	72,200
Petrol	29,268	25,6047	16,150	172,588
Roadmaking Materials	5,007	60,911	435	2,891
Salt	11,586	58,694	2,748	11,264
Tallow	3,745	32,426	14	124
Timber	2,306	20,590	92	564
Wheat	—	—	16,131	62,755
Other Traffic	42,912	552,093	40,468	320,292
TOTAL	381,637	2,655,355	150,678	1,131,804

GOODS TONNAGES AND REVENUE EXPORT TRAFFIC 1979

COMMODITY	VIA BORDER		VIA NAYUCI	
	Tonnes	Kwacha	Tonnes	Kwacha
Beans and Peas	9,976	27,758	2,608	7,967
Cassava	1,617	5,329	1,733	3,364
Cotton Lint	1,572	3,845	93	516
Groundnuts	—	—	15,080	27,119
Maize	2,813	19,133	18	158
Molasses	27,922	58,254	—	—
Miscellaneous Produce	3,880	16,418	4,616	15,148
Rice	3,686	23,082	585	1,950
Stockfeed	158	1,374	1,349	3,655
Sugar	78,910	303,458	—	—
Tea	16,212	162,723	19,172	125,067
Tobacco	26,007	325,484	22,793	169,541
Tobacco Scrap	8,315	74,601	3,237	24,592
Other Traffic	2,716	16,218	3,818	29,962
TOTAL	183,784	1,037,677	75,102	409,039

GOODS TONNAGES AND REVENUE LOCAL TRAFFIC 1979

COMMODITY	Tonnes	Kwacha
Beans and Peas	1,544	12,391
Beer and Minerals	6,240	77,909
Cement	11,267	65,860
Clinker	84,462	171,874
Coal	30,267	89,166
Cotton Lint	7,075	50,118
Cotton Seed	10,933	36,302
Cotton, Unginned	14,310	197,209
Fertilizer	45,946	231,653
Groundnuts	10,240	120,195
Livestock	777	8,403
Maize	49,830	261,813
Miscellaneous Produce	2,545	17,722
Returned Empties	4,279	31,098
Rice	11,130	61,370
Salt	3,090	20,804
Sugar	16,698	261,328
Tobacco	18,122	258,883
Other Traffic	87,215	316,747
Diesel	26,821	280,270
Paraffin	1,254	13,224
Petrol	5,328	51,476
TOTAL	449,373	2,635,815

When Malawi attained independence in 1964 the only rail link to the sea was a single track to the Mozambique port of Beira. The first part of the line had been opened between what is now Nsanje and Blantyre in 1908. An extension southwards to the north bank of the Zambezi river was opened in 1915, and the through all-rail route to the sea was opened in 1935, following the completion of the Zambezi bridge in Mozambique. In the same year the extension of the line from Blantyre to Salima was opened and this extension gave a connection to the lake service at Chipoka. The lake service had been operated by the government since 1931, but when the rail link was completed operations became the responsibility of the then Nyasaland Railways.

There were no further extensions to the system until 1970 when a line was opened from Nkaya, 88 kilometres north of Blantyre, to Nayuchi, on the Mozambique border, to connect with the Mozambique Railways line to Nacala.

Meanwhile studies had been carried out on alternative routes which could be followed from different points on the existing line to reach Lilongwe. It was finally decided to extend the line from Salima. A loan agreement was made with the Canadian government in February 1974 and initial construction commenced later that year.

The official opening of the 111 km extension from Salima to Lilongwe took place in February 1979 and in the remaining 10 months of that year 103,000 tons of goods and 21,000 passengers were carried on the new line.

At the same time that agreement was reached with the Canadian government to construct the line to Lilongwe, plans were also made for the completion of the rehabilitation of the line between Balaka and Salima with Canadian finance. Rehabilitation had already been commenced by Malawi Railways' own engineering staff and with its own finance, and the upgrading of this section has now been completed.

The loans which had been received by government for the construction of the line to Lilongwe and the rehabilitation of the line between Balaka and Salima were subsequently converted into grants, and a further grant was later made to enable the line to be extended from Lilongwe to Mchinji, near the Zambian border. Construction of this third extension actually commenced before the line to Lilongwe was completed, and it is now operational.

Malawi Railways now has a route distance of 785 km compared with 465 km in 1964.

In addition to the rehabilitation of the line north of Balaka much work has had to be carried out between Blantyre and the southern border. British aid has been received for this since 1971 and the worst parts have been improved, but much work remains to be done to bring this, the busiest section of the system, up to the standard of the new lines.

The rolling stock at the beginning of 1983 comprised 650 freight cars, 98 container flat wagons, 5 car transporters, 90 tank cars, 6 cattle wagons, 127 service wagons and vans, 30 third class and one second class passenger coaches.

Mozambique Railways took delivery of 7 car transporters at the same time and these 12 wagons run as one unit, a new departure for both the Malawian and Mozambican railways systems. The unit is employed principally on the Beira line. On the Nacala route experimental use of container block trains has shown that there is good potential for such a system in improving the efficiency of container traffic. Block trains of coal from Moatize (near Tete) in Mozambique have also met with some success.

Thirty engineering wagons were also delivered to the Railways Development Project in 1979 for use in the construction of the line to Mchinji. With the completion of construction they have now been transferred to Malawi Railways.

Locomotive power was strengthened with the introduction, during 1980 of 20 MLW Bombardier locomotives for mainline movements. This is in addition to the 14 AEI locomotives already in service. There are also an additional 19 locomotives which are used for placing and pulling out wagons from the more than 120 private sidings operating in Malawi.

The growth in the use of containers for the import and export of commodities has been impressive since 1978 with over 11,000 containers moved by rail into and out of Malawi during 1982. There are a total of seven container terminals in the country; two each in Blantyre, Lilongwe and Luchenza, and one at Balaka, and more are planned for the future.

With the expansion of the rail network, there has been a steady increase in the number of passengers carried. In 1982 more than 1,500,000 passengers were carried. During this expansion a new factory to manufacture concrete railway sleepers was constructed at Salima. Concrete sleepers have a considerably longer life than those made from timber and this durability reduces the cost of track maintenance.

The Lake Service.

Since 1935, Malawi Railways has operated a service on Lake Malawi. The fleet is based at Monkey Bay in Mangochi District and at the time of writing comprises six vessels. These are the Motor Vessels (M.V.s) *Mtendere*, *Ilala*, *Chauncy Maples*, *Nkhwazi*, *Karonga* and *Mpasa*. Of these, the *Mtendere* and *Chauncy Maples* are passenger vessels, the *Ilala* is a mixed passenger and cargo vessel whilst the remainder are cargo carrying vessels. The *Mpasa* is a fuel carrying vessel and a new 350 tonne capacity tanker is expected to enter service by the end of 1983.

The *Mtendere* and *Ilala* each operate a weekly service commencing from Monkey Bay, via Chipoka, where connections with the rail service are made, and then between them the two vessels serve Nkhotakota, Likoma, Nkhata Bay, Usisya, Ruarwe, Mlowe, Chitimba and Chilumba whilst the *Mtendere* alone also serves Kambwe and Kaporo. The round trips take six days. The cargo vessels operate to all ports as required.

To handle the increased volume of traffic in recent years, several lake ports, notably Chipoka, have been enlarged and redeveloped.

LINKING THE COPPERBELT
Malawi National Atlas Co-ordination Committee, 'Rail and Lake Transport', 1983.

This map from the *National Atlas of Malawi* shows the innovative cartography to which transport can lead, not least when it had to be summarised for other purposes. The respective role of lake and rail services, and of passenger and freight transport, were presented in schematic forms with the complementary nature of the two services readily apparent. Thus, the lake system to the north both supplied and benefitted from its train counterpart.

Malawi became independent in 1964 and, as part of post-colonial expansion, a line opened in 1970 from Nkaya to Nayuchi on the Mozambique border in order to connect with the Mozambique Ports and Railways line to Nacala. Container trains were used on the route, but rail links to and in Mozambique have been greatly affected by instability and conflict there.

In 1974–79, Malawi co-operated with the Canadian International Development Agency to build 70 miles of new track from Salima to Lilongwe, while an extension from Mchinji to Chipata in Zambia opened in 2010. In 1999, Malawi Railways was privatised in favour of the Central East African Railways consortium led by Railroad Development Corporation, which is an American holding company. This operation was sold in 2008 to an investment group based in Mozambique.

Similarly, the Tanzania–Zambia Railway Authority (TAZARA) line, which runs between Dar es Salaam and Kapiri Mposhi, was a major Chinese aid project. China sent about 50,000 people to work on the railway, which had in part to be constructed across difficult terrain – notably the escarpment between Lmimba and Makambako, which subsequently caused operating problems because of inadequately powerful engines. Also called the Tan-Zam railway, or Uhuru Railway, it was built in 1970–75 as part of China's pursuit of international political goals, notably as a way to advance its claim to lead the cause of national liberation, and thus in pursuit of its view that the Soviet Union lacked the necessary credentials. The two communist states were competing for leadership in southern Africa. That was the context for a railway that more specifically was intended to anchor Zambia in the cause of opposition to the pro-Western states in southern Africa by providing it with a means to avoid being dependent for transit on the white-minority ruled states of Southern Rhodesia (now Zimbabwe) and South Africa. In particular, this route would provide an opportunity for access from Zambia's part of the mineral-rich Copperbelt region it shares with the DRC. The destination was Kapiri Mposhi. As such, the railway was seen as a way to keep Zambia strongly against those colonial states. In 2023 studies supported by the United States and the EU were launched for the construction of a railway linking the Zambian Copper Belt with the Congolese Katanga region and the existing line to the Angolan port of Lobito.

However, in general it is the extent to which the earlier colonial legacy has been sidelined that is notable. In particular, the emphasis has been on freight moved by lorries. The railway became less attractive once the political situation changed in southern Africa and the fall of apartheid regimes in Zimbabwe, Namibia and South Africa has provided more outlets for Zambia. As another instance of Cold War rivalry, the United States funded the Tanzam Highway, built from 1968 to 1973. The management of the railway, once completed, proved inadequate and the cargo that was transported was well below the level planned. The rail system in southern Africa should, in theory, have provided volume, speed, regularity and predictability, but it did not do so, due in part to poor management that compromised the existing capability.

Further north, the deterioration from the 1980s to 2010s of the Ethiopia–Djibouti railway, built in 1894–1917, led to the Chinese-funded and managed construction of a replacement – a new, 472-mile electrified line that opened in 2018. However, revenue in 2019 did not match the operating costs.

MODERNISING ASIA
Quail Map Company, 'China', 2008.

Provincial boundaries, not physical geography, provide the background for this map (right) from the *China Railway Atlas*, which captures the density of China's rail network in the early twenty-first century but provides no details of usage. Under the communists, who gained power in the 1946–49 civil war, the railway – in accordance with both Marxist doctrine and the Republican-Nationalist tradition of Chinese rail politics from the 1900s on – was seen as a crucial tool of national defence and economic modernisation. The People's Liberation Army's railway corps provided a large, disciplined and inexpensive construction workforce, while captured Japanese engineers were used, and treated better than Chinese prisoners would have been by the Imperial Japanese. The end, in 1953, of the Korean War in which China had intervened in force, with major consequences for rail priorities, freed up more resources for rail expansion. As more generally for China with technological transfer, Soviet technical assistance was important, which replaced the earlier reliance on the West, notably Britain and the United States. By 1951 more than 13,000 miles had been restored, and in 1952, after difficult negotiations, the Soviet Union handed over the system in Manchuria.

Alongside the repair of railways damaged in warfare in 1937–49, came significant expansion of the network, notably beyond the core areas of earlier railways, which had focused on Manchuria. Instead, much effort was put into an east–west line from Lanzhou, capital of Gansu province on the Yellow River, into the Xinjiang Uygur Autonomous Region, northwest China, begun in 1952 and reaching Urumqi a decade later (see inset map). Tracklaying on a line from Chengdu, Sichuan province, to Kunming, Yunnan province, a line deep into the southwest, started in 1958 and was completed in 1970. In addition, a line completed in 1956 connected Fujian province in the southeast with the interior.

Internal disruption from the late 1950s, in the shape first of the Great Leap Forward of 1958–62 and then the Cultural Revolution of 1966–76, led to serious difficulties.

There was an emphasis on planning by exhortation, rather than implementation by practicality. Thus, freight shipment, which was close to 300 million tons in 1957, was supposed to reach 900 million tons by 1959 and three billion tons by 1972, with rail mileage reaching almost 75,000 miles. As a typical instance of a failure to address practicality, freight loads on all major lines were increased without taking account of the necessary power for the locomotives.

In the event, the freight load only reached 345 million tons in 1962, and there was widespread chaos.

The Cultural Revolution saw the railways taken over by radicals, which compromised performance, only for further reorganisation to follow from 1978 onward with the reaction against the radicals. Indeed, there was then a revival of growth, albeit one dependent on very heavy investment due to significant operating deficits. This funding was provided from the 1990s, and that led to the last mainline steam-powered service being taken out of service in 2005 despite the easy availability of coal, and at a controlled price. At the same time, there was scant attempt to place a realistic price on the opportunity costs of investment or indeed the running costs. Instead, for the plans of government to focus on rail development, railways were regarded as appropriate ideologically as well as economically. As earlier with the Soviet Union, rail in China was part of a strategy focused on heavy industry with no focus at that stage on large-scale passenger traffic.

However, in recent decades there has been a stress on high-speed trains. China Railway High-Speed was introduced in 2007 with the first passenger-dedicated line opened in 2008 from Beijing to Tianjin. The map depicts these lines in red, including the 819-mile-long Beijing–Shanghai line that opened to service in 2011 and moved over 210 million passengers in 2019. The maximum speed was 186mph (300kmh) and the train took just under five hours. The second map from the *China Railway Atlas*, 'Shandong, Jiangsu and Shanghai', shows the Shanghai Metro system, which by 2019 had almost 3.9 billion rides on a system with 498 miles of track and 408 stations on 19 lines. The system was built between 1986 and 1993, with new lines coming into operation from 2003, notably in preparation for the Expo 2010.

The railway network was extended westwards to Kashgar in 1999 and to Lhasa in Tibet in 2015. After the disruption caused by the Covid-19 pandemic, most of the world's high-speed rail services had returned to a more normal operating pattern by early 2023, although many of the best timings had been eased slightly. China retained its lead in the world speed survey.

In 2021, a 250-mile rail line carrying Tibet's first bullet train line connected Lhasa with Nyingchi, giving all 31 provincial-level regions of China access to high-speed train travel. The situation in China underlines the centrality of politics to much railway history. During the 2022 border confrontation with India in the Himalayas over the frontier in Arunachal Pradesh, the Chinese benefitted from the extent to which the new line to Nyingchi ran to within ten miles of the border, which speeded up China's ability to deploy troops.

Meanwhile, in Taiwan, a high-speed line opened in 2007 and in 2023 more trains were ordered so as to improve the service.

6. RAIL DEVELOPMENTS 1970–PRESENT   233

A DENSER NETWORK
John Yonge, 'London Transport: Railway Track Map', 1978.

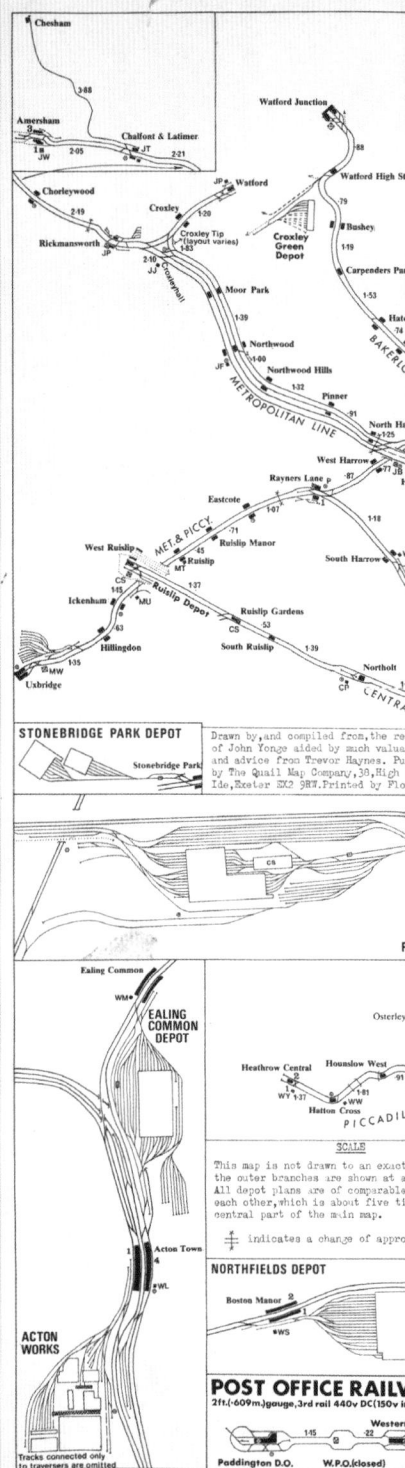

This is not the map displayed in 'Tube' stations or mentioned, like that of Harry Beck, in reverential tones. Instead, this track map sought to display geographical accuracy. The map is very different to the standard London Transport network map, although there was a shared departure from uniform scale. As is noted in the 'Scale' box: 'This map is not drawn to an exact scale and some of the outer branches are shown at a reduced scale.' The inclusion of the depot plans, all of comparable linear scale, and of the Post Office Railway, are other departures from the topological character of the official network map; as is the depiction of abandoned platforms, such as Brompton Road (west of Knightsbridge) and Lords (north of Marylebone). (Those are features of the system of which there are no signs in the official map for passengers.) As with other track maps, the map of the London system provides a mass of information for enthusiasts that was not made available or deemed relevant for travellers. Relevance, however, was not an automatically clear-cut category. The map differentiates track that is electrified from track that is not, includes the route of the Post Office Railway (depicted in an insert), shows platforms, abandoned platforms, washing machines, cleaning sheds, bridges, tunnel mouths and the mileage between open stations. The inserts on the depots offer material not readily available elsewhere.

This goal of accuracy had a demand of its own, but also, from the 1960s, the London Underground expanded considerably. As a result of Green Belt legislation from 1938, the key element in new lines was no longer the expansion of suburbia seen in the interwar years but instead new capacity within the already built-up area. Thus, the expansion to the Northern Line earlier planned from Edgware to Watford was abandoned. That new capacity was a product of increased commuter demand within this area, which in part reflected the movement of people from Inner London as a result of slum clearance and wartime bombing. Thus, London County Council had built large housing estates at Becontree, St Helier, Downham and Watling (Burnt Oak) in the interwar years. This pattern continued in the post-war years, so that Londoners from the East End were rehoused in new council estates, which included those at Debden, Hainault (both near the eastern part of the Central Line) and Harold Hill (not near a line).

The Victoria Line, opened in 1968–69 and the first automatic underground railway in the world, was followed by the Jubilee Line, completed in 1979, and the Piccadilly extension to Heathrow, opened in 1977. The latter was an aspect of the more general tendency in major urban areas to extend urban rapid transit systems to reach airports. In contrast, more distant airports, such as at Gatwick, Luton and Stansted, were served by conventional overland rail services. A mass transit system was to be extended to London City Airport, which opened in 1987.

The expansion of the Underground led to neglect of the smooth running of the rest of the system. Insufficient attention was given to making it efficient, user-friendly, reliable and comfortable, and lines like the Northern became notorious for poor service. Public access to the service was rarely free of steps, which reflected the problems of redesigning the existing system. Indeed, maps such as this, as well as the more famous Beck map, suffered from being two-dimensional. No guidance was provided on how to use the system. For passengers that meant how easy it would be to transfer between lines, and whether there was a need to use lifts or escalators. For managers this choice had clear consequences for capacity as far as station movements were concerned because the provision of lifts, or alternatively escalators, had consequences for a station's ability to handle the movement of passengers.

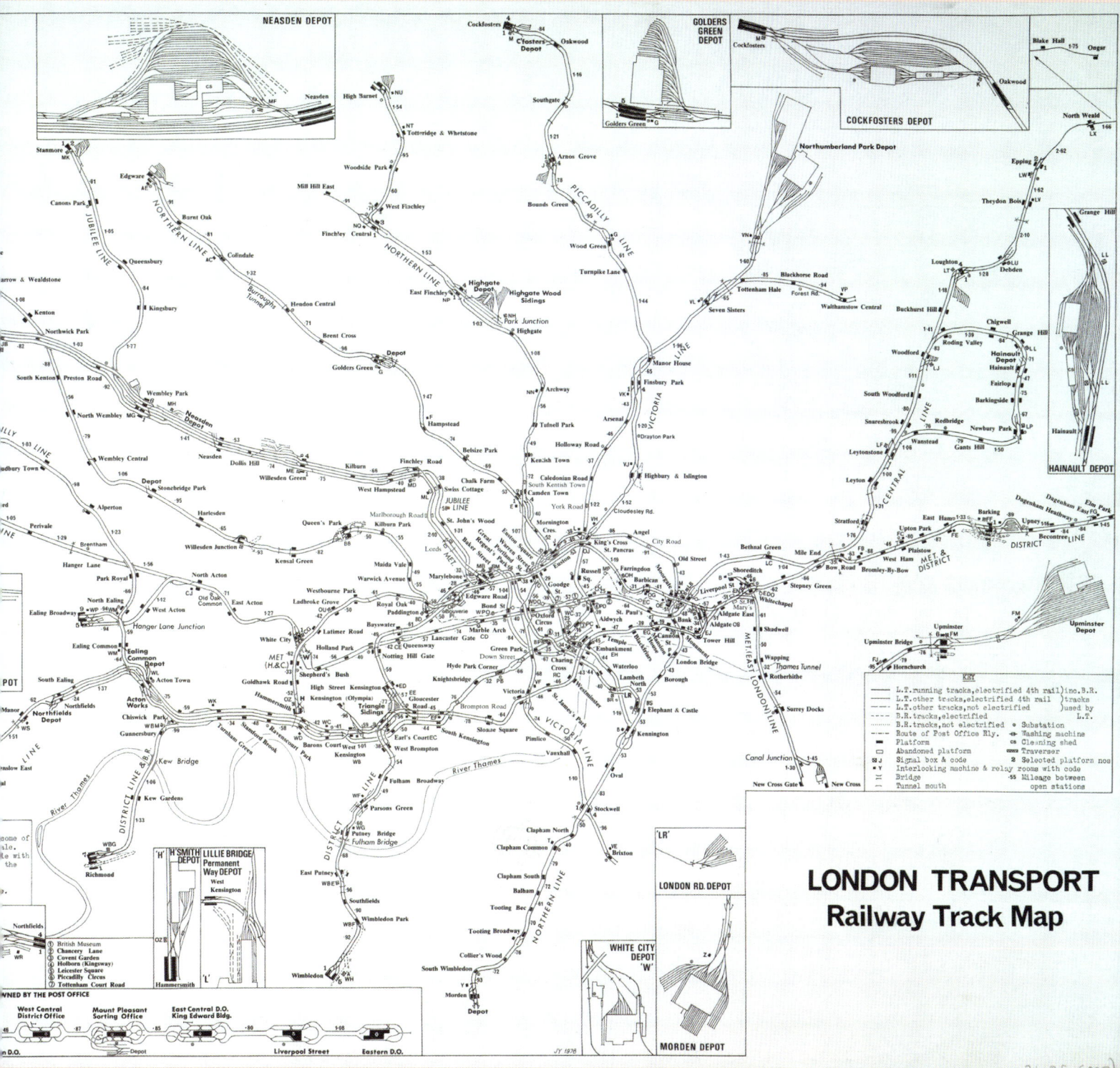

6. RAIL DEVELOPMENTS 1970–PRESENT 235

THE SINGAPOREAN CITY STATE
Land Transport Authority, 'Rail Network: This is the rail life', 2022.

This colourful and well-organised map of Singapore's rapid transit system was based by the operator on design ideas from a student called Samuel Lim. The system is divided between six lines on the Mass Rapid Transit (MRT), opened from 1987, and two lines on the Light Rail Transit (LRT), opened from 1999. The latter has automated guideway transit lines. Colour helps to organise the response of the viewer, while providing appropriate guidance on junction stations, and thus on the interchanges of lines necessary to make a system, as well as connections to other modes of transport forming parts of the public network (buses, cruise ships, cable cars, and so on).

The MRT network has 134 miles of track, more than 140 stations and is due for further expansion. The Circle Line was fully completed in 2012, the Downtown Line in 2017 and the Thomson-East Coast Line in 2020. The LRT network has 17 miles of track and more than 40 stations.

As is normal with the efficient corporatism of Singapore, it is well managed and supported by the appropriate revenue and investment that permits expansion as well as effective maintenance. In part, this effectiveness reflects a concentrated housing pattern and a high use of the system (daily ridership of over three million on the MRT; 200,000 on the LRT), given only limited car ownership. Indeed, citizens are required to bid on, and win, a state-issued certificate of entitlement before being able to purchase, operate and own a car for a ten-year period, after which another certificate has to be purchased. The cost of cars is very high. Whereas in 2022, among Americans car ownership was nearly 80 per cent, for Singaporeans it is about 11 per cent. Currently, a cross-border metro linking Singapore and Johor Bahru is planned.

Singapore, a key nodal point in the British Empire, had had a long railway history, with two main factors playing a role in shaping its future public transport development: serving the docks and helping move people in a crowded environment. The first saw the use of trains to help more goods and people in and from the docks, beginning with a shunting railway on the wharfs opened in 1877. Far more significantly, a railway to serve the New Harbour began operation in 1903, with an extension in 1907. This, the Singapore-Kranji Railway, used British-built locomotives, with trains reaching a top speed of 17 miles per hour. Electric tramways also developed in the 1900s, with services begun in 1905, the same year in which a power station was built.

In 1923 the opening of the Causeway meant that there was a continuous rail link between Singapore and Penang. As a result, the ferries required since 1909 to cross the Straits of Johor, when railheads had reached both sides, were no longer necessary. In time, this inherited system became redundant. Steam locomotives were phased out in Singapore in 1972, and in 2011 the main rail station at Tanjong Pagar was closed.

In July 2023 the Malaysian Ministry of Finance invited proposals for reviving the Kuala Lumpur–Singapore High Speed Rail project using a public-private partnership model.

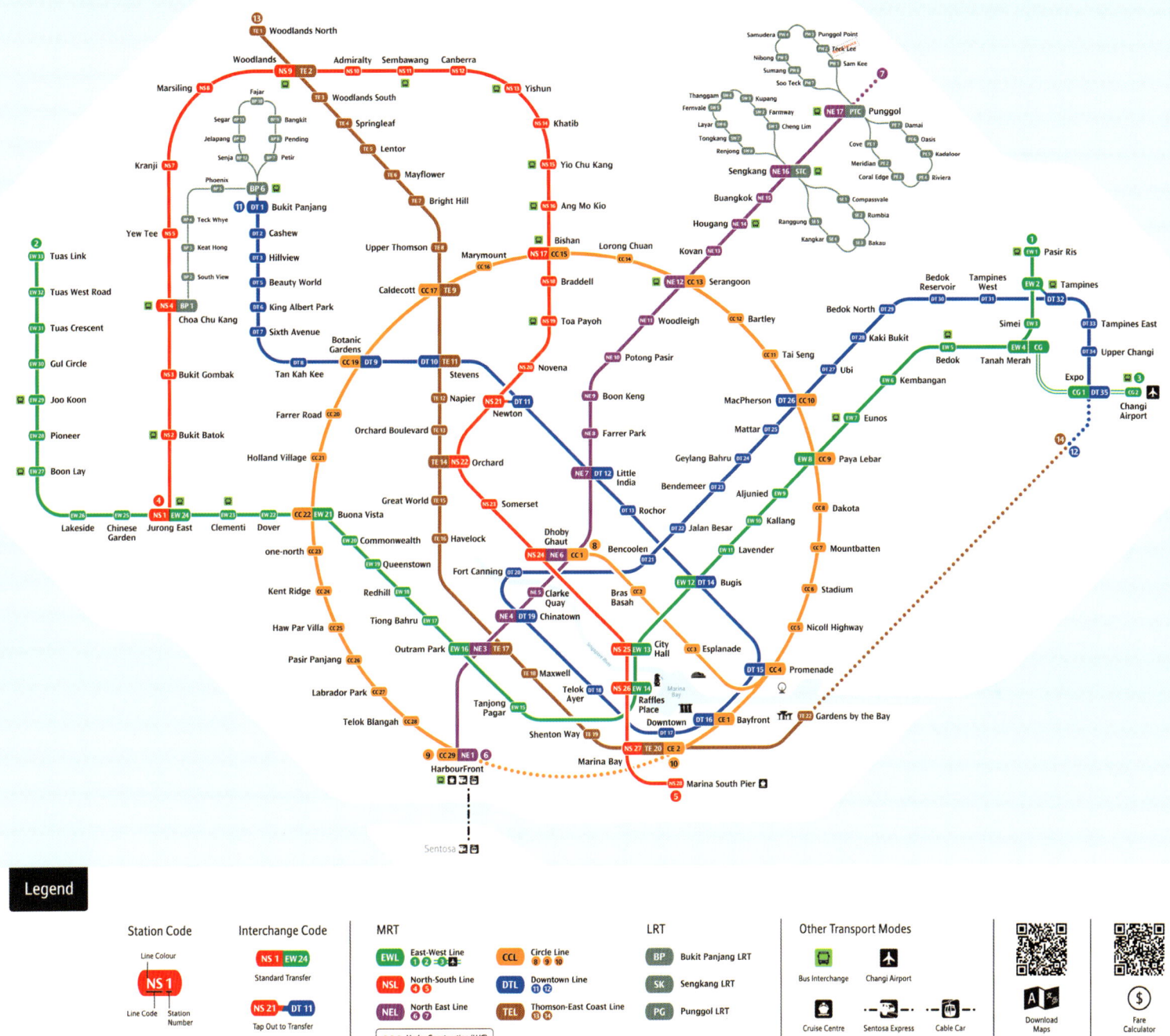

6. RAIL DEVELOPMENTS 1970–PRESENT 237

NORTH AMERICAN NETWORK
Amtrak, 'Amtrak's National Rail Passenger System', 1993.

With its effective use of colour, Amtrak's 1993 map of the national rail passenger system in the United States (see following pages) includes the lines for the Amtrak Thruway Bus Service as well as information about connecting rail services in Canada and Mexico. The background of national parks and terrain was appropriate for a federally funded map, and the identification of seasonal train services is useful for leisure travellers. As was normal in American maps, the organisational context for the rail lines (for ease of comprehension) was that of state boundaries, which had no relevance for running. The map also has details of urban areas served in various locations in the Mid-West and on the East Coast and West Coast.

The second map (right) was produced by the Illinois Department of Transportation in 1985 to try to reduce road congestion. This somewhat drab map contains very little information and is essentially a matter of a fit-for-purpose map. There is no sense of a wider network beyond, which underrates the role of Chicago and also of links via St. Louis. The map shows that Amtrak services in Illinois at the time are organised in four inter-city passenger rail corridors with 36 stations, including Chicago. My own experiences, of train from Champaign and Bloomington to Chicago, were of sad services that were much delayed. I have had better results on the Washington–Boston Northeast Corridor route and on the Keystone Corridor, which connects to Harrisburg, the state capital of Pennsylvania.

Rail was of particularly minor importance in travel to cities in the southern United States, both east and west. Downtown Phoenix's passenger train service ended in 1996. In contrast, although trains were stopped for a while in 2005 by Hurricane Katrina. New Orleans is served with three Amtrak routes, the City of New Orleans from Chicago, the Crescent from New York and the Sunset Limited from Los Angeles. Savannah is served by the New York–Miami Silver Service route. Moreover, in recent decades a number of major Southern cities have seen commuter services instituted, including Dallas, Houston and Miami.

The 'Amtrak Connects US' plan announced in March 2021 was conceived in part to create inter-city corridors. In the United States, as elsewhere, the return from Covid-19-blighted services provided an opportunity to regroup and to reorganise transport priorities, but it also reawakened political contention. Less contention surrounded the regional train lines, such as: SEPTA (the Southeastern Pennsylvania Transportation Authority), NJ Transit (New Jersey Transit), MARC (Maryland Area Rail Commuter) and VRE (Virginia Railway Express). As a result of such services, it is possible to find less-expensive alternatives to Amtrak – for example, by SEPTA and NJ Transit from Philadelphia to New York.

The availability of commuter operations along the Northeast Corridor that provide a frequent, lower-cost alternative to Amtrak's rail services between major cities is concentrated north of Trenton, New Jersey. There is NJ Transit between Trenton and New York; Metro-North Railroad between New York and New Haven, Connecticut; Shore Line East (SLE) between New Haven and New London; Connecticut Rail (CT Rail) between New Haven and Springfield, Massachusetts; and MBTA (Massachusetts Bay Transportation Authority) Commuter Rail between Providence and Boston. Together, they provide far more frequent service than Amtrak, but tend not to be shown on the same map. This is a more general problem with rail mapping, one that looks back to the nineteenth-century example of competing companies (when each mapped only its own service).

The ownership of existing tracks by freight-focused companies helps make it difficult to introduce more passenger services. Thus, the only long-distance passenger service to Worcester, Massachusetts, is an Amtrak Boston–Chicago service. There are daily commuter trains from Worcester to Boston and plans to run commuter services from Springfield to Boston, and then eventually from Pittsfield to Boston, mimicking the construction of the old Western Division. However, the cost would now be billions

Rail Passenger Routes

May, 1985

Special Family, Group, Senior Citizen, and Handicapped Fares

For Information Call
312-558-1075 (Chicago)
1-800-USA-RAIL
or Your Travel Agents

Illinois Department of Transportation

as well as the disruption of numerous environmental pacts hearings.

Deregulation in the 1980s and 1990s helped the railways to profit as freight carriers. The threat of an American freight train strike in late 2022 led to a wider realisation of the continuing economic significance of rail in the most car-obsessed country in the world. Food, energy, carmaker and retail groups urged Congress to intervene in the rail talks, noting that a freight shutdown could freeze almost one-third of American cargo, pushing up inflation, hitting food and fuel supplies, and costing the economy an estimated US$2 billion daily. About one-quarter of American grain shipments are made by rail. In December 2022 President Biden signed legislation to impose a contract deal tentatively agreed earlier.

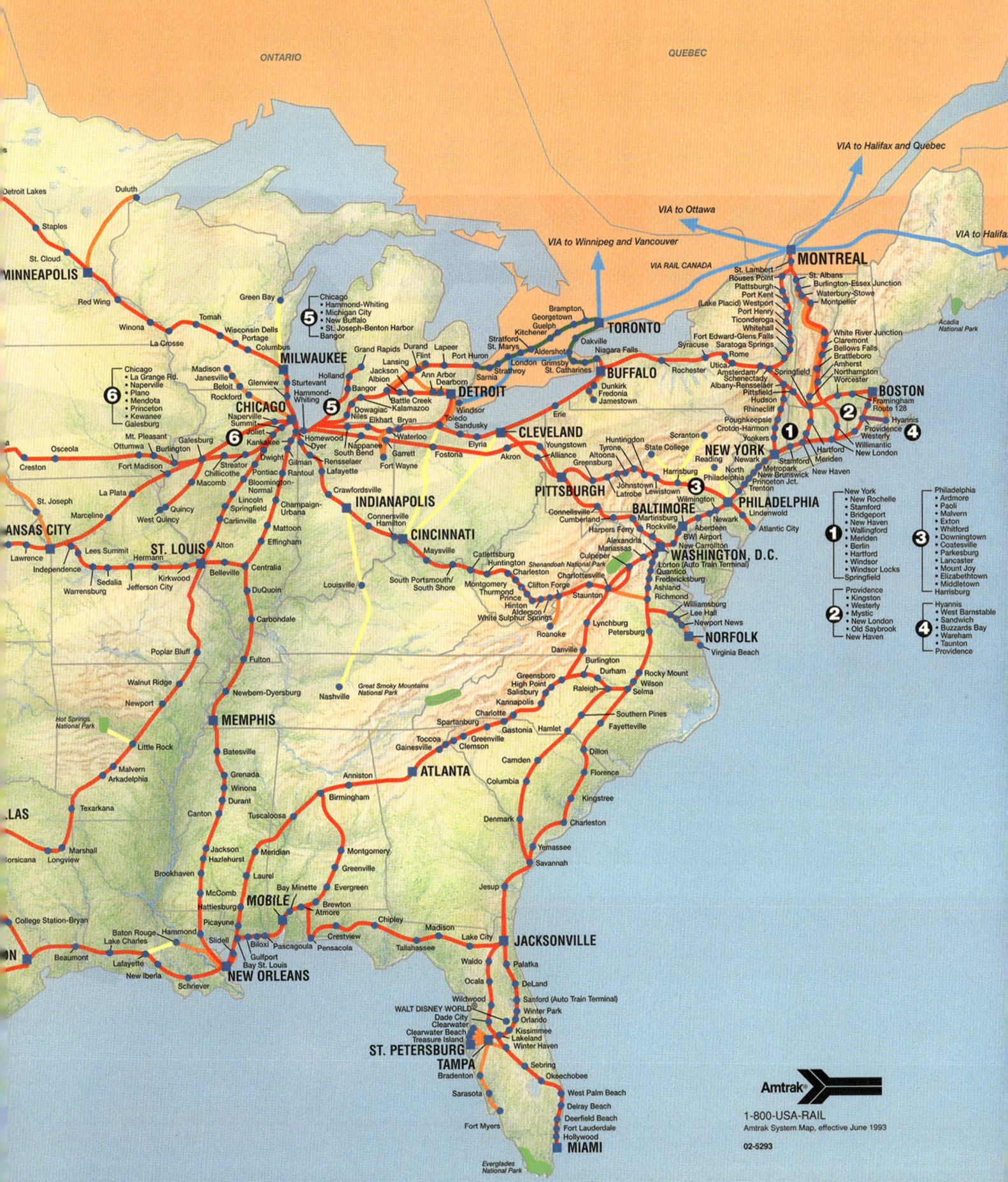

HIGH-SPEED RAIL ARTERIES
Jinbunsha Co. Ltd., 'New Railway Map of Tokyo and Vicinity', 1975.

The 2022 film *Bullet Train* provided a madcap adventure caper, but it also offered many in the audience a view of cutting-edge Japanese trains, including their aerodynamic character and their very short stops. These trains draw on a long history. Tokyo, which had had the first horsecar line in Japan in 1882 and opened Tokyo Central Station in 1914, with four platforms, was and remains a hub for national services.

This map of the Greater Tokyo area is a skilful topological map that fits in the lines run by the Japanese National Railway (JNR, the state-owned entity that operated the country's national network from 1949 to 1987) and the many private lines while also leaving the subway clear, although the two-dimensional nature of the map is poor in depicting the three-dimensionality made possible by subway and overhead monorail, and also in showing the subway junctions.

Tokyo is also Japan's largest urban centre and thus an important market in its own right, with rail services necessary to prevent the city breaking down because of congestion. As a result, the Ginza Line, the oldest underground service in Asia, was built in 1925–27, the inspiration being the London Underground. The project was delayed by financial problems, but completed in 1934 and then in 1941 merged with the Tokyo Rapid Railway, which had begun to operate in 1938. However, the Tokyo Metro Marunouchi Line, planned in 1925 and begun in 1942, was abandoned in 1944, in part due to a lack of resources. The Allied wartime attacks on Japan's rail infrastructure and on major urban areas resulted in the heavy bombing of Tokyo and much damage. The project for the Metro Marunouchi Line was revived in 1951, and the first part opened in 1954, with other sections following to 1996.

Meanwhile, the start of Shinkansen super expresses ('bullet train'), high-speed services, initially between Tokyo and Osaka, led to the opening of two new platforms at Tokyo Central Station in 1964. This was the year of the Tokyo Olympics and the trains, with their aerodynamic noses, were seen as a symbol and cause of national revival and

MAP OF TOKYO AND VICINITY

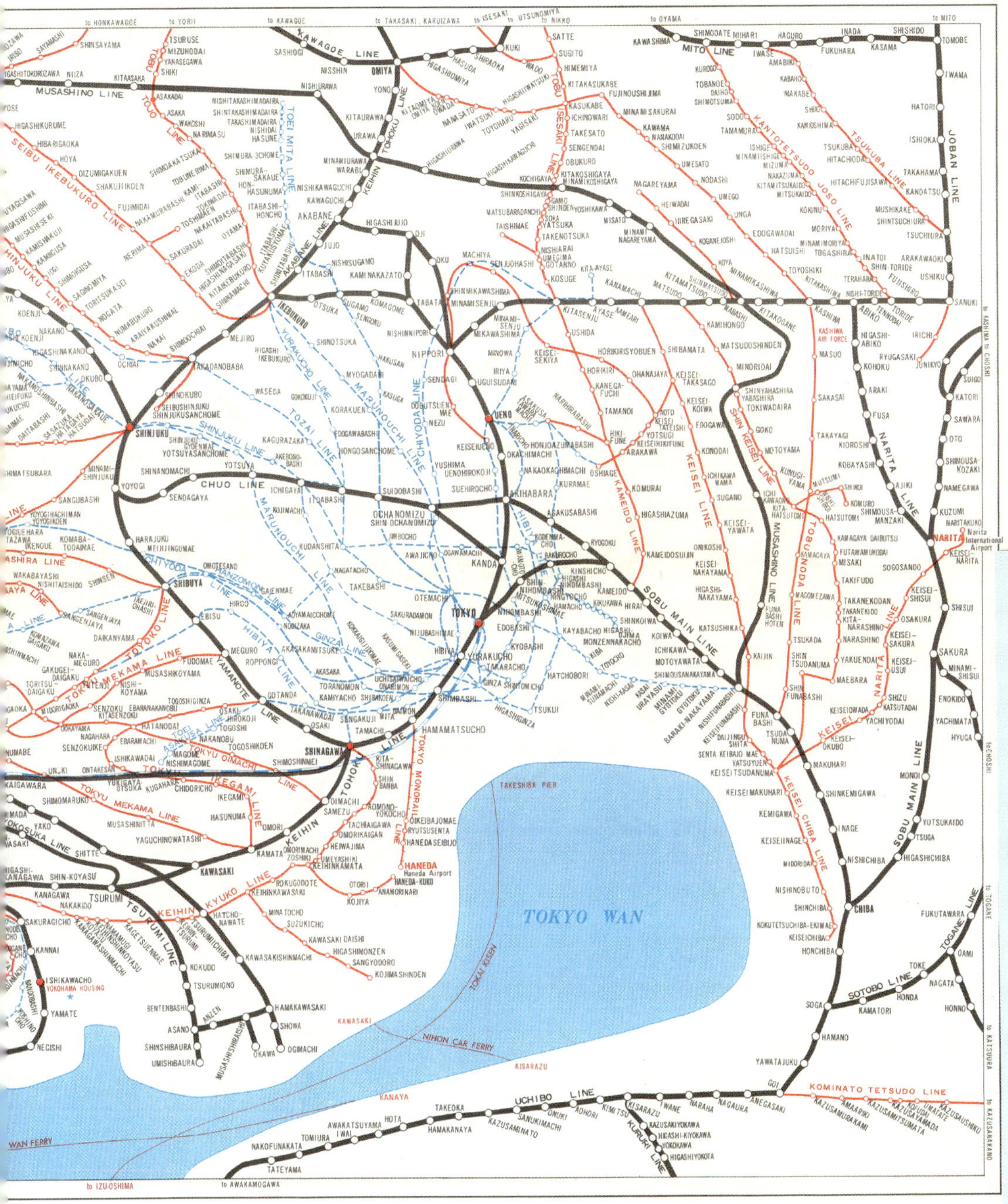

Published by JAPAN GUIDE MAP CO., LTD.
Compiled and Printed by JINBUNSHA CO., LTD.

"UKIYOE"
Painted in the beginning of Meiji Era (1872)

6. RAIL DEVELOPMENTS 1970–PRESENT 243

pride, with particular importance attached to Japan's successful embrace of the possibilities of technology. The Shinkansen system is restricted in its scope and the high-speed line to link Tokyo Central Station to Narita International Airport was stopped in the 1980s because of problems in obtaining the necessary land, although the Narita Express was to follow in 1991.

The 'bullet trains' are very impressive, not least for their punctuality. The main operator of inter-city services is the Japan Railways Group (seven for-profit companies that took over most of the assets and operations of JNR in 1987), but there are also other, private, railways – for example, Keisei Electric Railway Company. The opening of links between operators has been important to improving services. Thus, in 2023 the opening of a six-mile Sōtetsu–Tokyo link that connects the Sōtetsu and Tokyo railways has enabled the running of services through to Tokyo.

This 1975 map, compiled and printed by Jinbusha Co. Ltd. and published by Japan Guide Map Co. Ltd., appeared in a period of flux, with steam locomotives being withdrawn that year from JNR services, while economic expansion and urbanisation had increased commuter pressure in the Greater Tokyo area. That process of socio-economic change continued over the following decades. There was a considerable population transfer from rural areas to the major cities, in part as a consequence of agrarian mechanisation. This move increased the issues of scale in urban congestion and commuting. The map folds into covers (right) and also has a subway station guide, railway station guide and indexes to points of interest. The accompanying black-and-white sketch (see page 243) is from 'the beginning of Meiji Era (1872)' and labelled *ukiyoe*, which refers to a woodblock print of that period (literally, 'pictures of the floating world') depicting everyday scenes. In this case the steam train almost certainly refers to Japan's first train line, opened in 1872 (two decades after a Russian steam locomotive was shipped into Nagasaki), between Shimbashi (Tokyo) and Yokohama, which symbolised Japan's efforts to Westernise.

The centrality of rail to modern Japanese cities was captured in Riku Onda's novel *Eugenia* (2005):

'... if a station wasn't added later for a new bullet train line or to provide airport access ... old regional cities ... develop outwards from the station ... prefectural capitals all tend to look alike. At the front of the station you'll find a traffic circle surrounded by department stores and hotels.'

Tokyo

1:150,000

東京

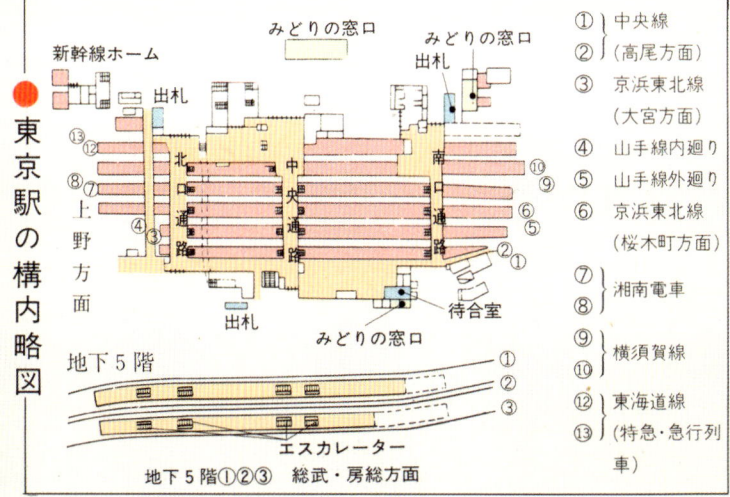

東京駅の構内略図

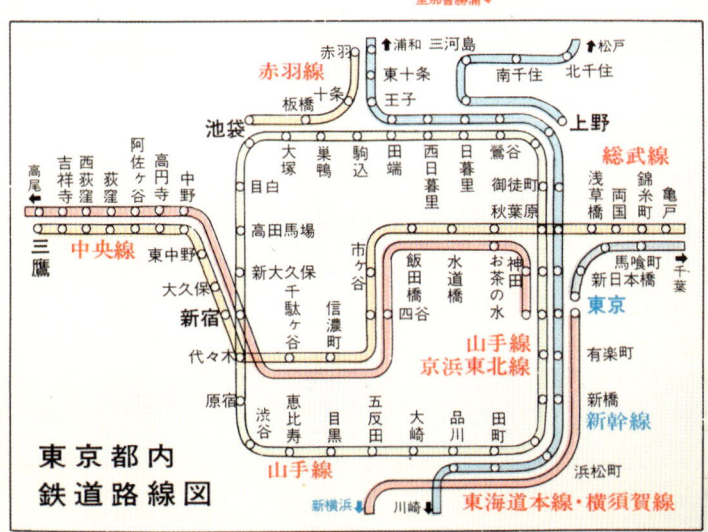

東京都内鉄道路線図

DEUTSCHE BAHN
Headquarters of the German Federal Railways, 'West German Railway Route Map', 1974.

A highly instructive feature of this map is the territorial border of West Germany, which was very much a creation of the decade post 1945 after defeat in the Second World War had led to the division of Germany. Aside from the losses of territorial gains from 1938, beginning with Austria, there was also the annexation of pre-1939 German territory (Pomerania, Silesia and southeast Prussia by Poland; northeast Prussia and Königsberg by the Soviet Union) and the creation of Allied military occupation zones in what is now Germany. The French and Soviet zones each had their own railways, and in the Soviet zone some train lines were dismantled so as to be relaid in the Soviet Union. In contrast, the American and British zones had a joint railway organisation from 1946 and when in 1949 the Federal Republic of Germany (West Germany) was formed from the British, French and West German zones, the French zone's railway was incorporated into this organisation.

The emphasis in West Germany was on repair and this saw the existing infrastructure brought back to the standards of the 1930s, with new steam engines built. The last was delivered in 1959, while the last steam engine ceased regular service in 1977. However, the diesel and electric West German railway faced considerable competitive pressure, in both freight and passenger services, from road. As West German levels of car ownership rose, so did the opportunities provided by a rapidly expanding autobahn system. Partly as a result, rail travel acquired the image of being the choice of failures, or at least of the poor, the old and students.

There was a parallel focus in freight, with the 'old' industries of coal, iron, steel and shipbuilding increasingly of less relative economic significance as light engineering and other industries rose in relative importance. Those sectors tended to rely on lorries for freight and cars for labour mobility, and were generally located in greenfield sites near roads, and increasingly in southern Germany.

There was also a political dimension – with rail associated, at least subliminally, with the state-directed heavy industry of communist East Germany. In West Germany, in contrast, rail suffered from the dominance of the Christian Democrats in government from 1949 to 1969. There was a parallel with the shift in the United States under the Republicans in the early 1970s.

This 1974 map, which shows the railway lines and topographical relief with shading and spot heights, contrasts with the French one of 1976 by Rudi Meyer (see page 13), where the focus was exclusively on the system, which was presented in a geometric form. The role of rail in Germany was actually lessened in maps, because those of the country – and indeed of its transport system – gave prominence to the terrain. Moreover, maps of Germany and also of German transport put an emphasis on the autobahns. As air services developed – the rival to rail – they were helped by the distance between major cities such as Hamburg and Munich.

In 1994, following reunification in 1990, the railway systems of West and East Germany united to form the German Railway Corporation (Deutsche Bahn, or DB). This provided the context for the extension of high-speed lines, which had begun in 1991 with the Hamburg–Mannheim and Stuttgart–Munich services. Hanover–Berlin, opened in 1998, provided a key link across former East Germany, as did the new line from Frankfurt to Berlin. The corporation moved its headquarters from Frankfurt to Berlin in 1996. By the mid-2010s, some of the rail services could reach speeds of 190mph. However, in yet another instance of the mismatch between slick-looking maps and disappointing services, it proved difficult to fulfil expectations. In 2021, only 61 per cent of long-distance trains arrived on time and there was an estimated EUR50 billion backlog of repairs. In March 2023, the Federal Audit Office strongly criticised Deutsche Bahn and the federal government for failing to deal with the railways' 'permanent crisis', its ongoing financial failings and its inability to cope with operational problems.

THE CHANGING SYSTEM: BELGIUM
Société nationale des chemins de fer belges, 'Network Card B: Belgian National Railway Company', 1969 and 1988.

These maps, each produced for internal purposes and with impressive inserts of areas of major rail concentration, show the rapid change in European railways in the late twentieth century with continued electrification and the abandonment of steam linked to a prioritisation of what were seen as major lines. These were very much those between the leading urban centres such as Antwerp, Brussels and Liège. In contrast, much of the earlier network was declared redundant, in particular routes to and between small towns. The 1969 map (right) includes lines being closed and lines that the Société nationale des chemins de fer belges (SNCB), the Belgian National Railway Company, could and could not use, and the 1988 map (below) drives home the amount of closure. For example, by 1988 there was no longer a service west from Jemelle to link the north–south line to the west. Nor was there a service from Marienbourg southwest to the French border.

These maps make no concessions to the physical geography, with no depiction of terrain or rivers, which is a characteristic of many modern rail maps and also of transport maps as a whole. It is as if the process of construction and operation has subordinated the terrain, which has therefore become inconsequential, and this characteristic is seen with the depiction of all long-distance transport systems. Thus, the modern mapping of rail systems, with the emphasis on the system as a whole, is affected by the existence of motorways and of air services.

The Belgian system emphasises the linkage between the cities that are its nodes and provides a form of hierarchy between them. The 1988 map clearly differentiates stations in a hierarchy of significance, and it separates major lines between those that were electrified and those not. Visual clarity is crucial, not least for potential customers who have to assess what means and route to take. Some of the secondary lines are shown as solely for industrial purposes and others as not in use. In the later map there is more electrification (red lines), but far fewer lines overall. That made it easier to present a map of the system.

So also to a degree in France, but again politics play a role in the extent and content of the system. From within the rail industry there can be strong pressure questioning the maintenance of current capacity, notably of underused rural lines. In part, this pressure is driven by regulation and the demand for uniformity; safety requirements alone ensure the need for significant investment if many of these lines are to remain open. The social investment of support for uneconomic lines is generally a frozen-case scenario of sustaining past commitments, not a progressive one of deciding that such lines should be constructed to face present and future concerns, a choice that reflects political assumptions. Thus, in France, lines in rural and upland areas, such as the Massif Central, which are used by relatively few, are sustained rather than using the money to construct similarly uneconomic, but new, schemes in expanding urban areas where public transport pressures are acute. That contrast reflects the significance of legacy issues, in infrastructure, institutionally, financially and in state and public assumptions, when looking at future provision and its contexts.

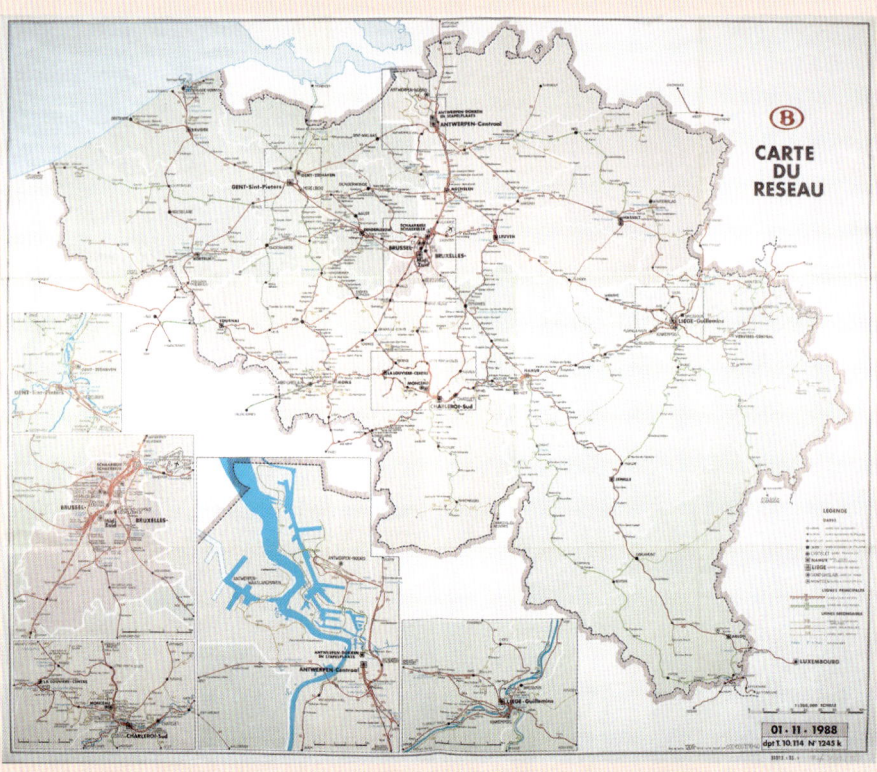

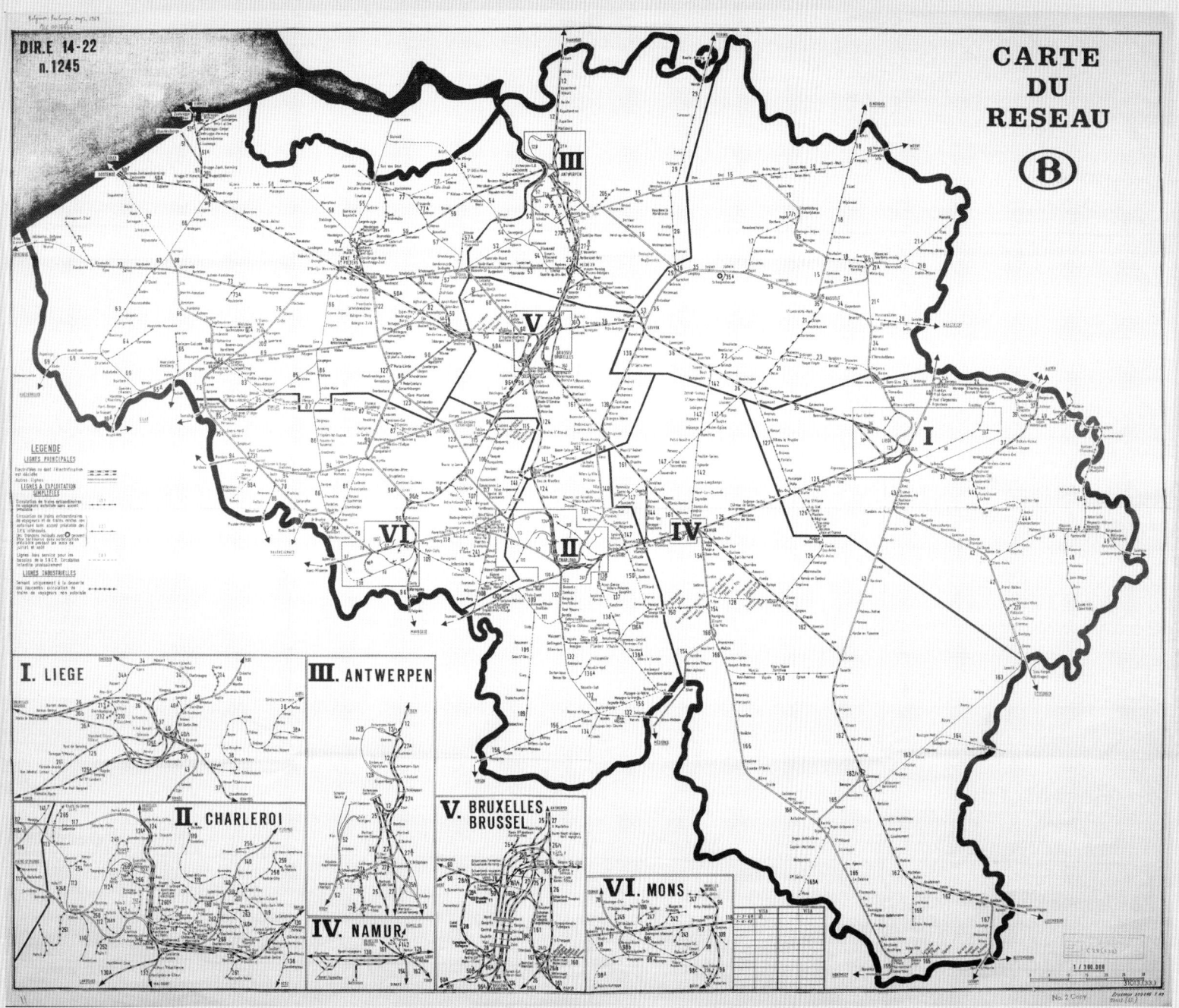

6. RAIL DEVELOPMENTS 1970–PRESENT 249

THE CHUNNEL
'Channel Tunnel Terminal White City', 1974.

Plans for a railway tunnel under the English Channel have not always envisaged the route currently followed, and this has also been the case with the connecting railways and the location of the metropolitan termini.

After earlier project proposals and alongside a detailed geological survey in 1964–65, Britain and France agreed in 1964 to build a state-funded tunnel. Work began on both sides of the Channel in 1974, only to be cancelled by the British government the following year. The 1974 plan was for twin tunnels on either side of a service tunnel. On the British side, the Channel Tunnel portal was between Folkestone and Dover.

As this map indicates (see inset), much of the new rail link was along an existing link, albeit with Ashford, Oxted and East Croydon bypassed, the last by a new rail tunnel. However, during the parliamentary debate in July 1974 concerns were expressed about the volume of traffic, with estimates that by 1990 there would be 280 trains a day. Furthermore, it was suggested in Parliament that additional rail tracks would be necessary for the foreseeable increase in rail traffic, because under the treaty with France an adequate rail link had to be provided.

The map does not indicate the likely disruption caused by the London route but it did entail a crossing of the Thames, which is not marked in the diagram. On the other hand, White City, the envisaged internal passenger terminal, was not as centrally located as the termini eventual built (first Waterloo and then St Pancras). White City was chosen in preference to two others on the shortlist: Victoria (the existing terminus for boat trains, but a crowded site) and Surrey Docks. The Greater London Council (GLC) backed the others as a means to economic regeneration, but they were more expensive because BR owned the White City site and Surrey Docks required more expensive additional rail and road links. White City, as the enclosed plan indicates, was close to existing transport links. Moreover, from there it would be possible for trains to go further into England.

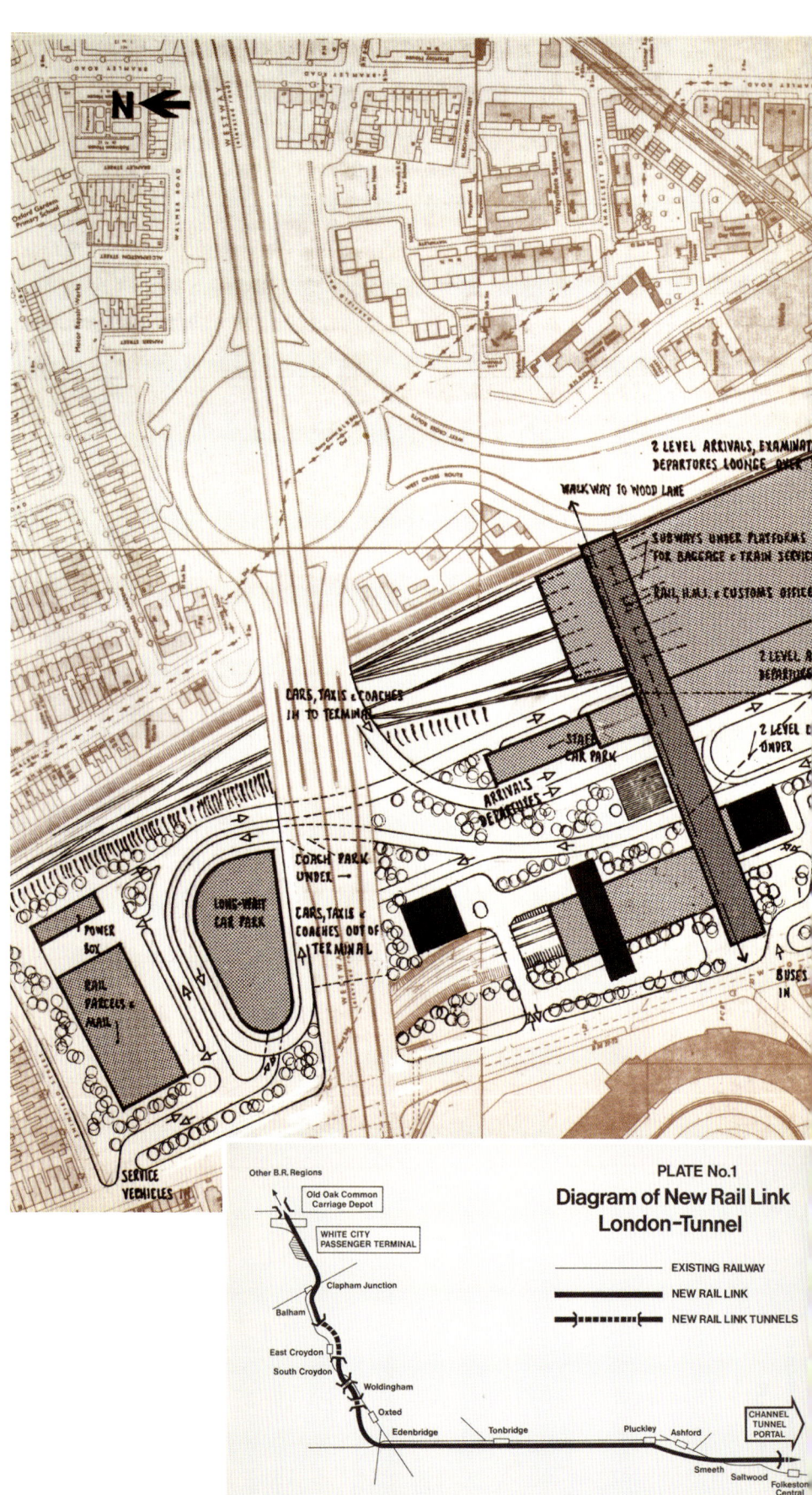

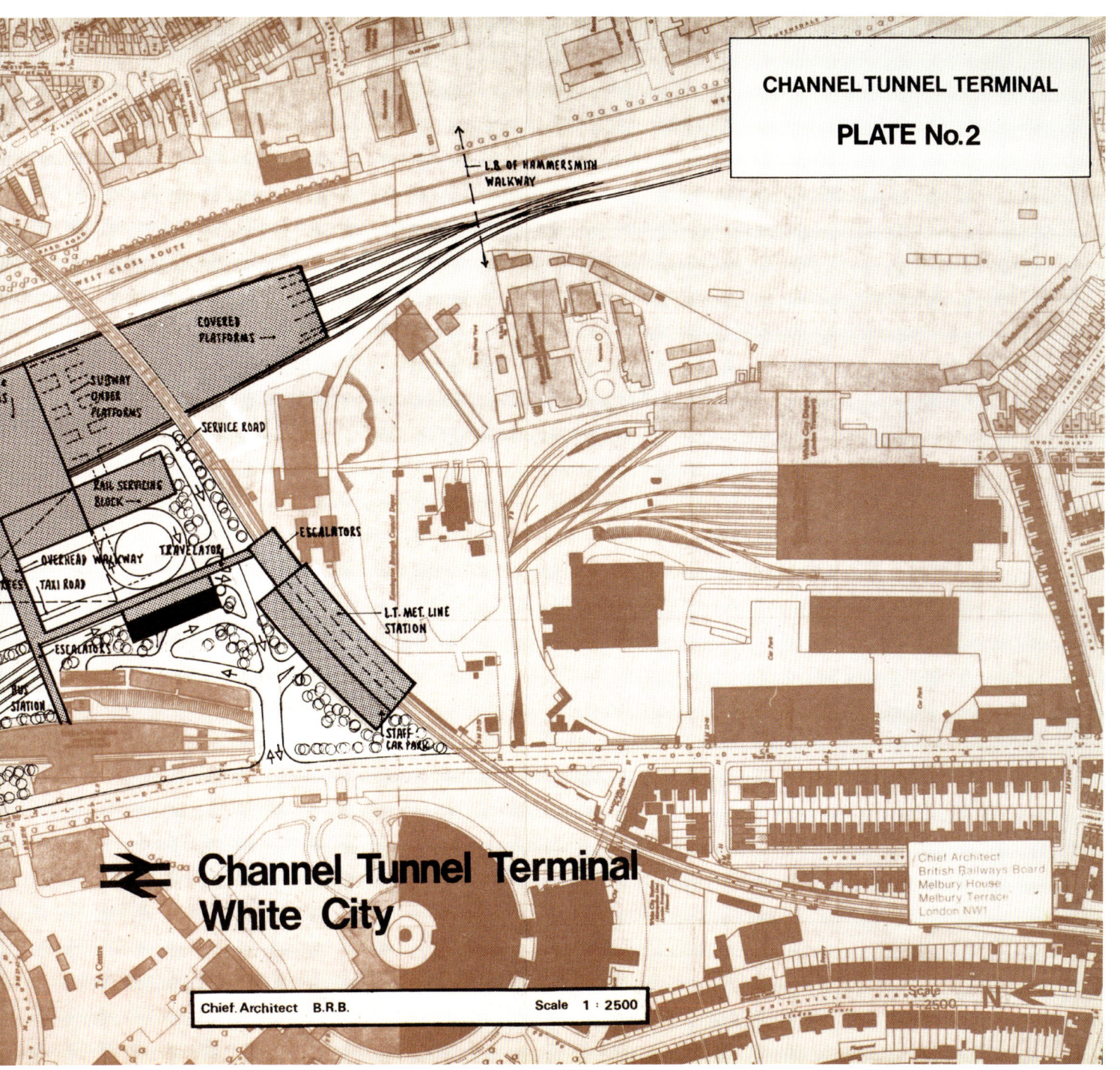

6. RAIL DEVELOPMENTS 1970–PRESENT 251

7

FACING THE FUTURE

In March 2023, the Surface Transportation Board (STB) in the United States approved an application, filed in 2021, from Canadian Pacific and Kansas City Southern to merge their two railways (in practice an acquisition of the latter) and create a single integrated Canadian Pacific Kansas City network connecting Canada, the United States and Mexico. This US$31 billion merger was the biggest rail news of early 2023, but it was largely ignored outside the specialist press and, more generally, some North American commentary. Certainly, commentators in Europe devoted insufficient attention.

Nevertheless, the scale of the new merger was instructive, as were the arguments for it, which included moving traffic from road to rail, thus improving safety and cutting carbon emissions. The acquisition, the first major American one since the 1990s, combined the sixth- and seventh-largest railways operating in the United States, which cut across the tradition of wariness about consolidation. However, the acquisition was defended by regulators on the grounds that it would not reduce competition, but would encourage investment and economic growth as well as shareholder value, the last an important element in encouraging investment. Specific economic benefits were discerned, including the flow of grain from the Midwest to the Gulf of Mexico and Mexico, and the movement of goods between Chicago and Dallas, and then between the United States and Mexico. The regulatory environment was clear, the STB declaring that it would require the new railway to justify 'rate increases over a certain level' on some movements.

Such developments indicated the vitality of North American rail. So, differently, did the private sector inter-city train operator Brightline's extension of its Miami–West Palm Beach service to Orlando in 2023.

Furthermore, given the need to confront economic change and the scale of investment required, mergers might well increase. To maintain that vitality, it was preferable for change to take place within a regulated commercial model rather than to be directed by the state and dependent on state funding. Yet it is precisely the latter that tends to dominate the world of mapping, not least because of the important part played by political actors. This is a crucial aspect of the rhetoric of mapping notably because political purposes are bolder in conception and also in display (see pages 254–255). Any understanding of the functional aspect of railway maps needs to be extended to include this rhetoric.

The interplay of rhetoric, functionality and geographical place was seen in 2023, when, as a sign of continuity, Derby became the head office of the newly established Great British Railways. This is envisaged in the rhetoric as the creation of a centrally controlled 'single guiding mind' for train contractors and infrastructure managers, under the control of the Secretary of State for Transport. Great British Railways is to be the successor of Network Rail, the headquarters of which is at Euston Station. Thus, there was a move of headquarters from London. An online referendum in 2002 had led to 205,000 votes cast. Derby is home to the historic Litchurch Lane Works and presents itself as the 'geographical and strategic heart of Britain's railways', having beaten off claims from Birmingham, Doncaster, York (home of the National Railway Museum), Newcastle (which claimed to be the birthplace of the railways because of its links to the Stephenson family) and Crewe (home to the industry's supply chain). As an instance of politics, Milton Keynes (where the Network Rail Operations Centre, the Quadrant building, employs 5,000) did not made the shortlist, which was defined in effect by ministers in terms of the political desire of helping the North by 'levelling up'.

In practice, a map of the choices would provide no indication of the problems currently facing rail in Britain. Income from fares has fallen dramatically, in large part due to the changes in commuting as a result of the post-pandemic growth of hybrid working; in April 2021–22, although £13.3 billion came from the government, fares provided only £6.5 billion compared to £10.4 billion a few years earlier in the 2018–19 financial year. Moreover, by 2022 more money was being spent on servicing Network Rail's accumulated debt than on maintaining track infrastructure. Alongside strikes there have been terrible levels of service, and concerns about the distribution of investment. There were particular concerns about the quality of regional and local services within the North, while the respective regional value of rail was suggested by data that revealed daily arrivals in London by train were over eight times greater than in the next city, Birmingham. In Britain, the birthplace of rail, the rail industry is in serious trouble.

More generally, the worldwide economic and fiscal problems of the early 2020s, including rising interest rates and high rates of inflation, as well as reductions in economic growth, have affected rail plans because it has become clear that ambitions can not be fulfilled, either in terms of existing services or new routes. This is seen in a variety of countries, including Australia and Britain, as well as in the failure to proceed with rail plans across much of Africa.

The dependence of rail projects on public funding leaves them vulnerable to problems with government budgets (see pages 258–259), not least accumulated debt and other expenditure priorities. Political weight plays a role in the resulting failures and successes. For example, in the United States in March 2023 the Southeastern Pennsylvania Transportation Authority (SEPTA) paused the development of the King of Prussia Rail project after it was not recommended for federal New Starts Program funding. The same month in Southeast Asia (see pages 256–257) saw the opening of Kuala Lumpur's Putrajaya Line. The contrast between global regions in the extent to which there is an expansion of networks and services provides a key context within which optimism based solely on the most successful routes should be critically assessed.

A CONNECTED EUROPE
Rail Baltica, *Linking people, nations and places*, 2022.

Planned in the 2010s, and confirmed in 2021 with the formal signing of an agreement by transport ministers from Estonia, Latvia and Lithuania, Rail Baltica is the product of the post-Cold War geopolitics of the Baltic states. From the early age of rail until 1917 and then again in 1940–41 and 1945–91, the three were part first of Russia and then, from 1940 and, more firmly still, 1945, of the Soviet Union. The rail routes of the region ran not north–south but from Russian centres out to Baltic ports, such as Klaipeda and Riga. Rail Baltica is designed to offer a different axis, one shown in these maps (below) from the 2021 news conference held in Tallinn, the capital of Estonia, when the first contracts were signed. This axis is a north–south one: from Warsaw, the capital of Poland, a stalwart opponent of Russian expansion, via Kaunas in Lithuania, with a main spur to its capital Vilnius and a lesser spur to the Baltic port of Klaipeda, north via the Latvian capital of Riga to Vilnius and then with a possible tunnel to Helsinki in Finland. The map underplays the Russian Baltic exclave of Kaliningrad (formerly Königsberg), which is deliberately bypassed by the new route.

The other map (opposite), from the official Rail Baltica brochure *Linking people, nations and places*, proclaims 'Connected Baltics in a Connected Europe', with the new line shown linking the Baltics to Poland and Finland depicted in a bolder colour than the rest of Europe, from which post-Brexit Europe is excluded. A 540-mile-long track is projected with electrification and standard-gauge track rather than Russian gauge. The initial intention was for completion in 2026, but this has been pushed back to 2030. As the map indicates, a straight route was preferred in order to produce efficiencies in construction and operation, notably greater speed. The EU has paid about a one-quarter of the costs. The Russian invasion of Ukraine in 2022 underlined the geopolitical rationale for the line.

This invasion ensured the destruction of part of Europe's rail network at the same time that plans such as Rail Baltica were in place. Air, missile and drone attacks, as well as ground action and bombardment, all posed problems. As a result, in September 2023 Ukraine State Railway responded to Russian drone attacks by imposing restrictions on the export transport of cargo to the Danube port of Izmail. The previous month, the construction of a standard-gauge 'backbone' was proposed to integrate the rail networks of Ukraine and Moldova with that of the EU.

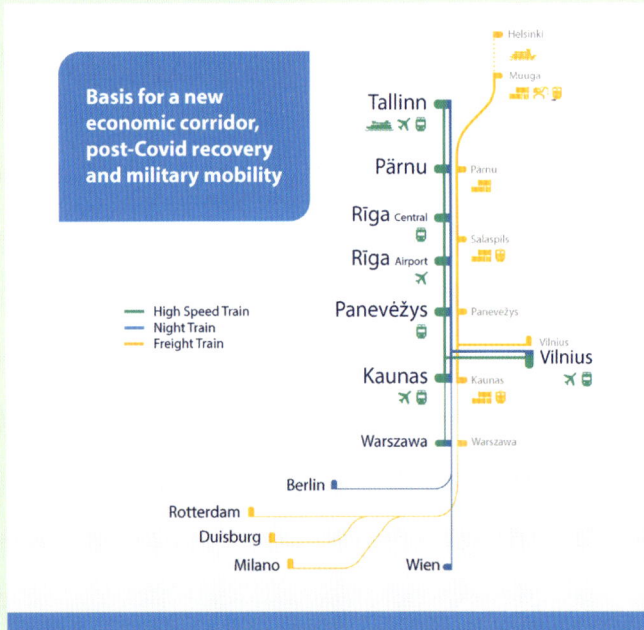

Rail Baltica

Connected Baltics in a Connected Europe

We are delivering seamless mobility for people, goods and services to accelerate social and economic development in the Baltics and beyond.

7. FACING THE FUTURE 255

THE NEW SILK ROAD

Mercator Institute for China Studies, 'Chinese Belt and Road Initiative', 2018.

The largest state-driven rail plans for the present and into the future are those of China. This is somewhat ironic as China's economy is more dependent on maritime links. Nevertheless, geopolitical attention centred on China's attempt to develop rail connections into the Asian interior and beyond. This represented a particular type of geopolitical speculation (that of rail-based links), but also the focus is on routes rather than the nodes of harbours and their related loading and unloading facilities. There is a rail equivalent to those, which would mean putting the stress on junctions and the related marshalling facilities, but neither has ever attracted the attention devoted to railway lines.

In November 2022 Mongolia opened a new rail link to China, between Zuunbayan and Khangi, in order to move products from mines, including Rio Tinto's Oyu Tolgoi project, to industrial hubs such as Baotou in the Chinese autonomous region of Inner Mongolia. This is one of three new rail links from Mongolia to China that were intended to boost Mongolian export capacity. Rail was presented as less expensive and quicker than the use of lorries, while the latter had been interrupted frequently due to Chinese concerns that drivers would transmit the Covid-19 virus.

The 2018 map shows the existing China–Russia network clearly, as well as the Chinese plan to expand it in Central and Southwest Asia to provide a through route to Istanbul and on to Budapest. The latter provides an alternative route for China to avoid Russia.

President Xi Jinping's Belt and Road Initiative (BRI) is an important component of his strategic vision for China, by which international aid for infrastructure is linked to the development of 'rail corridors' furthering Chinese strategic interests that are at once geopolitical and economic. For example, in April 2015 Xi launched the China–Pakistan Economic Corridor, a bilateral infrastructure project that envisaged modernisation of Pakistan's road, rail, air and energy transportation systems, including connecting the deep-sea Indian Ocean port of Gwadar by an overland rail route to China. As yet, the Pakistan line has not been built whereas the shorter China–Laos railway to Vientiane, the capital of Laos, was opened in 2021, and is seen as a stage in a route to Singapore, which is regarded as part of the Trans-Asian railway network adopted by the meeting of the Economic and Social Commission for Asia and the Pacific held in Jakarta in 2006. In practice, however, this network shows few signs of coming to fruition. In July 2023 officials from Afghanistan, Pakistan and Uzbekistan agreed the route for the proposed Trans-Afghanistan Railway, a joint initiative to promote regional connectivity and provide Central Asia with access to Pakistan's ports.

Meanwhile, the intended Chinese route westwards through Central Asia is a matter of pressure, diplomacy and investment. In 2022 China signed an agreement with Kyrgyzstan and Uzbekistan that fixed the route of a proposed new railway to European markets without going through Russia, as did existing links via Kazakhstan. Sadyr Japarov, the president of Kyrgyzstan, emphasised this as an assertion of his country's independence from Russia. For landlocked states like Kyrgyzstan and Uzbekistan, rail links, respectively to China and across Afghanistan to Pakistan, offer economic and political possibilities, but face serious political, security and financial issues. In October 2023, at the third Belt and Road Forum in Beijing, Kazakhstan agreed to buy up to 200 main line and shunting locomotives for its national railway.

Earlier, in 2016, Nepal and China agreed on a high-speed railway from Kathmandu to the border with China, part of the Chinese plan to win over Nepal from India. Nepal, however, cannot afford the cost, which will be high due to very mountainous terrain.

The Chinese railway boom has much to do with high-speed lines' potential to carry passengers, troops and freight, but most freight – indeed, a growing percentage of freight – within China is moved by lorry. The boom is driven by politics as much as economics (for example, the line to Lhasa in Tibet). There is the potential for political disorder to affect rail infrastructure and

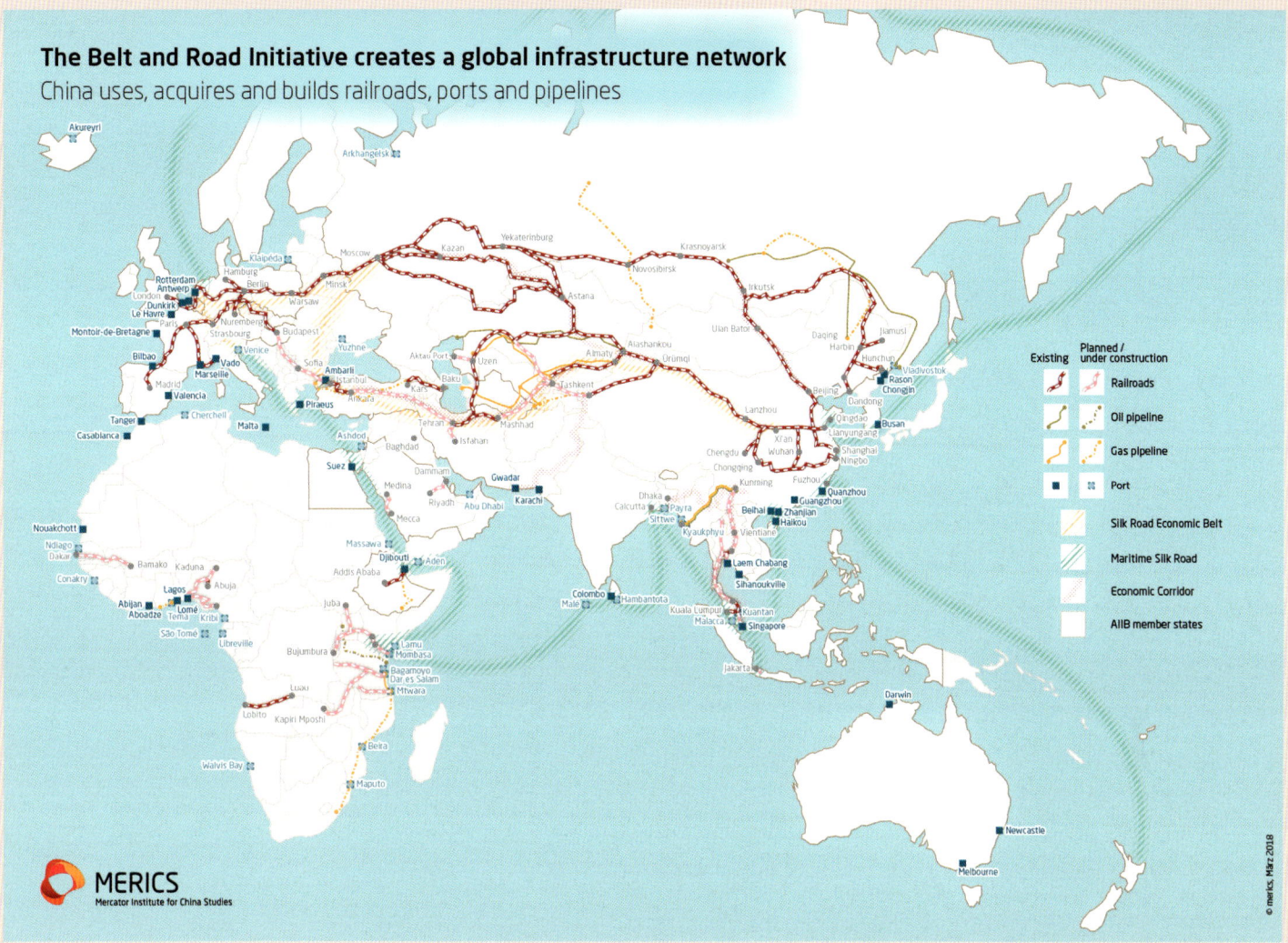

The Belt and Road Initiative creates a global infrastructure network
China uses, acquires and builds railroads, ports and pipelines

options, as with the war in Ukraine that ended BRI route plans via Ukraine as well as those across southern Russia. Yet, any option to the south was affected by wider regional instability and the potential for more, notably in the Kurdish areas of Iran, Iraq and Turkey, as well as in disputes within the Caucasus, particularly between Armenia and Azerbaijan, as well as regions of Georgia.

Yet, geopolitics aside, overland trade from China to Europe has not prospered primarily for economic reasons. Railway transport costs remain higher than shipping; indeed, containers have widened the gap. The financial problems facing the BRI plans led to a decline in enthusiasm in the early 2020s, not least as Chinese growth rates fell. Chinese railways, old and new, provide few links to Europe. Indeed, in 2022 the port of Hamburg and its Chinese state-owned partner Cosco, which owns the Greek port of Piraeus and shares of those at Antwerp, Bilbao, Hamburg and Zeebrugge, looked at setting up joint shipping projects in Europe.

The climate change that makes polar links possible as a result of the loss of Arctic ice may challenge the relative economic appeal of rail, in both North America and North Asia, while also more specifically affecting individual routes, not least due to permafrost melting – a particular issue across Siberia. Container travel by sea and rail faces similar issues in terms of local factors and trans-shipment requirements, but rail has the additional issue of being a fixed route with associated costs, as well as vulnerabilities to attack. The ship offers adaptability that can reduce the risk dimension of the choice for sea over rail.

Despite the emphasis on rail links, China depends on maritime ones. China has one of the lowest amounts of per capita arable land. Its energy and food imports largely come by sea, including half of its oil imports and its gas. These essential imports will continue to arrive by sea.

RAIL BRITANNIA

High Speed Two Ltd, 'HS2 Service Map', 2023 (prior to cuts announced in autumn 2023).

From the Beeching Report of 1963, a theme in Britain had been the future of rail. This principally involved debates over privatisation, difficult labour relations, the introduction of new trains and, much more rarely, ideas for new routes. There were also new map ideas, such as that of 'Inter-City Services' in May 1982. In that map, which drew on Meyer's 1976 map of the French system, the country was depicted, without a coastline, shaped by the rail network and the metropolitan focus was enhanced by the depiction of 'Outer London pick-up set-down points', such as Reading and Luton.

Following on the new – limited and successful – rail link (HS1) from London to the Channel Tunnel, the HS2 project was proposed in 2009 as a new route from London to Birmingham, with new branches onward to Leeds and Manchester. It was argued that projected passenger demand made this new route essential. Alongside this demand-led commitment, there was also the belief that such a link was necessary to affirm the government's commitment to 'the North', a commitment that was to be summarised by the term 'Northern Powerhouse'. After review, this plan was accepted by the Conservative–Liberal Democrat coalition. The relevant legislation took several years, and faced increasing criticism, but the independent Oakervee Review of HS2 in 2019–20 recommended implementation and, in 2020, construction companies were allowed to begin work. Rapidly, there was an impact. The line between London and Birmingham involved deforestation of part of the Chilterns as the route was prepared.

By July 2023 HS2 had an estimated cost of £85 million per kilometre excluding land purchases and stations, compared to a cost of £13 million in Continental Europe in order to build a high-speed railway. The difference was largely explained by the viaducts and tunnels added in Britain to the plan in order to overcome local objections. This map of the HS2 network as envisaged in early 2023 reflects the major cuts made in 2022 to meet the costs of Covid-19, cuts that ended HS2 East running northward from East Midlands Parkway to Yorkshire.

The financial and commercial viability of the HS2 high-speed rail link from London to Birmingham and 'the North' was always dubious but it has been additionally undermined by the impact of Covid-19, while in 2023 there were strong suggestions in *The Sunday Times* of the suppression of information about poor cost management and knowingly unrealistic financial predictions. In late 2023, on 4 October, the plan was dramatically curtailed with the end of HS2 East and HS2 phases 2a and 2b. The announcement meant that of the 540-kilometre-long network initially planned, only the 190-kilometre-long Phase 1, linking London with Birmingham and with a junction with the West Coast Main Line, the section actually under construction, was going ahead. In his speech, Prime Minister Rishi Sunak claimed:

> '... a false consensus has taken root that all that matters are links between our big conurbations. This consensus said that our national economic regeneration should be driven by cities, at the exclusion of everywhere else. It said that the most important connections those cities could have was to London and not anywhere else. And it said that the only links that mattered were north to south, not east to west.'

Instead, Sunak proposed more investment in regional rail and road links, notably in northern England.

In late 2022, the total number of trips by all modes of transport in London remained 11 per cent down on pre-Covid-19 levels. This ensured not only the general crisis in government finances that had led directly to the abandonment of the Yorkshire branch, but also a growth in 'working from home' (and the hybrid work model) that led to questions of likely demand for rail services.

Prior to Covid-19, the future of the major railway termini looked promising, with redevelopment scheduled for the ugliest in London – Euston – and the prospect of rising passenger numbers and the increased office rents so important to redevelopment projects. The situation now looks less promising. Moreover, there has been a changing pattern of car ownership, with an increase in the registration of private cars in 2011–21, especially in cities outside London. The chance of a new system of London termini, with a small number of central stations, appears less likely and the long-established pattern will probably continue.

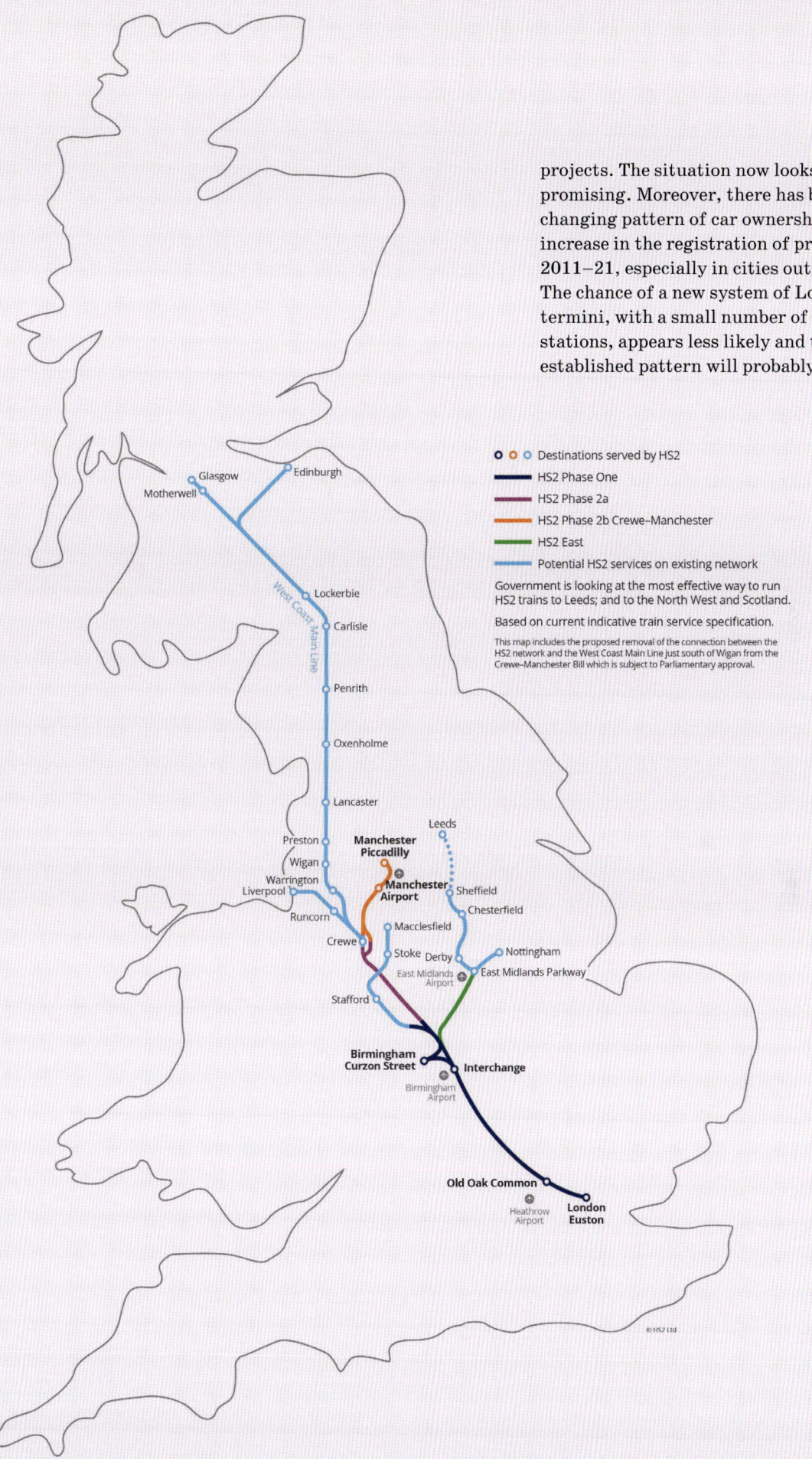

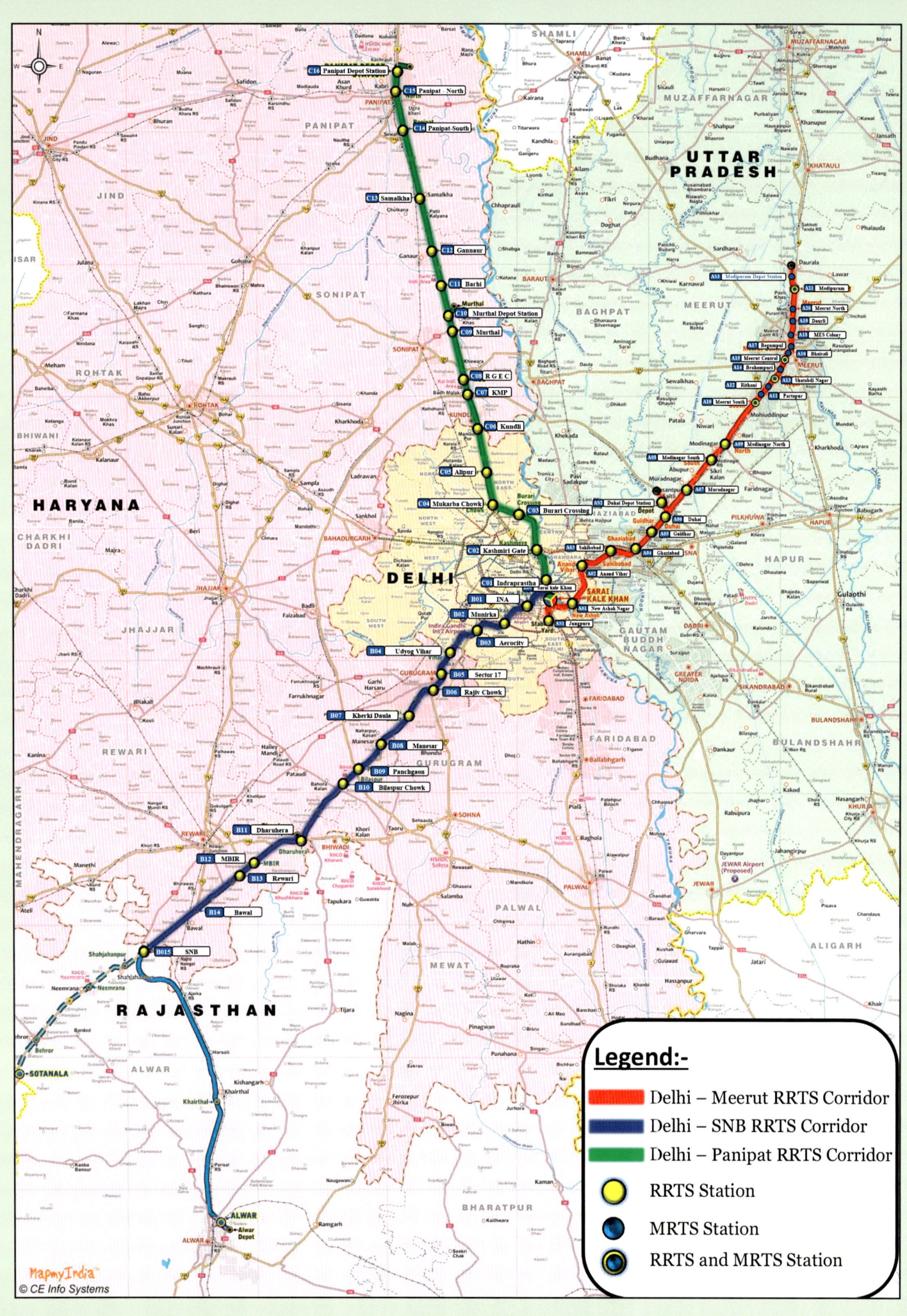

WORLD-CLASS COMMUTING

NCRTC, 'RRTS Network Phase 1', 2023.

In India, on 20 October 2023, Prime Minister Narendra Modi declared that the Delhi–Ghaziabad–Meerut Regional Rapid Transit System (RRTS) is a 'glimpse of India of the future', before the following day passenger services began on the initial section of the commuter railway. This map shows the RRTS's three priority corridors and its stations, as well as the crucial relationship with the metro.

The modern development of India's railways centres on Delhi as part of a commitment to develop the capital beyond the British pattern into what is presented as an appropriate post-colonial megalopolis (Prime Minister Modi referred in 2014 to the need of Indians to move beyond a 'slave mentality'), also seen in the building of a new Parliament. There is also the practical requirement to respond to a rapid increase in population, with India having passed China as the world's most populous country.

India's rising prosperity has encouraged bold plans for rail development, with rail in particular having significance as a means to move beyond the British legacy of transport infrastructure, which now appears dated and inadequate, as well as something that – Modi, as leader of the Bharatiya Janata Party (BJP), stressed – previous Congress Party governments had not replaced.

India has many plans for high-speed lines. For example, the National High Speed Rail Corporation Limited (NHSCRCL) was established in 2016 to connect Delhi, Mumbai, Chennai and Kolkata. The RRTS is designed to provide a commuter parallel in the Delhi region, and it is a product of the National Capital Region Transport Corporation (NCRTC) and a joint venture company of the government of India and the states of Delhi, Haryana, Rajasthan and Uttar Pradesh.

The system, with which Modi has been closely associated, is planned as a series of high-frequency corridors, providing fewer stops and higher speeds (of up to 180kph) than existing railways, and also extending farther and travelling faster than the metro. Based on a Task Force established in 2005 and a detailed project report that was commissioned in 2010, three corridors – from Delhi to Meerut, Alwar and Panipat respectively – were prioritised out of the eight identified. The first section of the line, from Delhi to Sahibabad, on the Delhi–Meerut corridor was begun in 2019 and inaugurated on 20 October 2023 by Modi, who had laid the foundation stone on 8 March 2019. Most of the 82km-long Delhi–Meerut line is elevated.

The RRTS network offers world-class commuter transit services and it ensures the convenience of quality last-mile connectivity, addressing the needs of all categories of travellers on the network. The network has been planned and spatially oriented to ensure seamless integration with the Indian Railways, Inter-State Bus Terminals (ISBTs), airports and the Delhi Metro.

The expansion of the Delhi Metro will be built with driverless technology – indeed, across the world driverless trains using automated systems have become more common. In London, the Docklands Light Railway has had automated trains since 1987. In December 2022 it was announced that a fourth line on the Paris Métro would go driverless, and in 2023 the first section of Montreal's automated light metro was inaugurated.

It is appropriate to end this account of the future with automation and light urban railways, because the bulk of the world's population lives and works in major cities, and railway relevance for most passengers rests on this capability. Moreover, an emphasis on such railways serves to underline the extent to which the grandstanding of geopolitically ambitious rail schemes, while impressive in terms of some mapping, does not capture the bulk of rail activity. However impressive, most rail transport, both freight and passenger, is not transcontinental, nor indeed international, it is instead a matter of shorter and more regular journeys, for both goods and people. The challenge in mapping the rail of the future is to explain and discuss these in often very crowded human spaces, notably cities and industrial belts. Nostalgia for the maps of the age of steam, however plentiful in historic collections, can only take us so far.

RAIL AND THE COLLECTIVE IMAGINATION

The art of rail was to be captured in many forms, but in two essential types: one by and for rail; the other inspired by rail. The functional aspect of art for the rail industry was seen, in particular, in posters and timetables that both displayed services and advertised them. Appealing to potential customers was a key aspect of the latter (see pages 268–271 and 276–279). Although there were cases of state monopolies that used imaginative and artistic display, that was very much not usually the case, as I discovered when travelling on Czech (1978), East German (1980) and Hungarian (1978) railways and when visiting India and Cuba, each twice, in the 2010s. Moreover, the episodic and unpredictable character of rail in Cuba helped to explain the way it was characterised in the outside world, with its frequent cancellations and unpredictable delays, even on the main line from Havana to Santiago.

In contrast, private enterprise sought to attract customers and designed material accordingly. This could be a simple and inexpensive form, as with the Great Northern Railway Company of 1889 to 1970, which ran the most northerly of the United States' transcontinental railways. It had a distinctive logo, a Rocky Mountain goat, which was based on a goat that William Kenney, president of the company from 1932 to 1939, had used to deliver newspapers in Minneapolis. Typical of its inexpensive advertising was the cover of the timetable for passenger service for 1948. This used only three colours, depicted the network as well as shared services and principal other services, and had the image dominated by five of the company's engines. The cover also referred to two of the named services, the Oriental Limited, which ran the Chicago to Seattle service as a secondary through train to the Empire Builder (the flagship train introduced in 1929, which

The Railway Station, by William Powell Frith, 1862, is set in London's Paddington Station, designed by Isambard Kingdom Brunel and built only a decade earlier.

took 45 hours and is still the name of an Amtrak service). The former was renamed the Western Star service in 1951, taking 58 hours, and both services were shown in a 1959 timetable. These services provided a more vivid colour for the map.

Different to those who were employed by the rail industry were those artists, musicians and writers who, inspired by it, addressed the full experience of rail, from sight to sound. Some illustrations were primarily descriptive, as in the interior of the train shed at Euston published by Ackerman and Co, London, in 1837 that made readily apparent the difference between covered and open stations. Others were more impressionistic, as in William Powell Frith's *The Railway Station* (1862), an engagement with a new and fascinating public space where different occupations and classes jostled each other. In this painting, the Victorian family and the arm of the law are both in evidence, and the station provides the opportunity for the storytelling that was so frequently involved in the artistic perception of rail.

Musical pieces engaged with the experience of train travel, but novels and pictures contained more frequent references. As far as literature was concerned, trains propelled characters into and through the narrative, but also the setting provided opportunities for conversation. Indeed, the carriage brought into close proximity characters of very different backgrounds and character. They were unchaperoned spaces.

Rail was part of the experience and vocabulary of the modern world, which even included what was not, in practice, part of the history of railways. Thus, the 'Underground Railroad', by means of which enslaved people fled northwards in the nineteenth century from the American South, was not a physical railway but a network of escape routes and helpers.

Rail itself has left a rich and fascinating range of museums, and many historic services, with revival societies often linked to the latter. There are also more specific memoires: in Santa Clara, Cuba, can be seen the derailed five wagons of an armoured train used by the forces of the dictator Fulgencio Batista, which crashed when the rebels under Che Guevara employed a bulldozer (also still on the site) to raise the tracks.

Yet, as an iconography for the twenty-first century, rail no longer has the imaginative potency it enjoyed in the late nineteenth. Whether that will change is part of rail's future and that of all of us.

A GEOGRAPHICAL BOARD GAME

John Jaques & Son, *The New Game or The Royal Mail or London to Edinburgh By L. & N.W. Railway*, c.1850.

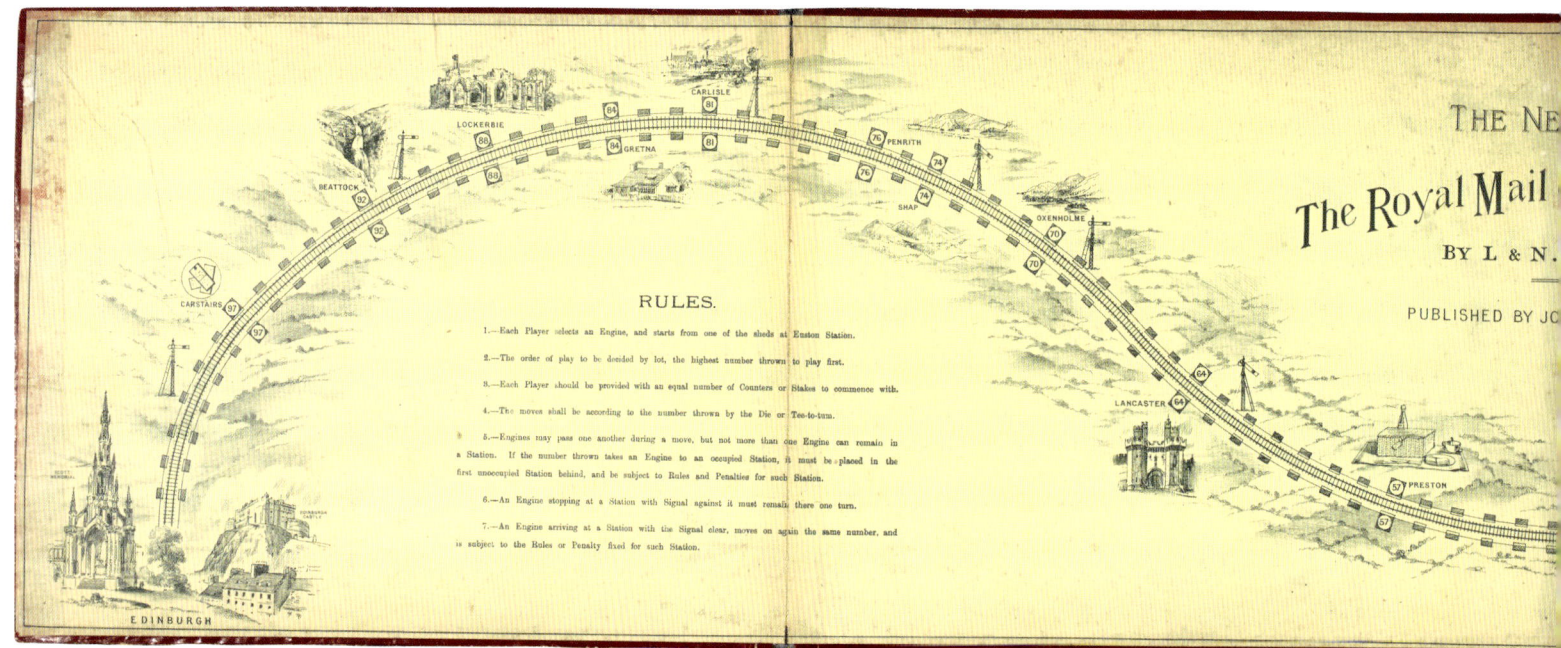

The game began at Euston, the headquarters of the London and North Western Railway (LNWR), where the Greek Revival entrance arch (built in 1837 and demolished in 1961) is depicted, and there are four engine sheds, each of which can provide the base for a competitor. Building on existing practices for board games, this was a dice-arbitrated race game on an illustrated board, which included the junction at Crewe – a town that was very much created by the needs of rail. There was information in the rules about the nature of rail travel, as in rule 6: 'An Engine stopping at a Station with Signal against it must remain there one turn.' Stopping at stations led to payments, some of which involved providing players with information about the railway, as in Wolverton and Crewe, where the railway works had to be visited, and Shap, where it was necessary 'to pay two stakes for an extra engine to take train over Shap Fells'. That was a pertinent introduction to the issues posed by terrain, also alluded to in the snaking route to Edinburgh, decorated with vignettes of places through which the line ran. These include views of Lichfield Cathedral, Crewe, the Lake District, Gretna and, finally, Edinburgh (with the Castle and the Walter Scott Memorial). The Westminster-based company was launched in 1795 by Thomas Jaques and continued to play a major role in publishing games, including playing cards incorporating views of English landmarks, under his grandson John Jacques the younger. Other rail-themed British puzzles included the *Great Western Railway*, The *Torbay Express* jigsaw puzzle of 1930 and its *Cornish Riviera Express* counterpart.

The construction of a railroad also played a central role in some board games, with competing lines advancing across the map

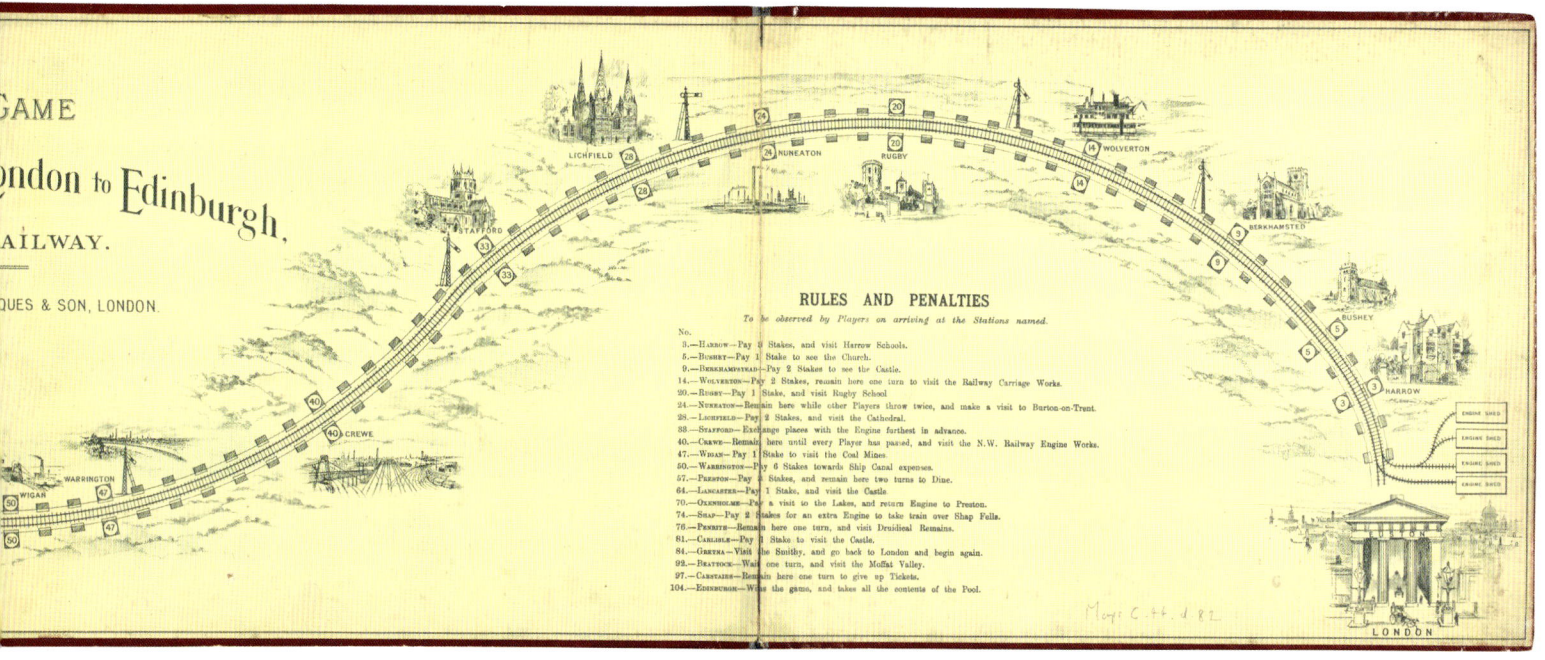

accordingly. When I was a child, I frequently played *Railroader* (1963) – 'A race, in the Wild West, to pioneer the first railroad from Junction City to Buffalo Creek', in the face of competitors who might employ dynamite, as well as landslides, hostile Native Americans, and other contingencies, which used an imaginary landscape, but, on the box, a real map of the United States. Such a combination was typical of many games.

A key later instance, from 1982, was *Empire Builder*, a railway-building game played on a board of the United States and southern Canada and requiring the building of a railway linking six out of the seven major cities. This was followed by new editions, which included Mexico, as well as by a sequence of new areas, notably in *EuroRails*, *Nippon Rails*, *British Rails*, *Russian Rails* and *Australian Rails*. Produced in 1994, the last of these versions involved constructing rail lines on the map (which wiped clean after the game), with the requirement to build a network including Perth and three out of Adelaide, Brisbane, Melbourne and Sydney. Terrain costs were provided, from mountains to rivers. In the case of the *Railroader* and *Empire Builder* games, there was competition between railway companies but no political dimension.

America is the setting for the original basic version of *Ticket to Ride*, which has become an award-winning series with a number of versions, including *Europe* (2005), *Germany* (2006), *Nordic Countries* (2007), *Switzerland* (2007), *The Heart of Africa* (2012), *Netherlands* (2013), *United Kingdom* (2015), *France* (2017), *Poland* (2019), *Japan* (2019) and *Italy* (2019), with the *Europe* version having a slant in terms of international competition. By 2022 there had been ten million copies of these games sold.

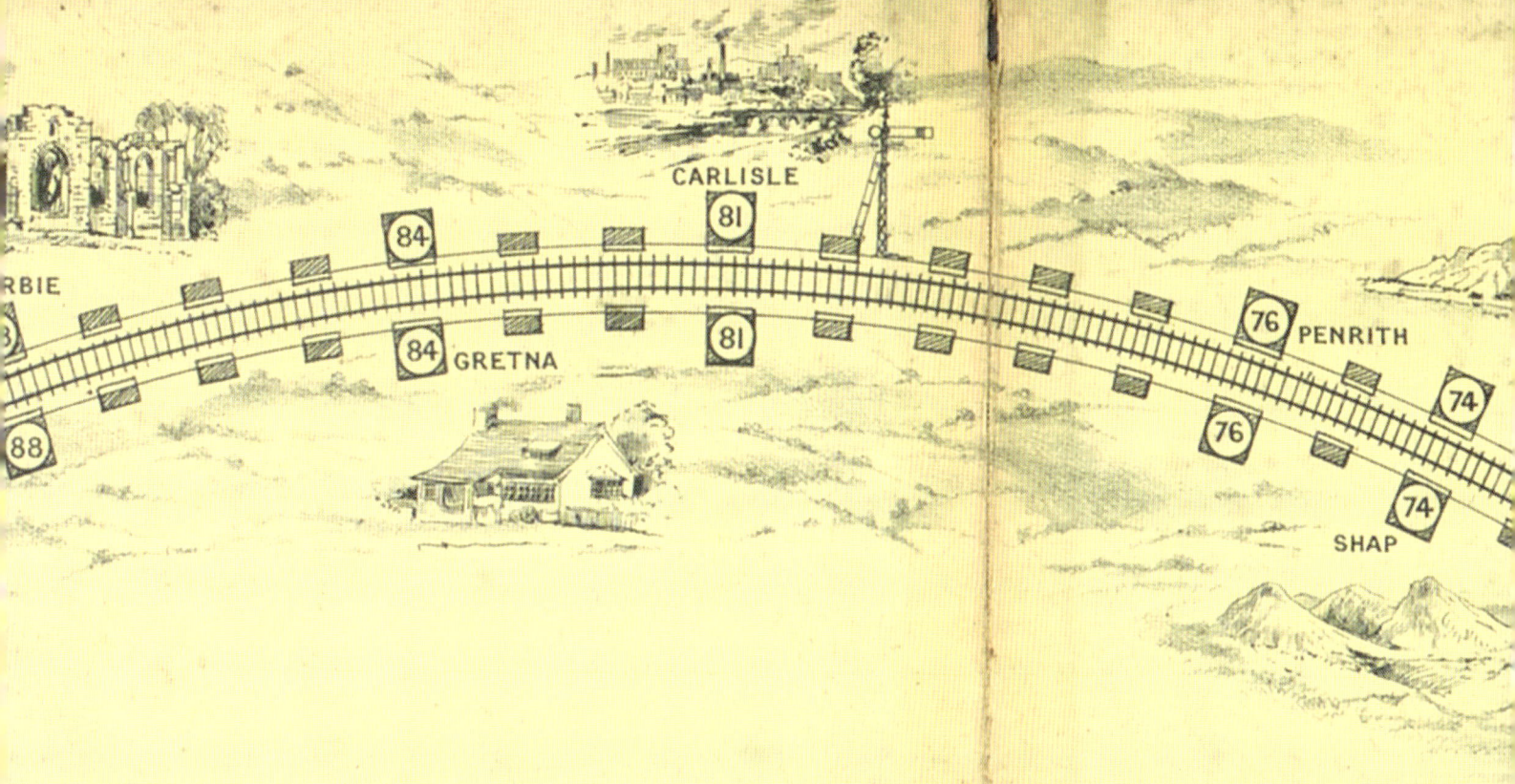

RULES.

.—Each Player selects an Engine, and starts from one of the sheds at Euston Station.

.—The order of play to be decided by lot, the highest number thrown to play first.

.—Each Player should be provided with an equal number of Counters or Stakes to commence with.

.—The moves shall be according to the number thrown by the Die or Tee-to-tum.

.—Engines may pass one another during a move, but not more than one Engine can remain in on. If the number thrown takes an Engine to an occupied Station, it must be placed in the occupied Station behind, and be subject to Rules and Penalties for such Station.

.—An Engine stopping at a Station with Signal against it must remain there one turn.

.—An Engine arriving at a Station with the Signal clear, moves on again the same number, and ct to the Rules or Penalty fixed for such Station.

EXOTIC EAST
Compagnie internationale des wagons-lits, 'Orient Express', 1889.

The romance of rail was to be suggested for long by the Orient Express. This poster indeed is a classic instance of the cartographic imagination, with the relationship being two-way: geographical ideas affected the imagination, while the imagination influenced the response to the cartography, both the facts of distance and the imagination of it. The Orient Express became associated with luxury and intrigue, the two combining in a sense of sexuality focused on the person of the *femme fatale*. Initially, literature provided a functional account of the train, with Bram Stoker in his novel *Dracula* (1897) having the count's opponents take that train in order to confront him. In contrast, more recent computer games have used the Orient Express to provide a background to games, as in the 1985 board game *Orient Express*, which employs the route as a backdrop.

This 1889 poster not only contributes to the appeal of the train it also provides a map that offers valuable details about the journeys. The route crosses the Rhine just east of Strasbourg and the Danube near Vienna, before skirting the latter en route to Budapest. Thence to Bucharest it stayed north of the Danube, but the route to Constantinople (now Istanbul) went via Belgrade and Sofia. Also included are several timetables in French and English: London–Paris–Vienna; Paris–Constantinople; and Paris–Bucharest. Used to far longer expresses, the modern reader might be most surprised by the size of the train, solely one locomotive and three carriages. By cutting out extraneous detail, the map focuses solely on the route, which is a standard device in transport maps – one that essentially captures their role as depictors of a system.

The Belgian company Compagnie internationale des wagons-lits was founded in 1874 on the example of the American Pullman night trains with sleeping cars. It was the sole group providing such luxury international travel, and its trains included the Paris–St. Petersburg Nord Express, established in 1896, and the Paris–Lisbon Sud Express, established in 1887 and extended to London in 1888.

Established in 1883, the Orient Express initially went via Bucharest to the River Danube port of Giurgiu, after which ferry, train and boat took passengers via the Black Sea port of Varna to Constantinople. The first direct train to Constantinople left Paris in 1889.

The opening of the Simplon Tunnel in 1919 ensured that the Simplon Orient Express via Venice also became possible (see pages 178–179). *Stamboul Train* (1932) by Graham Greene and *Murder on the Orient Express* (1934) by Agatha Christie were only two of many works set on the train or on similar luxury services.

The train has recently been revived as a Venice–Paris service, which deliberately sells itself on a suggestion of opulence. This indeed helps to characterise an age, and possibility, for rail that is read not only back more generally into the age of steam, but also extended to assumptions centred around luxury train travel elsewhere – for example, in Southeast Asia and southern Africa, as well as special excursion experiences in countries such as Britain and the United States.

Moreover, since 2023 Accor, the French hotel group, has run the Paris to Istanbul service that had ended in 1977. To this end Accor has restored 17 carriages dating to the 1920s. Accor also runs the London to Venice service, which is all that remains of the old Simplon Orient Express.

PRESENTING THE 'TUBE'
MacDonald Gill, 'In the heat of the summer...', 1922.

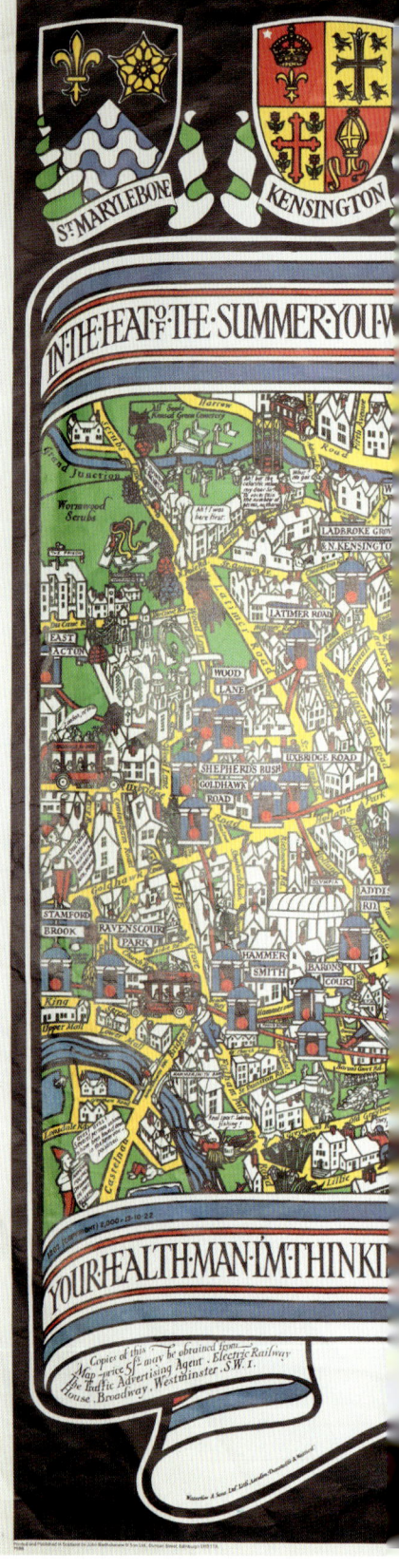

This map by MacDonald Gill (1884–1947) provides an absorbing pictorial depiction of central London with the 'Tube' lines also present, albeit in a somewhat confusing fashion, and bears the notice:

'In the heat of the summer you will find me cool, in the cold of the winter find me warm, Come down underground, You've bought your ticket? Your health man I'm thinking of. No longer 'twill stick it for cheapness, celerity. What else can compare. You are fed up above. Feed below on your fare.'

The map was actually an advertisement that could be purchased for five shillings from Electric Railway House. It includes a note near the bottom that the scale is six inches to the mile.

Gill was a skilful artist and graphic designer who provided striking images. At the time the recent expansion of the 'Tube' had incurred heavy costs and it was necessary to tackle construction and operating expenditure, not least in order to ensure fresh investment. In doing so, there was a wish to add to the somewhat predictable commuting revenue a further source of use and revenue from encouraging leisure uses. To do this, it was appropriate to emphasise the appeal of central London to day-travellers. Moreover, attracting travel to the 'Tube' was made more serious by the competition posed by other means, notably trams, trolleybuses, buses and cars. The last two benefitted from a flexibility of route not seen with tubes, trams and trolleybuses.

The map shows buildings that still exist as well as others that do not, notably the Great White City, in Shepherd's Bush, a grand exhibition complex opened in 1908 but one that was not used for exhibitions after 1914, although the White City Stadium, which seated 66,000, remained and was to be used as a greyhound racing track from 1927.

Gill's map built on his earlier 'Wonderground Map of London Town' (1914) that gave the underground system a strong brand. Although Gill had elements of whimsy, he shared the characteristics of initial maps of the 'Tube' in being accurate in terms of distance and direction. Within a decade Harry Beck had abandoned scale and used the typological method to create, in 1931, the present tube map, which was issued as a pocket edition in 1933. It made Gill's map appear dated.

The 'Tube' was especially significant in the perception of urban railways because London had the oldest and largest underground railway system in the world.

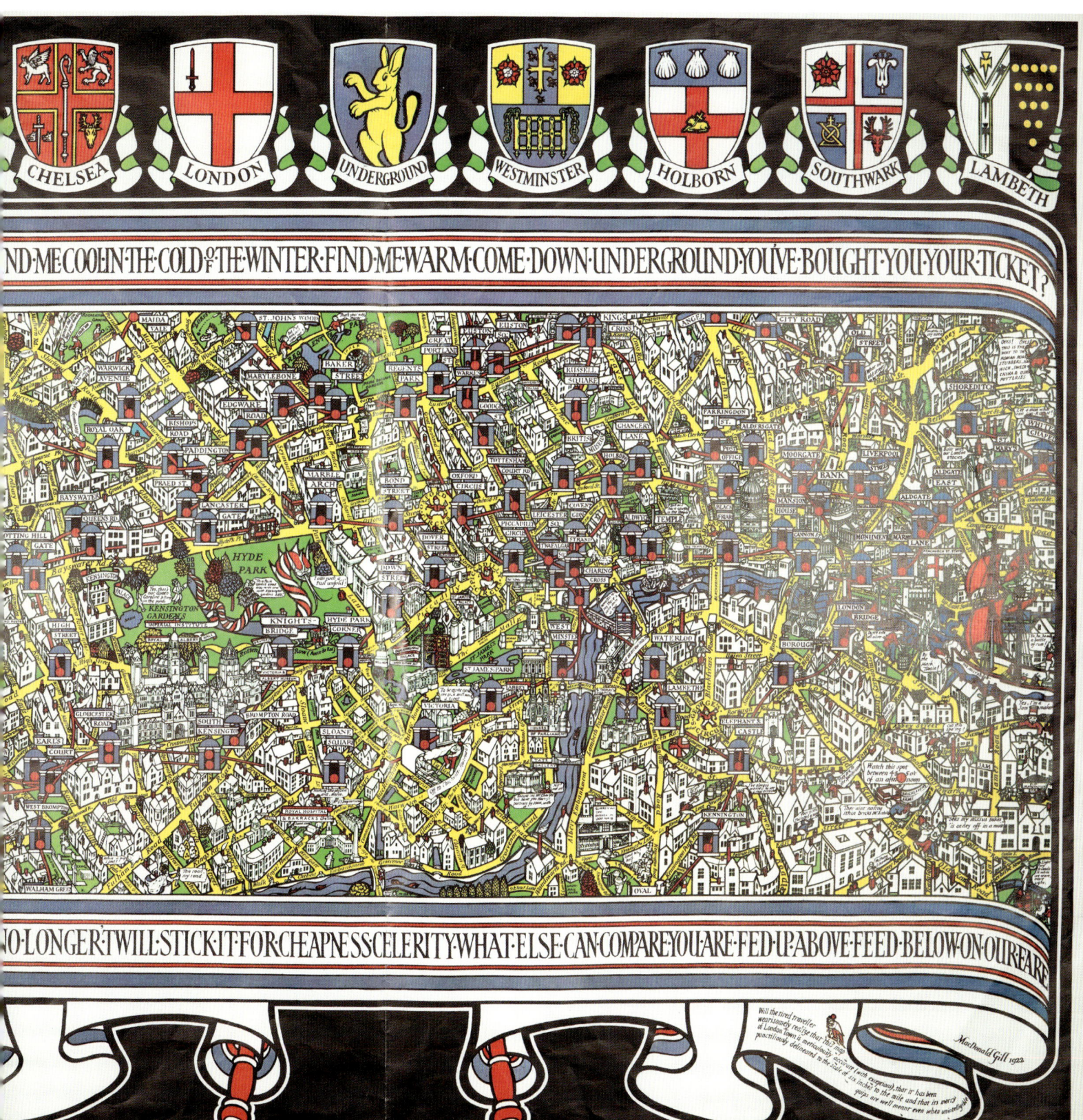

8. RAIL AND THE COLLECTIVE IMAGINATION

RAIL MYSTERIES

Dorothy Sayers, 'Map of Galloway for use with "The Five Red Herrings"', 1931.

Many detective novels, notably Golden Age ones of the 1920s and 1930s, were set on and around trains and stations. For example Miles Burton's *Death in the Tunnel* (1936), Agatha Christie's *Murder on the Orient Express* (1934) and *4.50 from Paddington* (1957). Although most were not illustrated with maps, several did offer them – often of fictional settings and thus in the form of imaginary plans. However, others sought to use real settings, as with Freeman Wills Crofts' *Sir John Magill's Last Journey* (1930) and Dorothy L. Sayers' *The Five Red Herrings* (1931). In her foreword, Sayers begins: 'Here at last is your book about Gatehouse and Kirkcudbright. All the places are real places and all the trains are real trains, and all the landscapes are correct.' So also with finding information: '…the stationmaster at Gatehouse, the booking-clerks at Kirkcudbright, or any of the hundred-and-one kindly people who so patiently answered my questions about railway tickets and omnibuses.'

The map shows the rail and sea routes referred to in the book, which Sayers was familiar with because she had visited frequently with her husband, an enthusiastic fisherman. Timetables are printed in the text, and the plot relies on them and on accurate distances. Some of the criticisms of the books rested on the details, which could be tedious. The solution to *The Five Red Herrings* depends in part on the nature of punching railway tickets. Modern rail travellers in Britain will respond with amazement to the idea that clues rely on trains observing rail timetables. Somewhat differently, in *The House with the Green Shutters* (1901), George Brown's critical novel about his native Ayrshire, carting was ruined by the railway.

Murder on the Orient Express does not need a map because the location of the snowdrift blocking the line is not material to the plot. In Christie's *The Mystery of the Blue Train* (1928) there was also no need for a map, and if one would have contributed to the interest of the plot it might have had to involve a time sequence as well.

In Christie's British stories, the railway acted as an integrated network, with branch lines feeding into long-range systems, and cross-country lines providing further integration, the whole helping her characters move, but also providing opportunities for villains. *4.50 from Paddington* provides the idea of a system: 'The train now arriving at Platform 1 is the 5.38 for Milchester, Waverton, Roxeter, and stations to Chadmouth. Passengers for Market Basing take the train now waiting at No. 3 platform. No. 1 bay for stopping train to Carbury…', all of which were fictional places.

The ability to make trains and also connections was very important to plots of many detective novels. Villains, victims, witnesses, detectives and human red herrings, all moved by train. As J. Jefferson Farjeon writes in his *The Z Murders* (1932):

'If you really wish to test the depths of atmospheric depression, visit Platform No. 3 at Euston Station on an early autumn morning…. The earlier stages had not been enlivened by the murky Lancashire platforms through which they had glided. Shadows crept over the rowed-up milk-cans at Carnforth…. At Preston, there had been a tedious change … three hours for the Glasgow-London train … it had been packed… Euston, that doubtful Mecca, reached at five o'clock; and people at their worst were dribbling out on the ill-lit platform.'

In Alfred Burrage's short story 'Oberon Road' (1924), the protagonist, Michael Cubitt, a lawyer living in Southeast London and commuting to work in the City, 'subject to railway strikes and minor alterations in the timetable … went up by the same train every morning, and came back by the same train every evening'.

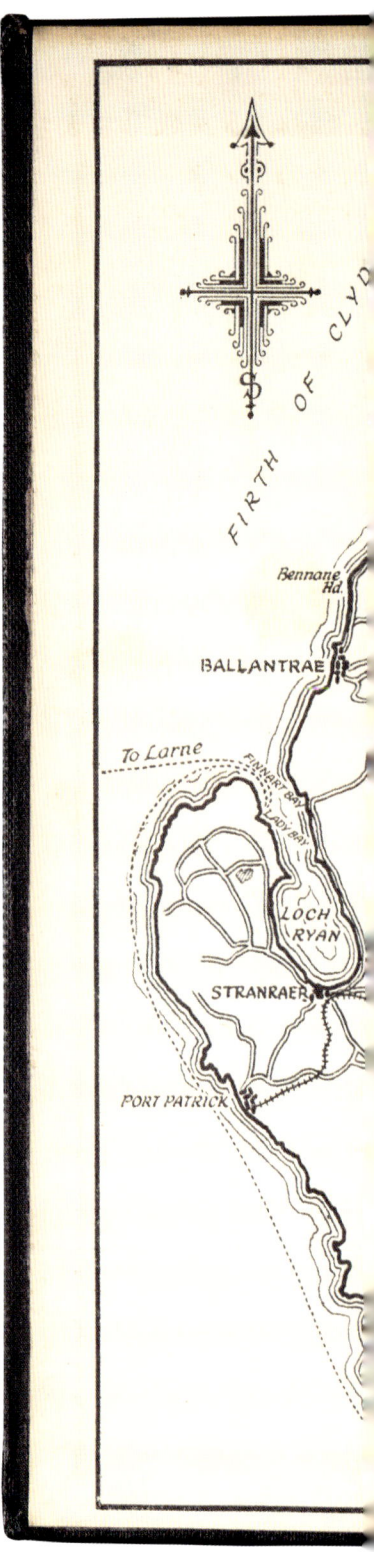

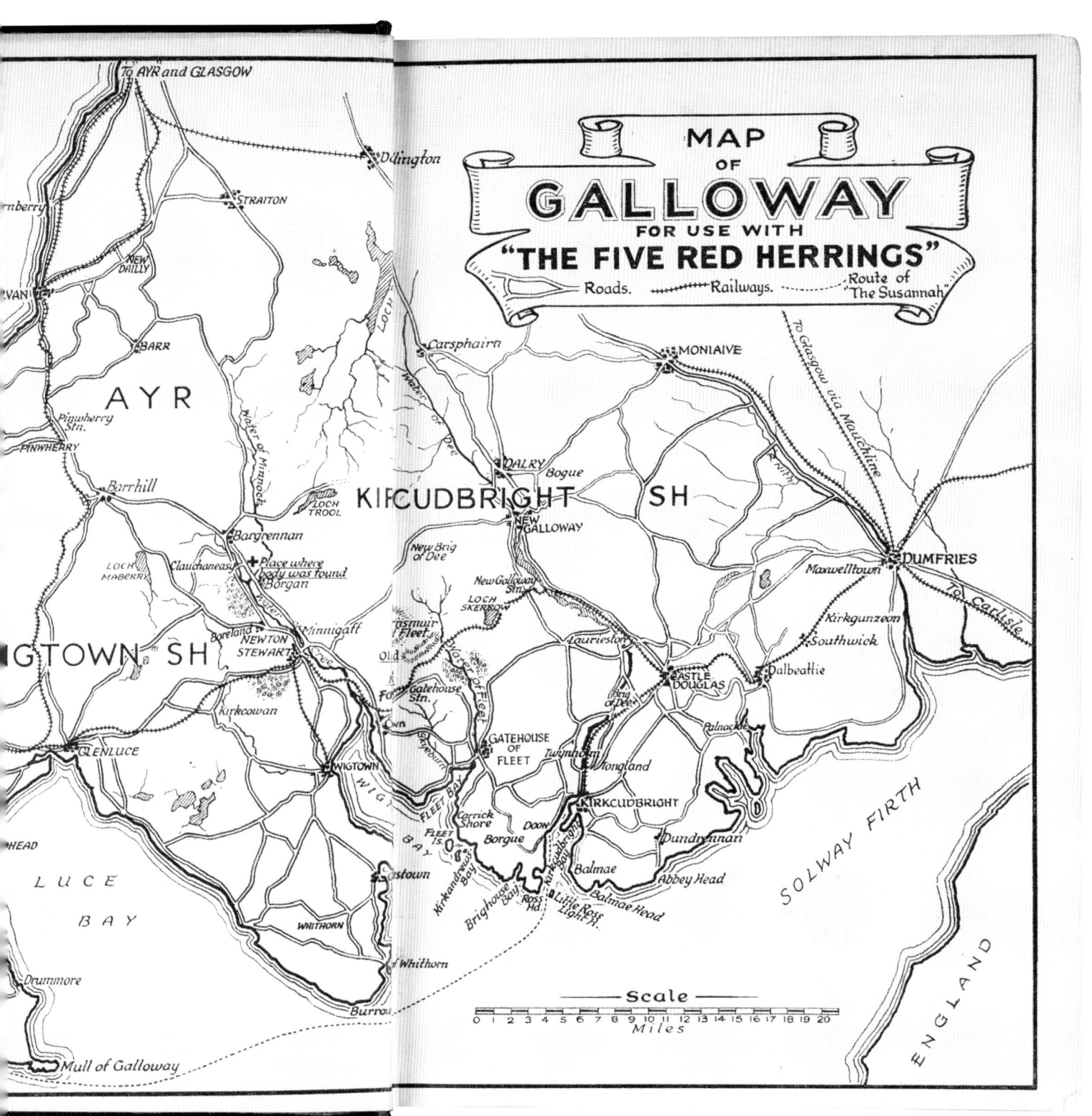

THOMAS THE TANK ENGINE
Peter Edwards, 'Railway Map of the Island of Sodor...', 1945.

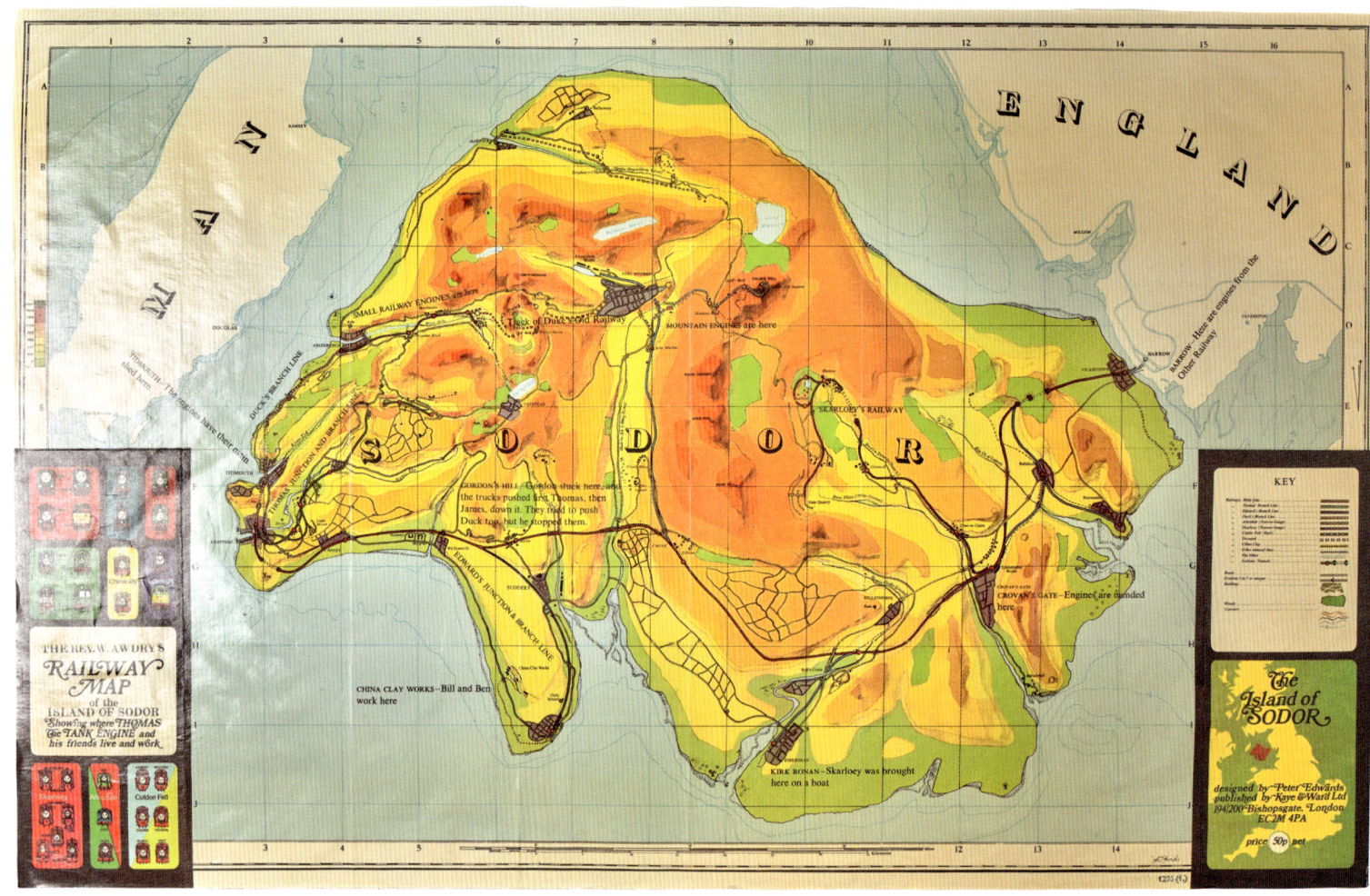

Published from 1945, when *The Three Railway Engines* appeared, the Thomas the Tank Engine stories are benign and were first designed to comfort the son of the author, Wilbert Awdry, who was suffering from measles. The map, which is subtitled 'Showing where THOMAS The TANK ENGINE and his friends live and work', provides both the imaginary island, including its terrain (in height) and towns, as well as the main line and branch lines. The map also locates various episodes mentioned in the text, as in: 'GORDON'S HILL. Gordon stuck here, and the trucks pushed first Thomas, then James, down it. They tried to push Duck too, but he stopped them.' and 'CHINA CLAY WORKS – Bill and Ben work here.' The key would have been familiar to anyone used to British Ordnance Survey (OS) maps, with its depiction of woodland, contours, roads, gradients, buildings and stations. The rail lines are coded using colour and differentiated by gauge. The maps offer an alternative to the many pictures in the stories. It is possibly only on the imaginary island of Sodor that any post-war British cartographer could map roads alongside railways and make the former far less prominent. In most modern maps, whether OS maps or their French equivalents, railways are less prominent. Conversely, transport in the United States was such that, from about 1850 until the 1920s, general maps depicted railways but not roads.

The second book, *Thomas the Tank Engine* (1946), was illustrated by Reginald Payne

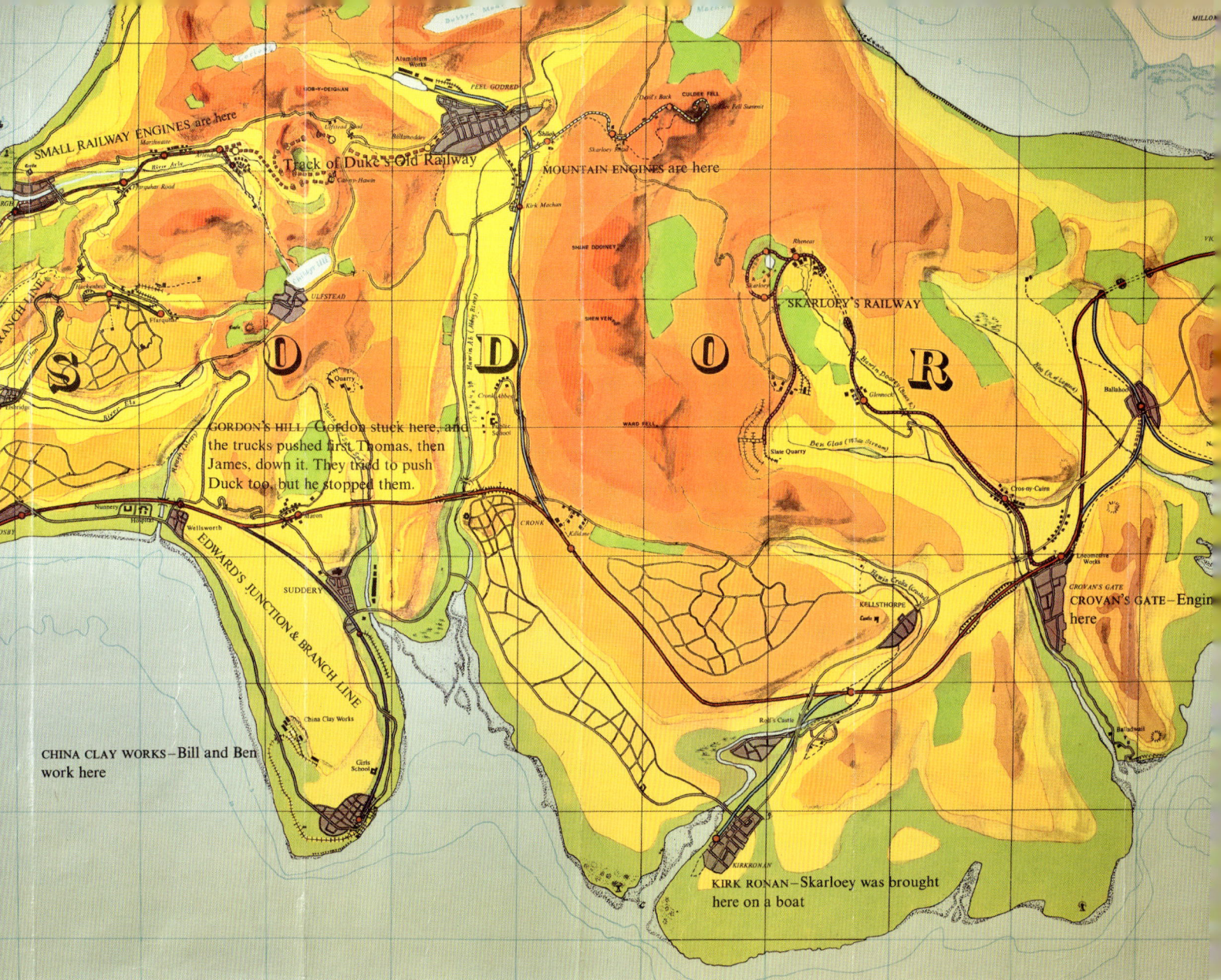

who was shown by Awdry, as a model, a photograph of a 0-6-0 E2 Class engine of the London, Brighton and South Coast Railway. These locomotives were indeed then in service, but there was a historic dimension as they had been built between 1913 and 1916. The locomotives were to be withdrawn from service and scrapped between 1961 and 1963, which was further to date the stories. Diesel was the villain.

In the chronology of the stories, Thomas arrives on the fictional Sodor swiftly after he is built in 1915, and seeks a major role rather than his part as 'a tank engine'. Eventually, he is given his own branch line. Wilbert Awdry wrote the stories from 1945 to 1972, and his son Christopher followed suit in 1983–2011. Awdry's notes were the basis of *The Island of Sodor: Its People, History and Railways* (1987), which provided background details to aid consistency. In Awdry's stories, refugee engines arrive from Britain with tales of closed lines and stations. Set in 1963, episode 67 of *Stories of Sodor Wiki*, aired on 28 January 2022, has Beeching visit Sodor and close most of its lines. The Thomas characters were referred to at length in the 2022 film *Bullet Train*, which was an interesting instance of the ability to use fictional models as references, and between very different moods of fiction. Maps in other children's stories in which railways feature, such as the 1994 edition of Kenneth Graham's *The Wind in the Willows*, do not always include the railways.

8. RAIL AND THE COLLECTIVE IMAGINATION 275

THE ART OF RAIL
Railway companies, 'Paris' and 'East Coast Route', c.1900–1920s.

A particular form of rail map was that provided in advertising material, most significantly rail posters, which became a major art form, and one that took forward the insistent attempts to advertise services, as well as the important role of rail travel in nineteenth-century art, both Realist and Impressionist, from Turner to the numerous French painters who handled topics such as the Gare Saint-Lazare in Paris, for example Claude Monet (see pages 6–7) or Gustave Caillebotte. Poster art covered a variety of aspects of rail, as in Edward McKnight Kauffer's 'Power: The Nerve Centre of London's Underground' (1930), which dealt with the electricity generation that drove the London Underground. More conventionally, there were posters of trains, such as '"The Coronation" Crossing the Royal Border Bridge Berwick-upon-Tweed' (1937) by Tom Purvis, or 'Trains of our Times' (1949) by Vic Welch, which showed a diesel and an electric.

The advertising campaigns of rail companies encouraged travel for pleasure, which provided a source of revenue, not least of First Class passengers. In Britain, the Great Western Railway (GWR), which was very much the system used by Agatha Christie, helped invent the idea of a Cornish Riviera: the Cornish Riviera Express train began its service in 1904, running non-stop from London to Plymouth, before pressing on to Penzance in Cornwall.

Posters served to present an imaginative account of space easily overcome and sights ready to be glimpsed. The prime market was the discretionary one in the shape of the leisure tourist, as seen (right) in the 1920s' poster 'Paris' by Midland Railway, where a stylishly dressed lady, accompanied by luggage, is considering a visit to Britain by means of the Midland Railway Company, which provides the address of its two offices in the French capital.

The earlier, c.1900, 'East Coast Route' poster (following pages), by the Great Northern Railway (GNR), North Eastern Railway (NER) and the North British Railway (NBR), proclaims the 'Shortest & Quickest Route' from London to Scotland, and thereby challenges the routes offered by both the London and North Western Railway (LNWR) and the Midland Railway. The 'East Coast Route' posters also offers the facilities for comfortably-off travellers wishing to visit the Western Highlands, with dining cars, sleeping cars and lavatories. The views offered – Lincoln, York, Scarborough, Durham, Edinburgh, the Forth Bridge, Stirling, Loch Lomond and Fort William – are designed to tempt. So also with the LNWR poster (following pages) from a similar era, which includes views of Euston Station, Lichfield Cathedral and Edinburgh.

Journey speed was an issue, both in the Cornish Riviera Express service and in the rival routes to Scotland. This competition dramatically cut travel time from London, and improved the predictability and regularity of rail travel. The same was also true in other countries. The Dutch rail service from Amsterdam and The Hague via Paris to Marseille, Nice and Monte Carlo was advertised in a 1913 poster by Jan Willem Sluiter in terms of a man and woman exchanging a warm greeting at the seaside.

Poster art became less florid and crowded as the twentieth century progressed. Less information was conveyed, and there were fewer subordinate illustrations. This was the case both for countries where such posters had earlier been prominent and for others where they became newly significant. However, posters continued to be important, and could be dramatic; but photographs became more so. In particular, instead of rather self-consciously posed shots, principally of trains or obviously staged workers, came photographs that were closer to the mixed experience of rail, as with the photographs of Colin Gifford from the 1950s to the 1970s.

By the twenty-first century, the image of rail travel remains important to sales, but practical information is more accessible by Internet sources. This also affected the character of mapping, and thus the representation of the railway, past, present and future, through maps. It is easy to neglect Internet sources when considering maps, but that is mistaken. The maps of the past are not a guide to those of the future.

8. RAIL AND THE COLLECTIVE IMAGINATION 279

FURTHER READING

Akerman, James R. (Editor) *Cartographies of Travel and Navigation*, Chicago: University of Chicago Press, 2006. (Jerry Musich chapter 'Mapping a Transcontinental Nation: Nineteenth- and Early Twentieth-Century American Rail Travel Cartography'.)

Bailey, Michael R. *Built in Britain: The Independent Locomotive Manufacturing Industry in the Nineteenth Century*, Market Drayton: Railway & Canal History Historical Society, 2021.

Bradley, Simon, *The Railways: Nation, Network and People*, London: Profile Books, 2016.

Broch, Ludivine, *Ordinary Workers, Vichy and the Holocaust: French Railwaymen and the Second World War*, Cambridge: Cambridge University Press, 2016.

Buchanan, Angus, *Brunel: The Life and Times of Isambard Kingdom Brunel*, London: Hambledon, 2002.

Carter, Ian, *Railways and Culture in Britain: The epitome of modernity*, Manchester: Manchester University Press, 2001.

Carter, Ian, *British Railway Enthusiasm*, Manchester: Manchester University Press, 2008.

Cobb, Michael H. *The Railways of Great Britain: A Historical Atlas*, London: Ian Allan, 2003.

Coleman, Terry, *The Railway Navvies: A History of the Men Who Made the Railways*, London: Pimlico, 2000.

Elleman, Bruce A. and Kotkin Stephen (Editors), *Manchurian Railways and the Opening of China: An International History*, London: Routledge, 2010.

Freeman, Michael J. and Aldcroft, Derek H. (Editors) *Transport in Victorian Britain*, Manchester: Manchester University Press, 1988.

Freeman, Michael, *Railways and the Victorian Imagination*, New Haven, Connecticut: Yale University Press, 1999.

Gwyn, David, *The Coming of the Railway: A New Global History 1750–1850*, New Haven, Connecticut: Yale University Press, 2023.

Haywood, Richard M. *Russia Enters the Railway Age, 1842–1855*, New York: Columbia University Press, 1998.

Holland, Julian, The Times *History of Britain's Railways: From 1600 to the Present Day*, London: Times Books, 2015.

Katsumasa, Harada, *Japan's Discovery, Import and Technical Mastery of Railways*, Tokyo: United Nations University, 1979.

Koyagi, Mikiya, *Iran in Motion: Mobility, Space, and the Trans-Iranian Railway*, Stanford, California: Stanford University Press, 2021.

Modelski, Andrew M. *Railroad Maps of North America: The First Hundred Years*, Washington, DC: United States Government Printing Office, 1984.

Nigrin, Tomáš, *The Rise and Decline of Communist Czechoslovakia's Railway Sector*, Budapest: Central European University Press, 2022.

Oliver, Lizzie, *Prisoners of the Sumatra Railway: Narratives of History and Memory*, London: Bloomsbury Academic, 2019.

Ovenden, Mark, *Underground Cities: Mapping the Tunnels, Transits and Networks Underneath our Feet*, London: Frances Lincoln, 2020.

Ovenden, Mark, *Transit Maps of the World: Every Urban Train Map on Earth*, London: Penguin, 2015.

Ovenden, Mark, *Railway Maps of the World*, New York: Viking Press, 2011.

Ovenden, Mark and Ashworth, Mike, *Transit Maps of the World: The World's First Collection of Every Urban Train Map on Earth*, London: Penguin Books, 2007.

Richards, Jeffrey and MacKenzie, John M. *The Railway Station: A Social History*, Oxford: Oxford University Press, 1986.

Roberts, Maxwell J. *Underground Maps After Beck: The Story of the London Underground Map in the Hands of Henry Beck's Successors*, London: Capital Transport Publishing, 2005.

Rolt, L.T.C. *George and Robert Stephenson: The Railway Revolution*. Stroud: Amberley Publishing, 2016.

Schivelbusch, Wolfgang, *The Railway Journey: The Industrialization of Time and Space in the Nineteenth Century*, Oakland: University of California Press, 2014.

Simmons, Jack, *The Victorian Railway*, London: Thames and Hudson, 1991.

White, John H. *American Locomotives: An Engineering History, 1830–1880*, Baltimore: The Johns Hopkins University Press, 1997.

Wolmar, Christian, *Railways & the Raj: How the Age of Steam Transformed India*, London: Atlantic Books, 2017.

Wolmar, Christian, *To the Edge of the World: The Story of the Trans-Siberian Express, the World's Greatest Railroad*, New York: Public Affairs, 2013.

Wolmar, Christian, *Engines of War: How Wars Were Won & Lost on the Railways*, London: Atlantic, 2010.

Yenne, Bill, *Atlas of North American Railroads*, St Paul, Minnesota: Motorbooks International, 2005.

Zayarnyuk, Andriy, *Lviv's Uncertain Destination: A City and Its Train Terminal from Franz Joseph I to Brezhnev*, Toronto: University of Toronto Press, 2019.

LIST OF MAPS

Note: All maps are from the collections of the British Library unless otherwise stated.

Page 1 Gabriele Chiattone, 'Gotthard-Bahn, Laghi Como, Maggiore & Lugano', poster (Milan: Stab. D'Arti Grafiche, 1902). Collection Arjan den Boer/retours.eu.

Page 2 Map of the New York Central and Hudson River Railroad and its principal connections (Chicago: Rand McNally and company, 1876). Library of Congress, Geography and Map Division.

Page 10 London Overground map – Autumn 2024 (Transport for London, 2024).

Page 13 Rudi Meyer, SNCF Rail Network (SNCF, 1976). Archives SNCF.

Pages 18–19 J. Gibson, 'Plan of the Collieries on the Rivers Tyne and Wear also Blyth, Bedlington and Hartley; with the Country 11 Miles round Newcastle' (London: R. Ward, 1788). Maps K.Top.32.44.

Pages 21, 22–23 'A Geological, Railway & Canal Map of England and Wales, and Part of Scotland' (C. Smith & Son, 1843). Maps 1180.(4.).

Pages 24–25, 26–27 James Wyld, 'Plan Shewing the Proposed Line of the London and Greenwich Railway'. (J. Wyld, Geographer to the King, Charing Cross, 1832). Maps Crace Port 19.56.

Page 28 J.B. Wallace, 'Drake's Map of the London and Birmingham Railway, and Drake's Map of the Grand Junction Railway'. (Birmingham: J. Drake, 1839). Maps Crace Port 19.57.

Pages 30–31 J.B. Wallace, 'Drake's Map of the Grand Junction Railway'. (Birmingham: J. Drake, 1839). Maps Crace Port 19.57.

Page 33 Charles Fowler, 'Fowler's Map of the Railways in Great Britain', published in *Fowler's Railway Traveller's Guide*, second edition, 1841 (Leeds: C. Fowler, 1841). 1609/1392.

Pages 34–35 'Chemin de Fer de Paris à Marseille. Partie compires entre Avignon et Marseille', from a collection of pamphlets and maps relating to the construction of the railways between Lyons, Avignon and Marseille, 1840–46. Maps 16.b.25.

Page 36 Grundress und Profil der Sächs Bäyerisch (Staats-Eisenbahn von Leipzig, K. Bayersche Grenze, undated). Maps 18.c.2.(51.).

Page 37 Die Eisenbahn von Innsbruck nach München eröffnet am 24. November 1858 (Innsbruck Redlich, 1859). David Rumsey Map Collection Cartography Associates.

Pages 38–39 Henry Schenck Tanner, 'Map of the Canals & Rail Roads of the United States' (Philadelphia, 1830). Library of Congress, Geography and Map Division.

Page 41 'Map of Routes for a Pacific Railroad' (New York: Office of P.R.R. Surveys, 1855). Maps 2.d.23.(13.).

Page 47 'Map of Alabama, Describing the line of the South & North Alabama Railroad. And its Connections' (New York: GW & CB Colton & Co., undated). Maps 2.d.32.

Pages 48–49 'Map showing the Lake Superior and Mississippi Rail Road. A portion of its tributary country, the lines of railway with which it connects and the extent of its land subsidies' (Philadelphia: Sheble, Smith & co., 1869). Maps 2.d.32.

Page 51 'The Great Rebellion of 1861–4', Ensign, Bridgman & Fanning's 'New Military and Rail-Road Map of the United States, showing the Depots and Stations' (Ensign, Bridgman & Fanning, N.Y., 1865). Maps 71490.(109.).

Pages 52–53, 54–55 'Birds Eye View of the Seat of War: arranged after the latest Surveys' (Boston: L.Prang & Co., 1862). Maps 71495.(69.).

Pages 56–57 'Watson's New Township & Rail Road Map of the Dominion of Canada: With Plans of the Cities of Montreal, etc.' (Boston: Perry & Spaulding, 1873). Maps 70615.(32.).

Pages 58–59 'Map of part of Manitoba and the Canadian Northwest Territories: Alberta, Assiniboia and Saskatchewan shewing system of land survey and the lines of the Canadian Pacific Railway' (Winnipeg: Burland Litho., Co., 1890). York University, Canada.

Pages 60–61 Admiralty Chart of the 'Isthmus of Panama, showing the proposed Panama Canal and the Railway' (London, 1885). Maps SEC.10.(657.).

Pages 62–63 'Uruguay Central and Hygueritas Railway of Montevideo' (London: Whitehead, Morris & Lowe, undated). Maps 18.c.2.

Pages 64–65 F.I. Rickard, 'Plan of the City and Suburbs of Buenos Ayres Shewing the Projected Line of Tramway of the Buenos Ayres Street Railway Company Limited' (Buenos Aires: publisher unknown, 1870). National General Archive, Argentina. AR AGN MAP01–II191.

Page 66 'Plano del Gran Ferro-Carril de Buenos-Ayres al Rosario' (Buenos Aires: Mariano Billinghurst & Ca., 1885). National General Archive, Argentina. AR AGN MAP01-II126.

Page 67 'Mapa de las Lineas Ferreas de la República Argentina' (Buenos Aires: Guia Kraft, 1889). National General Archive, Argentina. AR AGN MAP01-114.

Pages 68–69 Martin Wanner, 'Übersichtskarte der Gotthardbahn' (Bern: M. Wanner, 1880). Maps 24480.(20.).

Pages 70–71 'Military Map of Great Britain: Distribution of Troops 1st May 1883' (London: War Office, Intelligence Branch, 1883). Maps MOD IB.293.

Page 72 Alf A. Langley, 'Midland Railway. Map of Lines and Stations' (Derby: Engineer's Office, 1884). Maps 199.a.4.

Pages 74–75 Kartor öfver utsträckningen af Sveriges jemte Norges och Danmarks Jernvägar vid slutet af hvarje år under tjugufemårs-perioden 1856–1880 (Stockholm: Gen. Stab. Lit. Anst, 1880). Maps 19.c.36.

Pages 76–77 Spoorweg-Kaart van het Koningrijk der Nederlanden ('s Gravenhage: J. Smulders & Ctie, 1877). Maps 32065.(6.).

Page 78 'Map of Portion of North Island Affected by the Wellington & Manawatu Railway' (Wellington & Manawatu Railway Company Ltd, 1883). NZ National Library, Alexander Turnbull Library.

Pages 84–85, 86–87 'Stanford's Map of the Siberian Railway, the Great Land Route to China and Korea' (London: E. Stanford, 1904). Maps 49590.(3.).

Pages 88–89 'Mombasa-Victoria (Uganda) Railway and Busoga Railway' (London: Geographical Section, General Staff, 1916). Maps 66430.(56.).

Pages 90–91 'Railway Map of South Africa with the Routes to the Diamond & Gold Fields' (Cape Town: Dennis Edwards & Co, 1895). Maps 67031.(9.).

Pages 92–93 'Map of Railway Freetown-Baiima, Sierra Leone, to illustrate Messrs. Shelford & Son's report, June 1904' (London: War Office, 1904). Maps 65645.(3.).

Page 94 Die Eisenbahn-Konzessionen in der Asiatischen Türkei im Jahre 1914, nach den Verträgen entworfen von Generalleutnant z. D. Imhoff (Gotha: J. Perthes, 1915). Maps 46986.(5.).

Pages 96–97 Map to illustrate *The Short Cut to India* by David Fraser (Edinburgh, London: William Blackwood & sons, 1909). 010057.f.56.

Page 98 Al Hijāz. A post-card map of the Hijāz Railway (Cairo: 'Alī Rida Mu'īn, 1910). Maps 49185.(7.).

281

Page 99 Die Hedschas-bahn (Gotha: Justus Perthes, 1906–1908). Tr.969(b).

Page 100 'Map of the Nitrate Railway and Nitrate Works of Tarapacá, Chile' (London: W. Morrison & Sons & Mallet, 1890). Maps 86565.(2.).

Pages 102–103 J.H. Trott, 'Map of Railway Systems in India, Burma and Ceylon. Compiled for the Railway Department … India' (Bilaspur: J.H. Trott, May 1909). Maps 52437.(6.).

Pages 104–105 'Map of New Zealand, Shewing Railways Open for Traffic, March, 1901' (Wellington: Department of Lands and Survey, 1901). NZ National Library, Alexander Turnbull Library.

Pages 106–107 Map 23 from D.A. Sanborn *Insurance Map of Boston*, Volume 1 (New York: D.A. Sanborn, 1867). Library of Congress, Geography and Map Division.

Pages 108–109 Augustus Koch, 'Panoramic View of the West Bottoms, Kansas City, Missouri & Kansas showing stock yards, packing & wholesale houses' (publisher unknown, 1895). Library of Congress, Geography and Map Division.

Pages 110–111 'All elevated trains in Chicago stop at the Chicago Rock Island and Pacific Railway Station, only one on the Loop' (Chicago: Poole Bros., 1897). Library of Congress, Geography and Map Division.

Pages 112–113 'The Heart of New York: Grand Central Terminal, Only Railway Station on the Subway, Elevated and Surface Lines' (New York, Chicago: Rand, McNally & Co., 1918). David Rumsey Map Collection Cartography Associates.

Pages 114–115 Max von Brandt, map of licensed and planned railway in East Asia from *Industrielle und Eisenbahn-Unternehmungen in China* (Berlin-Charlottenburg: Deutsche Kolonialgesellschaft, 1899). Universitätsbibliothek, Bremen.

Pages 116–117 'The Railway Map of Japan', 1900 (publisher unknown, undated). Boston Public Library, Norman B. Leventhal Map Center.

Pages 118–119 Ceramic tile map of 'North Eastern Railway', *c*.1900, produced by Craven Dunnill & Co. Ltd. National Railway Museum/Science & Society Picture Library (SSPL).

Pages 120–121 North Eastern Railway poster. Percy Home, 'The Humber, the Company's Docks at Hull & The Railway Connections' (North Eastern Railway, 1900–15). National Railway Museum/ SSPL.

Pages 122–123 J.P. & W.R. Emslie, 'Official Railway Map of South Wales' (London: Railway Clearing House, 1910). Maps 6098.(18.).

Pages 124–125 J.P. & W.R. Emslie, 'Official Railway Junction Diagram: Cardiff, Cogan, Penarth & Taffs Well (London: Railway Clearing House, 1905). Maps C.44.d.86.

Pages 126–127 'The "District Railway" Map of London' (London: W.J. Adams & Sons, 1892). Maps 3485.(105.).

Pages 128–129 J. P. & W.R. Emslie, 'Official Railway Map of London and Its Environs' (London: Railway Clearing House, 1899). Maps 3487.(20.).

Pages 130–131 'London's Most Healthy Residential Area Served by the Great Central Rly. Co.' (London: Great Central Railway Co, 1910). Maps 3487.(34.).

Pages 132–133, 134–135 F. Walseck, 'Neueste Eisenbahn-Karte von Deutschland und den angrenzenden Ländern …' (Cologne: F. Walseck, 1891). Maps 20.a.30.

Pages 136–137, 138–139 'Artaria's Eisenbahnkarte von Österreich-Ungarn' (Vienna: Artaria & Co., 1913). Maps 27723.(52.).

Pages 140–141 Jules Décor, 'Carte Kilométrique Officielle des Chemins de Fer Suisses' (Geneva: J. Chanel & Cie, 1888). David Rumsey Map Collection Cartography Associates.

Pages 142–143 D.J.M. Serra, 'Plano de Barcelona y sus alrededores' (Barcelona: Foto. Lit Thomas & Ca, 1891). Institut Cartogràfic i Geològic de Catalunya.

Page 145 'Mappa dos caminhos de ferro portuguezes em 1 de Janeiro de 1895 no Continente e no Ultramar' (Lisbon: des. Goullard e Nogueira, *Gazeta dos Caminhos de Ferro de Portugal*, 1895). Biblioteca Nacional de Portugal.

Page 146 'France, Chemins de Fer' from *Atlas général Vidal-Lablache* (Paris: Armand Colin & Cie, 1894). Maps 48.e.46.

Pages 146–147 'Nouvelle Carte de la Belgique indiquant la situation et le resort de chaque bureau de poste anise que toutes les communes situées dans un rayou de trente kilometres d'un bureau d'origine. Déterminé Chemins de fer, Stations, Routes, Canaux, etc.' (Brussels: C. Granzella, 1851). Bibliothèque Universitaire Moretus Plantin. RCA 6/006.

Pages 148–149 Federico Sauer, 'Strade Ferrate Italiane' (Bologna: Sauer & Barigazzi, 1908). Maps 20603.(21.).

Pages 150–151 A. Combes, 'Railway Map of Part of Australia, Showing Through Connections from South Australia to Queensland and Proposed Strategic Railways' (Melbourne: Department of Lands & Survey, 1916). Maps 90053.(21.).

Pages 152–153 'Third Army Railways' (London: Geographical Section, General Staff, 1918). Maps C.14.g.(137.).

Pages 154–155 'Railway Map of the Western Theatre of War: Showing Broad-Gauge Lines Only' (London: Geographical Section, General Staff, 1916). Maps 14317.(146.).

Pages 158–159, 160–161 A.J. Clevely, 'Africa' map for *The Story of the Cape to Cairo Railway & River Route, from 1887 to 1922* by L. Weinthal (London: Pioneer Publishing Co., 1923–26). Tab.435.c.5.

Page 161 A.J. Clevely, 'Egypt and the Anglo-Egyptian Sudan showing the Valley of the Nile' map drawn for *The Story of the Cape to Cairo Railway & River Route, from 1887 to 1922* by L. Weinthal (London: Pioneer Publishing Co., 1923–26). Tab.435.c.5.

Page 163 'Map of the Argentine Railways' (London: Buenos Aires & Pacific Railway Co., 1925). Maps 87312.(10.).

Pages 164–165, 166–167 'Erie Railroad: Serving the Heart of the Industrial Empire' (New York: Whitney-Graham Company, 1927). David Rumsey Map Collection Cartography Associates.

Pages 168–169 'Map of the Canadian Pacific Railway: the Minneapolis, St. Paul & Sault Ste. Marie Railway; the Duluth, South Shore & Atlantic Railway; the Spokane International Railway; Northern Alberta Railways and connections' (Chicago: Pool Bros, 1931). Maps x.1600.

Pages 168–169 'Bird's-Eye View of the Laurentian Mountains showing roads, lakes & rivers within the area served by the Canadian Pacific Railway' (Map of the Gatineau Valley, etc) (Montreal: Canadian Pacific Railway Co., 1915). Maps 70720.(13.).

Pages 170–171 'Railway Map of India' (Edinburgh & London: W. & A.K. Johnston, 1938). Maps 52430.(50.).

Pages 172–173 M.G. Bouillard, 'Pe King' map from *Carte de la Chine* (Beijing: Ministére des Communications, 1925–26). Maps Y.674.

Pages 174–175 'Canton–Kowloon Railway Raids 14 October 1937–9 February 1938' (London: Air Ministry, United Kingdom, 1937–38). Maps MOD MSAM 2633.

Pages 176–177 'Situation Militaire en Mandchourie avant le 18 Septembre 1931' produced for *le Report de la Commission d'Etude de la Société des Nations* (Geneva: League of Nations, 1932). Library of Congress.

Page 177 'Shina Jihen shutsudō kinen. Sketch map of China, 1939, marking railways and showing Japanese advance into central China, 1937–8' (publisher unknown, undated). National Library of Australia. Bib ID 6977815.

Pages 178–179 Brochure for the 'Simplon-Orient Express Taurus-Express' (Paris: J. Barreau & Cie, 1930). Collection Arjan den Boer/retours.eu.

Pages 180–181 'Great Western Railway: Map and Index of Goods Stations' (London: Geographia, 1933). Maps X.10555.

Pages 182–183 Otto M. Müller, 'Exploring Switzerland' (Zurich: Office Central Suisse de Tourisme, 1939). Maps X.1337.

Pages 184–185 Heinrich Caesar Berann, 'Primiero: La Conca d'Oro delle Dolomiti' (Italian State Tourist Board, 1936). Maps 1060.(4.).

Pages 186–187 Wladyslaw Gorszek, 'Map of the Railway Network of the Republic of Poland' (Krakow: T-Wa Ruch SA, 1938). Maps Y.3620.

Pages 188–189 'Railway Map of the European Soviet Socialist Republics of the USSR' (Moscow: Union of Soviet Socialists Republics, 1938). Maps X.2047.

Page 193 'State of Damage to Rail Centres in Western Germany, 16 Feb. '45' (Great Britain: Office of the Assistant Chief of Staff, Supreme Headquarters Allied Expeditionary Force [SHAEF], 1945). Maps MOD SHAEF OACS 1933.

Page 195 'Burma - Siam Railroad Installations' (Washington: Department of State, Division of Map Intelligence and Cartography, 1944). NARA.

Pages 196–197 'Site Plan for Auschwitz-Monowitz', 1944. Fritz Bauer Institute, I.G. Farben Archives.

Page 199 Maps of 'Transport Engineering', 'Principal Land Communications' and 'Principal Rail Traffic Flows' in the USSR. (Great Britain: Ministry of Defence, Survey Production Centre, 1951). Maps MOD GSGS Misc 1521.

Page 201 'Railway Map of East Africa' (Kampala: Saben & Co., 1956). National Library of Australia. MAP G8401.P3 1956.

Pages 202–203 T. Filipeyan, 'Turkey–Iran railway link, plan and profile, Qureh Tapeh to Tabriz' (Tehran: publisher unknown, 1959). Maps X.13640.

Pages 204–205 'Commonwealth Railways Map of Australia, showing railway systems' (Canberra, A.C.T.: Division of National Mapping, Department of National Development, 1960). Maps 900053.(39.).

Pages 206–207 'The Pennsylvania Railroad Regional Map, effective November 1, 1955' (Philadelphia: Pennsylvania Railroad, 1955). Private Collection.

Pages 208–209 'Major Rail Network: 1966', map from *The National Atlas of the United States* (Washington, D.C.: United States Geological Survey, 1970). David Rumsey Map Collection Cartography Associates.

Pages 210–211 British Railways Board, 'British Railways Liner Train Routes and Terminals under Consideration', Map No. 11 from *The Reshaping of British Railways* (London: Her Majesty's Stationery Office, 1963). Maps C.44.d.97.

Pages 212–213 'Get Out and About on the Southern' (Great Britain: British Rail, 1968). Collection Arjan den Boer/retours.eu.

Page 215 'Rete Delle Ferrovie dello Stato' from the 1959 annual report of the Ferroviarie Stato (Rome: Ferroviarie Stato, 1959). Giorgio Stagni/www.stagniweb.it CC BY-SA 3.0.

Pages 216, 218–219 'Les Buffets de Gare' (Paris: SNCF, 1954). Archives SNCF.

Pages 220–221 'Mapa de los Ferrocarriles Españoles 1968' (Madrid: RENFE, 1968). Biblioteca Ferroviaria. Museo del Ferrocarril de Madrid-FFE.

Pages 222–223 'Densidad de Circulacion 1971' (Madrid: RENFE, 1971). Biblioteca Ferroviaria. Museo del Ferrocarril de Madrid-FFE.

Pages 224–225 'Carte des communications ferroviaires par Trans Europ Express' from brochure *Across Europe* (Paris: C.I.P.C.E., 1965). X.512/216.

Pages 228–229 Transport USSR (Moscow: Main Directorate of Geodesy and Cartography, 1978). Maps 35796.(160.).

Page 230 'Rail and Lake Transport' from *National Atlas of Malawi* (Blantyre: Naitonal Atlas Co-ordinating Committee, 1983). Maps Ref. G.5.(Africa).(2.).

Page 232 Index map from *China Railway Atlas* (Exeter: Quail Map Company, 2008). Maps 247.a.89.

Page 233 'Shanghai' map from *China Railway Atlas* (Exeter: Quail Map Company, 2008). Maps 247.a.89.

Pages 234–235 John Yonge, 'London Transport: Railway Track Map' (Exeter: Quail Map Company, 1978). Maps 3485.(307.).

Page 237 Singapore Transit System Map (Singapore: Land Transport Authority, 2022). Singapore Land Transport Authority.

Page 239 Amtrak 'Rail Passenger Routes: May, 1985' (Springfield, Illinois: Illinois Department of Transportation, 1985). Maps X.11396.

Pages 240–241 'Amtrak's National Rail Passenger System' map, effective June 1993. Amtrak.

Pages 242–243 'New Railway Map of Tokyo and Vicinity' (Tokyo: Japan Guide Map Co. Ltd., 1975). David Rumsey Map Collection Cartography Associates.

Pages 244–245 Shinkansen gaido mappu (Tokyo: Shobunsha, 1975). Maps 203.e.18.

Page 246 Eisenbahnstreckenkarte (Frankfurt: Hauptverwaltung der Deutschen Bundesbahn, 1974). Maps X.13235.

Pages 248–249 Société nationale des chemins de fer belges, 'Carte du resaux B' (Belgium: 1969–1988). Maps 31013.(33.).

Pages 250–251 'Diagram of the New Rail Link London-Tunnel' and 'Channel Tunnel Terminal White City', from British Railways Board, *Channel Tunnel* (London: British Railways Board, 1974). YA.1987.b.4274.

Pages 254–255 *Linking people, nations and places*, Rail Baltica brochure, 2022–24. www.railbaltica.org.

Page 257 'The Belt and road initiative creates a global infrastructure network', 2020. Mercator Institute for China Studies.

Page 259 'HS2 Service Map', 2023 (prior to cuts announced in autumn 2023). HS2 Ltd.

Page 260 Phase 1 of the RRTS network linking Indian Railways, inter-state bus terminals, airports and the Delhi Metro. www.ncrtc.in.

Pages 264–265, 266–267 *The New Game or The Royal Mail or London to Edinburgh* (London: John Jaques & Son, *c*.1850). Maps C.44.d.82.

Page 269 Poster for the Orient Express 'Londres. Paris. Constantinople', 1889. Bibliothèque nationale de France, Paris.

Pages 270–271 MacDonald Gill, 'In the heat of the summer... come down underground' (London: Traffic Advertising Agent, Electric Railway House, 1922). Maps 3485.(305.).

Pages 272–273 'Map of Galloway', – endpapers for *The Five Red Herrings* by Dorothy L. Sayers (London: Victor Gollancz, 1931). NN.17856.

Pages 274–275 Peter Edwards, 'The Rev. W. Awdry's Railway Map of the Island of Sodor: Showing where THOMAS the TANK ENGINE and his friends live and work' (London: Kaye & Ward, 1958). Maps 1296.(2.).

Page 277 Midland Railway poster. National Railway Museum/SSPL.

Page 278 'England & Scotland East Coast Route: Great Northern, North Eastern & North British Railways' poster. National Railway Museum/SSPL.

Page 279 London & North Western Railway poster. National Railway Museum/SSPL.

INDEX

(Figures in **bold** refer to maps; those in *italics* refer to other illustrations.)

accidents 68–9, 217
Adams, W.J. **126–7**
Adelaide 150, 204, 265
advertising 1, 73, 116, 130–1, 140, 157, 204, 206, 212, 217, 262, 270, **276–9**
Afghanistan 84, 102, 256
Alabama **46–7**
Alberta 58, 168
Alexandria 45, 52, 61, 88, 178
Algeria 34, 217
Alps 13, 36, **68–9**, 182, 185
Alta Velocidad Española (AVE) 220
American Civil War 46, 50, 52, 108
Amsterdam 76, 276
Amtrak 206, 227, **238–41**, 263
Amur Valley *81–3*, 85, 228
Angola 93, 144, 200, 231
Antwerp 76, 147, 248, 257
Argentina 56, 62, **64–7**, 80, 156, **162–3**, 190
Arles 34–5
Arnhem 76, 192
Artaria & Co. **136–7**
Ashford 12, 213, 250
Atlantic Ocean 12, 40, 48, 56, 60, 64, 93, 107, 217
Auschwitz **196–7**
Australia 56, 80, 91, 104, **150–1**, 191, **204–5**, 224, 253, 265
Austria **36–7**, 68, 132, 138, 182, 185, 224, 247
Austro-Hungarian Empire 132, **136–7**, 178, 187
Avignon 34–5
Awdry, Rev. W. 274–5
Azerbaijan 202–3, 257

Baghdad 81, 94–5, 138, 178
Balkans 95, 136–7, **138–9**, 192
Baltics 133, 152, 197, **254–5**
Baltimore 38, 52–3
Bangladesh 170
Barcelona 142–3, 223, 225
Barnet 130
Barreau, J. & Cie **178–9**
Bavaria 35–6, 132
Beck, Henry 130–1, 234, 270, 280
Beeching Report 11, 210–13, 258, 275
Beijing 84, 115, 172–3, 177, 233, 256
Belgium 76–7, 142, **146–7**, 154–5, 160, 192, 196–7, 224–5, **248–9**
Bengal 102, 170, 194
Berann, Heinrich Caesar 182, **184–5**
Berlin 81, 94–5, 115, 132, 138, 225, 247
Bianchi, Riccardo 148
Bilaspur 102, 132
billboards 102, 140, *157*
Billinghurst, Mariano 64
Birmingham 29, 46, 73, 253, 258
biscuits 73
board games *264–7*, 268
Boer Wars 91
Bolivia 101, 162
Bombay 102, 170

Boston, Mass. 38, 52, **106–7**, 168, 238
Bouillard, M.G. **172–3**
Bourne, J.C. *16–17*
Bradshaw, George 32
Brandling, Charles 18–19
Brassey, Thomas 64
Brazil 62–3
brewers 122
bridges:
  in Africa 159
  in Arabia 98
  in Argentina 64
  in Australia 151
  in Burma 194–5
  in Canada 168
  in Germany 133, 192
  in Italy 214
  in Russia 152, 229
  in UK 234
  in USA 50
Brighton 12, 32, 212–13, 275
Brisbane 150, 265
British Columbia 56, 59, 85, 168
British Rail (from 1965) 210, 212
British Railways (1948–1997) 210, 213
British Railways Board (1963–2001) 210
British Transport Commission (1948–1964) 210
British War Office **70–1**, **92–3**, 172, **174–5**, **198–9**
Brunel, Isambard Kingdom *263*, 280
Brussels 224–5, 248
Budapest 137, 256, 268
Buenos Aires 62, 64–6, 162–3
Buffalo 166–7
Bulgaria 137
bullet trains 233, 242, *244*, 275
Burma **102–3**, 115, **194–5**
Burrage, Alfred 272
Burton-on-Trent 73, 122, 129
Butlin's 180

Cairo 45, 89, 98, 158–60, 178, 200
Calcutta 102
Calgary 58
California 41, 50, 60–1
California Gold Rush 40, 60
Canada 40, **56–9**, 80, 91, 111, 168, 191, 228, 238, 252, 265
Canadian International Development Agency 231
Canadian National Railways 168
Canadian Pacific Railway 45, 58–9, **168–9**, 252
canals 9, 14, 18, 20–1, 23–5, 29, 32, 34, 38, 42, 46, 56–7, 122, 147
Canton 115, 172–3
Cape to Cairo railway **158–9**, 200
Cape Town 89, 91, 158–60, 200
Cardiff 18, 122–3, 211
Carlisle 73, 119
Carnac, Carol 185
Catalonia 142, 220
Caucasus 95, 257
Ceylon **102–3** (*see also* Sri Lanka)
Channel Tunnel **250–1**, 258
Channon, Henry 'Chips' MP 212

Charing Cross, London 126
Chesapeake Bay 52–3
Chiattone, Gabriele *1*
Chicago 40, 48, 109, **110–11**, 112, 166, 206, 238, 252, 262
Chile 12, **100–1**, 162
China:
  citizens 56, 60, 261
  railways 80, 95, **114–15**, 156, **172–5**, 198, 229, **232–3**
  world politics 84, 115–16, 172, 177, 227–9, 231–3, 256–7
Chinese Belt and Road Initiative **256–7**
Christie, Agatha 178, 268, 272, 276
chromolithography *43*, 102, 140
Clark, Estra 213
Cleveland, Ohio 166
Clevely, A.J. **158–61**
coal 9, 14, 69, 81, 191, 227
  in Africa 200
  in Australia 150
  in Belgium 147
  in Canada 58
  in China 233
  in France 35, 38, 155
  in Germany 76, 132–3, 155, 247
  in India 102, 170
  in Netherlands 76
  in Nigeria 159
  in Poland 187, 196
  in Portugal 144
  in Spain 142
  in UK 16–17, **18–19**, 20, 23, 29, 32, 35, 42–5, 73, 119–20, **122–3**, 155, 180
  in USA 46, 106, 206
  in USSR 198
Cobbett, William 17
Cold War 198, 231, 254
Cologne 132
Colton, J.H., G.W. and C.B. **46**
Combes, A. **150–1**
Commonwealth Railways, Australia 150, 204
commuters 32, 111, 129, 130–1, 212–13, 234, 238, 244, 260–1
Companhia dos Caminhos de Ferro Portugueses (CP), 144
computer games 268
computer technology 226
Congo 160, 200, 231
Constantinople 137–8, 268 (*see also* Istanbul)
Covid-19 223, 233, 238, 256, 258
crashes 148, 170, 263
Crewe 12, 29, 253, 264
Crimea 71, 84, 227
Cuba 12, 190, 262–3
Cumbria 18, 43, 73
Curzon, Lord 102
Czech railways 156, 262, 280

Daily Mirror 181
Daily Telegraph 8, 210
Darlington 20, 119, 213
de la Blache, Paul Vidal **146–7**
Décor, Jules 140
Delhi 261

Denmark 74–5, 132, 224
deportation 196–7
Depression, Great 156, *157*, 162, 166, 180, 187, 206, 217
Derby 12, 73, 253
Deutsche Bahn 246–7
diamonds 90–1
diesel
dining cars 178, 217, 276
Djibouti 231
Dolomites 182, **184–5**
Doncaster 12, 213, 253
Douro, River 144
Dover 20, 25, 71, 213, 250
Drake, James **28–9**
Duluth 40, 48, 168
Dunnill, Craven **118–19**
Durban 45, 91
Durham 19–20, 73, 119, 276

Ealing 32–3, 130
East Africa 88–9, **200–1**, 231
East Coast Route 210–11, 276, 278
Edgware 130–1, 234
Edinburgh 20, 84, 95, 210–11, 264–7, 276
Edmonton 58
Edwards, Dennis **90–1**
Edwards, Peter **274–5**
Egypt 71, 84, 88, 98, 160, 178
electrification:
　in Argentina 162
　in Baltics 254
　in Belgium 248
　in East Africa 231
　in India 170
　in Italy 214–15
　in Netherlands 76
　in Poland 187
　in Soviet Union 228
　in UK 126, 180–1, 211–12, 234
　in USA 206
Emslie, J.P. & W.R., **128–9**
engineering 12, 17, 20, 33, 42, 68, 81, 92, 95, 122, 180, 198
engineers 20, **24–5**, 32, 38, 40–1, 45, 64, 71, 73–4, 101, 116, 148, 152, 169, 172, 227, 232
engraving 11, *15*, 18, 29, 140, 157, 162
Ensign, Bridgman & Fanning **50–1**
environment 60, 130, 138, 225, 227, 236, 239
Estonia 254
Ethiopia 231
Euston Station, London 29, 253, 258, 263–4, 272, 276
Evelyn, John 15

fares 29, 122, 129, 131, 210, 253
Ferrovie dello Stato, Italy (FS) 148, 214–15
Ffestiniog Railway 122
Filipeyan, T. **202–3**
First World War 76, 81, 95, 98, 138, **152–5**, 156, 159–60, 162, 178, 180, 200
Fleming, Ian 198, 206
Florence 214
Florida 50, 206
Fowler, Charles **32–3**

France:
　in nineteenth century 24, 33, **34–5**, 68, 71, 76, 84, 92, 132, 142, 144, **146–7**
　in twentieth century 12–13, 81, 95, 142, 147, 150, 152, 154–5, 185, 192, 198, **216–19**, 224, 247–8, 250, 280
　in twenty-first century 197, 217, 223, 265, 268
　overseas 64, 84, 88, 91–3, 112, 115, 159–60, 162, 172, 200
Frankfurt 247
Fraser, David **95–7**
Fray Bentos 62
freight 8–12, 157, 190–1, 226
　in Africa 88, 231
　in Argentina 66, 162
　in Australia 151
　in China 232–3, 256
　in France 147
　in Germany 247
　in India 261
　in Italy 214
　in Middle East 178
　in Netherlands 76
　in New Zealand, 104
　in Panama 60
　in Poland 187, 196
　in Spain 143, 220
　in Thailand 194
　in UK 18, 20, 42, 45, 180, 210
　in USA 106, 108, 166, 206, 238–9
　in USSR 198
　in Wales 122
Frith, William Powell *263*

Gare Saint-Lazare *6–7*, 276
Gast, John *43*
gauge widths:
　in Australia 150, 204
　in Baltics 254
　in East Africa 200
　in France 147, 217
　in India 102, 170–1
　in Middle East 98
　in New Zealand 79
　in Panama 60
　in Poland 187
　in Portugal 144
　in Scandinavia 74
　in South Africa 91
　in South America 101
　in Spain 142, 220, 223
　in UK 274
　in USA 50, 166
　in West Africa 92
　in World War, First 152–5
　in World War, Second 192
General Staff (British Ministry of Defence) 74, 88, 95, 152, 154, 192, 198
General Strike 168, 180–1
Geneva 140
Gentleman's Magazine 24
geology **20–3**, 206, 250
Germany:
　citizens 52, 95, 98, 108, 137, 182
　railways 74, 95, **132–5**, 154, **192–3**, 196, 224, **246–7**, 262, 265

　world politics 36, 68, 74, 76, 80–1, 88, 115, 138, 142, 152, 154–5, 159, 187–8, 192, 196–7, 200, 202
Gibson, John *18–19*
Gill, MacDonald **270–1**
Glamorganshire Canal 18, 122
Glasgow 12, 126, 210, 272
gold 48, 60, 90–1, 104
Gotthard Railway *1*, 68–9, 140–1
Grand Central Station, NY 112–13
Grand Junction Railway (GJR) **28–31**
Grand Trunk Pacific Railway (GTP) 57, 59, 168
Great British Railways 253
Great Central Railway (GCR) 120, **130–1**
Great Depression 59, 107, 156–7, 162, 166, 180, 187, 206, 217
Great Eastern Railway (GER) 120
Great Indian Peninsular Railway (GIPR) 102, 170
Great Northern Railway (GNR) 73, 262, 276
Great Train Raid 52
Great Western Railway (GWR) 24–5, 32, 81, 122, 130, **180–1**, 212, 264, 276
Greenwich 24–5
Groszek, Wladyslaw **186–7**

Haarlem 76
Haj 98
Hamburg 247, 257
Hanover 132, 247
Harpers Ferry 38, 52
Hejaz line 98–99
heliozincography 92
high-speed trains 29, 197, 214, 217, 220, 223–4, 233, 242, 247, 256, 258, 261
Hitchcock, Alfred 206
Home, Percy **120–1**
horses 14, 16, 18, 20, 38, 52, 64, 71, 74, 137, 202
HS2 29, **258–9**
Hudson River 2, 112, 166
Hull 119–20
Humber, River 120
Hungary 132, 137, 196, 262

I.G. Farben 196–7
Illinois 50, 111, 238–9
India:
　before independence 61, 81, 84, 88, 94–5, **102–3**, 156, 162, **170–1**, 178, 194, 202, 280
　since independence 190, 203, 233, 256, **260–1**, 262
Indian Ocean 88, 159, 256
Indonesia 190, 194
Industrial Revolution 14, 45
Innsbruck *36*, *37*
InterCity 125 211
investment 12, 14, 17, 157, 190–1, 252–3
　in Africa 88, 156, 160, 231
　in Belgium 248
　in Canada 56
　in China 115, 156, 233, 256
　in India 170
　in Italy 148
　in Netherlands 76

INDEX 285

in New Zealand 104
in Portugal 144
in Romania 137
in Scandinavia 64
in Singapore 236
in South America 62, 64, 66, 101, 162
in Spain 142, 220
in Switzerland 68
in Turkey 95
in UK 18, 20, 25, 32, 42, 180, 210, 258, 270
in USA 60, 108, 166
Iran 98, 178, 202–3, 257, 280
Iraq 178, 202–3, 257
Irish labour 38
iron:
  in France 147
  in Germany 132–3, 247
  in Portugal 144
  in Spain 142
  in UK 6–7, 9, 16–18, 20, 23, 42, *44*, 73, 119–20, 129
  in USA 46, 48, 52, 166
  in USSR 187
Istanbul 178, 256, 268 (*see also* Constantinople)
Italian State Tourist Board
Italy 68, 95, 98, **148–9**, 182, **214–15**, 224, 265
Ivory Coast 160, 200

Japan:
  railways 80, 95, **116–17**, **242–5**, 265, 280
  world politics 81, 85, 115–16, 172, 177, 194, 232
Jaques, John & Son, **264–7**
Jinbunsha Co. Ltd. **242–3**
Johnson, Dr Samuel 126
Johnston, W. & A.K. 84, 94, **96–7**, 170

Kansas City *108–9*, 252
Kazakhstan 152, 202, 256
Kenya 88, 200
Khartoum 84, 88, 160
Kimberley 91, 159
Knapp, Drewett & Sons Ltd **130–1**
Koch, Augustus *108–9*
Korea 84–5, 116, 177, 198
  Korean War 198, 232
  North Korea 190, 198
Kowloon 172, 174–5
Kuala Lumpur 236, 253
Kwai, River 194
Kyrgyzstan 256

Lachambeaudie, Pierre 147
Lake Erie **164–5**, 166
Lake Superior 40, **48–9**, 56
Lake Victoria 88–9, 159
Lancashire 18, 29, 272
Landmann, Colonel George Thomas 24–5
Lange, Dorothea *157*
Langley, Alf A. 73
Latvia 197, 254
Lawrence, T.E. 98
League of Nations 176–7
Leeds 17–19, 29, 32, 73, 120, 258
Leningrad 188, 228 (*see also* St. Petersburg)
Liberia 92–3
Light Railways Act 119
Lincolnshire 120, 276

Lisbon 144, 268
lithography 11, 43, 52, 58, 60, 62, 71, 102, 108, 132, 140, 162
Lithuania 254
Liverpool 29, 73, 101
Liverpool and Manchester Railway 29
livestock 17–18, 62
locomotives 9, 157, 190, 227
  in Africa 88, 92
  in Argentina 64
  in Australia 151
  in Austria 36
  in Bolivia 101
  in China 232
  in France 6–7, 35, 112
  in India 102, 170
  in Iraq 178
  in Italy 148
  in Japan 244
  in Kazakhstan 256
  in Manchuria 177
  in Middle East 98
  in Poland 187
  in Singapore 236
  in Spain 143
  in Thailand 195
  in UK 12, *15*, 20, 25, 33, 275, 280
  in USA 52, 112, 166, 206, 280
  in USSR 198
London and Birmingham Railway 16–17, 29
London and North Eastern Railway (LNER) 180, 213
London and North Western Railway (LNWR) 29, 264, 276, 279
London and Southwestern Railway (LSWR) 32
London Overground **10**
London Passenger Transport Board 131
London termini, *see* individual station names
London Transport 210, **234–5**
London Underground **126–7**, 130–1, 226, 234, 242, **270–1**, 276
  Bakerloo Line 130
  Central Line 234
  District Line 126–7, 130
  Jubilee Line 130, 234
  Metropolitan Line 126
  Northern Line 130, 234
  Piccadilly Line 130, 234
  Victoria Line 234
London, Brighton and South Coast Railway 32, 275
London, Midland and Scottish Railway (LMS), 180
Lott, C.S. **58–9**
Louisville 46
Lowe, Robert 126
Luxembourg 154, 192, 224
Lviv 138, 280
Lyon 12, 34–5, 147

Madrid 142–4, 220–3
Malawi 160, **230–1**
Malaya 194
Manchester 29, 32, 211, 258
Manchester Guardian 89
Manchuria 84–5, 115, 172, **176–7**, 188, 232, 280
Manhattan **112–13**
Manitoba 58–9

Marseille 34–5, 196, 276
Maryland 38, 50, 238
Marylebone Station, London 129, 234
McIntosh, Hugh 24
meat transport 62, 66, 104, 108
Mecca 98
Medina 95, 98
Melbourne 150, 265
Merthyr Tydfil 18, 122
Mexico 40, 60, 156, 238, 252, 265
Meyer, Rudi 12, **13**, 217, 247, 258
Midland Railway **72–3**, 129, 276–7
Milan 148, 214
Minden 132
Minneapolis 48, 168–9, 262
Mississippi 9, 38, 40, **48–9**, 50, 111
Missouri 50, 60, 108, 166
Mombasa 88–9, 200
Monet, Claude *6–7*, 276
monorails 242
Mont Cenis Tunnel 68
Montevideo 62–3
Montreal 45, 56–7, 169, 261
Morden 130
Morel, Edmund 116
Morrison, W. & Sons & Mallet **100–1**
Moscow 188, 228
Mozambique 91, 144, 160, 200, 231
Müller, Otto **182–3**
Munich 36, 247

Nairobi 88, 200
Naples 148, 214
Napoleon I 35–6, 38
Napoleon III 35
nationalisation 157
  in Argentina (1948) 64, 162
  in Austria (1908) 137
  in Canada (1917) 59
  in China (1911) 172
  in France (1938) 217
  in Mozambique (1941) 200
  in Peru (1876) 101
  in Portugal (1975) 144
  in Spain (1941) 220
  in Switzerland (1901) 140
  in UK (1948) 190, 210
Native Americans 41, 58, 265
navvies 10, 14, *16–17*, 38, *83–4*, 280
Nepal 256
Netherlands **76–7**, 91, 192, 197, 224, 265, 275
Network Rail 253
New Haven 238
New Orleans 40, 50, 206, 238
New South Wales 150
New York 38, 46, 60, 80, 106–7, 112–13, 129, 166, 206, 224, 238
New York Central Railroad 2, 112
New York, Chicago and St. Louis Railroad 166
New York, Lake Erie and Western Railroad 166
New Zealand 56, **78–9**, 91, *104*, **105**
  Maoris 56, 79, 104
Newcastle, NSW 150
Newcastle, UK 18, 29, 42, 45, 119, 210–11, 253
Newcomen, Thomas 15
Nicaragua 60
Nigeria 88, 92, 159

Nile, River 84, 88–9, 160
nitrates 100–1
Nord Express 268
North British Locomotive Company 12
North British Railway (NBR) 276
North Eastern Railway (NER) 118–19, *120*, **121**, 276
Northumberland 18–19
Norway 74–5

Ohio 38, 50, 52, 166
Onda, Riku 244
Oporto 144
Orange and Alexandria Railroad 52
Orange Free State 88, 91
Ordnance Survey 92, 152, 154, 198, 226, 274
Orient Express 137, 178–9, 268, *269*, 272
Otago Gold Rush 104
Ottawa 45, 56
Ottoman Empire 81, 95, 98, 137–8, 152

Pacific Ocean 12, 40–1, 60–1
Pacific Railroad 40–1, 111, *157*
Paddington Station, London 32, 130, 211, *263*, 272
Pakistan 84, 202, 256
Pampas **64–5**, 66, 162
Panama 12, **60–1**
Paris 12, 34–5, 80, 126, 143–4, 147, 154, 196, 217, 224, 268, 276
Paris Métro 261
passenger numbers 66, 119, 211, 258
Pennines 119
Pennsylvania 16, 38, 50, 166, 238, 253
Pennsylvania Railroad 112, **206–7**
Pennsylvania Station 112, 129
Perry & Spaulding **56–7**
Persian Gulf 202
Perth 204, 265
Perthes, Justus **94–5**, **98–9**
Peru 101, 162
Philadelphia 38, 48, *191*, 238
Philadelphia Inquirer 191
pilgrimage, postcard 98
Poland 152, **186–7**, 247, 254, 265
Poole Bros. **110–11**, **168–9**
Port Elizabeth 91
Portugal 24, **144–5**
Portuguese colonies 91, 93, 115, 159–60, 200
Post Office Railway 234
postcard *98*
posters *1*, *120*, 138, 178, 192, *216–17*, 262, 268, 276, *277–9*
Potomac Railroad 38
Potomac River 52
Prang, Louis **52–3**
Preston 24, 272
printers 32, 52, 62, 204, 244
printing 11, 20, 191, 226, 244
Prussia 36, 115, 132, 147, 247, 253
Punch 69

Quail Map Company **222–3**
Quebec 45, 56, 168
Queensland 150–1

Rail Baltica **254–5**
Railway Staff College, Dehradun, India 170
Rand McNally 2, **111–13**

Red Sea **98–9**, 160, 178
RENFE 220–1
Rhodes, Cecil 89, 91, 159
Rhodesia 91, 200, 231
Rickard, F.I. **64–5**
Rocket 15, 213
rolling stock 42, 46, 64, 108, 148, 157, 166, 170, 172, 180, 190, 210, 227
Romania 137–8, 188
Rome 148, 214
Rosenheim 36
Rotterdam 76
Royal Geographical Society 24
Russia 81, **84–5**, 95, 102, 115, 137, 142, 152, 177, 188, 198, 202–3, 227–8, 244, 254, 256–7, 265, 280 (*see also* Soviet Union)
Russian Empire *81–3*, 187
Russian Federation 228
Russian Revolution 168, 172, 188

Saben and Co. **200–1**
Salzburg 36
San Francisco 40, 60, 101
Sanborn, D.A. **106–7**
Sauer, Federico **148–9**
Sayers, Dorothy **272–3**
Scandinavia **74–5**
Scarborough 32, 119, 276
Scotland 12, 20–1, 73, 88, 119, 122, 211, 276
Scott, William Bell 42, *44*
Seattle 108, 262
Second World War 198, 247
Selby 120
Semmering Tunnel 68
Serbia 137–8
Serra, D.J.M. **142–3**
Seville 220, 223
Shaw, W. James **206–7**
Sheble, Smith & Co. **48–9**
Siam **194–5** (*see also* Thailand)
Siberia **84–5**, 95, 198, 228, 257
Sicily 148, 214
Sierra Leone **92–3**
Silesia 187, 196, 247
Silverlink Metro Services 10
Simplon Orient Express **178–9**, 268, **269**
Simplon Pass 140
Simplon Tunnel 69, 268
Singapore 194, **236–7**, 256
slates 122
slavery 46, 50, 60, 92, 196, 263
sleeping cars 178, 204, 224–5, 268, 276
Smith, Charles **20–1**
Smulders, J. & Co. **76–7**
Société nationale des chemins de fer français (SNCF) **12–13**, 197, 217
South Africa **90–1**, 104, 231
Southampton 180, 213
Southern Pacific Railroad 40, *157*
Southern Railway (SR) 180–1, **212–13**
Soviet Union (*see also* Russia):
  railways 156, **188–9**, 192, **198–9**, **228–9**, 233
  world politics 172, 187, 192, 196–8, 202, 231–2, 247, 254
Spain 12, 24, 42, **142–3**, 144, 150, 156, **220–3**, 224
Spokane International Railway 168
Sri Lanka 8 (*see also* Ceylon)

St Pancras Station, London 73, 129, 250
St. Lawrence River 45, 57
St. Louis 48, 50, 166, 238
St. Petersburg 228, 268 (*see also* Leningrad)
Staffordshire 18
Stagg, John 29
Stanford's 84
steam power 9, 11, 14–17, 42, 60, 62, 81, 120, 133, 150, 191, 233, 268
steam trains:
  in Australia 150
  in Belgium 248
  in Chile 101
  in China 233
  in France 6–7, 217
  in Germany 192, 247
  in India 261
  in Iran 202
  in Iraq 178
  in Italy 146
  in Japan 244
  in Norway 74
  in Singapore 236
  in Switzerland 69
  in UK 18, 210, 213
  in USA 38, 112, 206
steamships 9, 38, 50, 60, 66, 74, 80–1, 88, 93, 101–2, 104, 107, 116, 146, 150, 162
steel 12, 35, 73, 119, 150–1, 155, 166, 172, 187, 224, 247
Stephenson, George 16, 45, 253, 280
Stephenson, Robert *15*, 42, *44*, 45, 74, 253, 280
Stockholm 74
Stockton and Darlington Railway 119
Stoker, Bram 137, 268
Sud Express 144, 268
Sudan 84, 88, 160, 200
Suez Canal 61, 81, 84, 98, 178
Sunday Times 8, 258
surveyors 29, 32, 40, 106
Sweden 74
Swindon 12, 180
Switzerland 68, **140–1**, **182–3**, 185, 224, 265
Sydney 150, 265

Tabriz 202
Taff Vale 18, 122, 180
Taiwan 116, 223
Tanganyika 159, 200
Tanner, Henry Schenck **38–9**
Taurus Express **178–9**
Taurus Mountains 95
Tehran 178, 202
telegraph 9, 15, 42, 44, 166
Tennessee 46
Texas 41, 50, 108, 206
Thailand **194–5**
Thames, River 24, 129–30, 250
Third Army 152–3
Thomas the Tank Engine **274–5**
Tibet 233, 256
Times, The 95, 172, 181, 280
Tokyo 116, 242–4
Toronto 56–7, 168
trade unions 122, 142, 198
Train à grande vitesse (TGV) 12, 217
Train Bleu 224
trams 18, 64, 236, 270
Trans Europ Express (TEE) **224–5**

Trans-Australian Railway 204
Trans-Iranian Railway 202
Trans-Siberian Railway *81–3*, 84, **86–7**, 95, 228, 280
Transport for London (TfL) **10**
Transvaal 84, 88, 91–2, 160
Treno Alta Velocità (TAV) 214
Trieste 68, 137, 214–15
Trott, J.H. **102–3**, 132
tunnels 9, 12, 14
 in Austria 36
 in Baltics 254
 in Canada 168
 in Iran 202
 in Spain 142, 223
 in Switzerland 68–9, 268
 in Turkey 95
 in UK 20, 126, 130, 234, 250, 258
 in USA 107, 112
 in USSR 229
Turin 214
Turkey 80, **94–5**, 152, 178, **202–3**, 257
Turkmenistan 202
Tyneside 18, *44*, 45, 119
typesetting 191

Uganda **88–9**, 200
Ukraine 227, 254, 257
United States of America 12, **38–41**, *43*, 50–1, 56, 58, 60, 110–11, 138, 156, 166, 168, 190–1, 194, 197, **206–9**, 224, 228, 231–2, 238–9, 247, 252–3, 262, 265, 268, 274, 280
Uruguay **62–3**, 66, 162
Utrecht 76

Van Sweringen brothers 166
Vanderbilt, Cornelius 60
viaducts 24, *36*, 64, 129, 258
Victoria, Australia 150
Victoria Station, London 129, 185, 250
Vienna 68, 132, 137, 187, 268
Vietnam 190, 198, 227
Virginia 38, 50, 52, 238
Vladivostok 85
von Brandt, Max 115
Wales 20, **122–5**, 147, 180, 211
Wallace, J.B. **28–9**
Walseck, Ferdinand 132
Walter, George 24
Wanner, Martin **68–9**
Warrington 29
Warsaw 187, 254
Washington, D.C. **52–5**, 238

Waterloo Station, London 250
Wear, River 18, 119
Weinthal, L. 159
Wellington, NZ 78–9
Wellington, SA 91
West Africa 88, **92–3**, 200
West Coast Gold Rush, NZ 104
West Coast Main Line, UK 258
Wharton, W.J.L. **60–1**
Whitehead, Morris & Lowe **62–3**
Whitney-Graham Co. **164–5**
Wilkinson, John 15
Winchester, US 38
Wolverton 12, 264
Worcester, Mass 38, 106, 238
Wyld, James **24–5**

Yemen 98
Yonge, John **234–5**
Yorkshire 32, 119–20, 213, 253, 258, 276
Yugoslavia 178

Zambezi, River 159–60
Zambia 159, 200, 231
Zaragoza 142, 223
Zeebrugge 257
Zimbabwe 200, 231 (*see also* Rhodesia)

PICTURE CREDITS

All images from the collections of the British Library except the following: **Pages 6–7** Mr and Mrs Martin A. Ryerson Collection, Art Institute of Chicago; **15** Wellcome Collection; **43** Library of Congress, Prints & Photographs Division, Washington, D.C.; **44** National Trust Photographic Library/Bridgeman Images; **82–83, 157** Library of Congress, Prints & Photographs Division; **191** Cornell University – PJ Mode Collection of Persuasive Cartography; **263** Royal Holloway, University of London/Bridgeman Images.

ACKNOWLEDGEMENTS

I would like to thank James Brownjohn, Bill Gibson, David Gwyn, Hilton King, Thomas Otte, Mark Ovenden, José Palma Sardica, Geoff Rice and Ulf Sundberg for advice on earlier drafts or particular points. This book owes much to the patient skill of the excellent Christopher W. It is dedicated to my grandson Fred, a prince among smilers.